Routledge Revivals

The Development of the British West Indies 1700-1763

The Development of the British West Indies 1700-1763

Frank Wesley Pitman

Routledge
Taylor & Francis Group
ROUTLEDGE REVIVALS

First published in 1945 by Archon Books

This edition first published in 2018 by Routledge
2 Park Square, Milton Park, Abingdon, Oxon, OX14 4RN
and by Routledge
52 Vanderbilt Avenue, New York, NY 10017, USA

Routledge is an imprint of the Taylor & Francis Group, an informa business

Publisher's Note
The publisher has gone to great lengths to ensure the quality of this reprint but points out that some imperfections in the original copies may be apparent.

Disclaimer
The publisher has made every effort to trace copyright holders and welcomes correspondence from those they have been unable to contact.

A Library of Congress record exists under ISBN:

ISBN 13: 978-0-367-14254-4 (hbk)
ISBN 13: 978-0-367-14259-9 (pbk)
ISBN 13: 978-0-429-03094-9 (ebk)

The Edinburgh Geographical Institute

WEST INDIES
AND
CENTRAL AMERICA
Scale - 1:14,300,000
English Miles
WEST INDIES
ATLANTIC OCEAN
PUERTO RICO
(Sp. till 1898)
DOMINGO
LEEWARD Is
OR Nth CARIBEES
WINDWARD Is
OR Sth CARIBEES
LESSER ANTILLES
Leeward Isles (of the Spanish)
SEA
VENEZUELA
GUIANA
BRITISH
GUIANA
(Dutch till 1803)
DUTCH
GUIANA
GEORGETOWN
New Amsterdam
Trinidad
Port of Spain
Tobago
Grenada
Barbadoes
St. Vincent
St. Lucia
Martinique
Dominica
Guadeloupe
Antigua
Barbuda
Montserrat
Nevis
St. Christopher
LEEWARD ISLANDS
WINDWARD ISLANDS
Scale - 1:3,000,000
Scale - 1:3,000,000
CARACAS
Ciudad Bolivar
S. Fernando
Cumana
Barcelona
Margarita
Orinoco
John Bartholomew & Co.

THE DEVELOPMENT OF THE BRITISH WEST INDIES

1700-1763

By

FRANK WESLEY PITMAN

ARCHON BOOKS
1967

[*Yale Historical Publications: Studies IV*]

Library of Congress Catalog Card Number: 67-19509
Printed in the United States of America

TO
MY WIFE

PREFACE

The West Indies have attracted, in recent years, an increasing interest from students of American colonial society. The more the history of that society is investigated the more do we realize the significance of the West Indies in the development and also in the disruption of the old British empire. For it was from the West Indies that England derived, perhaps, the greatest increment of wealth in modern times. The settlement of the West Indies seemed to the British and French in the eighteenth century the most vital of their outside interests. From the European point of view it is probably not too much to say that tropical colonies were then considered of far greater value than the temperate zone settlements in North America. To the British Northern Colonies, moreover, the West Indies were no less significant. For in the great slave communities to the southward Americans found the only great and permanent market for all their staples. It was the wealth accumulated from West India trade which more than anything else underlay the prosperity and civilization of New England and the Middle Colonies. The situation was analogous to that in the first half of the nineteenth century when the growth of our middle western states rested so largely upon the demand from the cotton planters of the lower south for foods and live stock, cheap access to whose market was made possible by the Mississippi River.

In this book, which in its original form was a doctoral dissertation, I have attempted an investigation of industrial and social conditions in the British West Indies in the effort to reach a better understanding of the part those islands played in the growth and dissolution of the empire. In the seventeenth century the British sugar islands served as an adequate market for northern produce and furnished tropical commodities in abundance for northern consumption and commerce. But the habit of free access to the foreign West Indies, especially in the first quarter of the eighteenth century, enormously stimulated the growth of North American industries and commerce. America was fast outgrowing, therefore, the empire to which mercantilists and West India planters desired to confine its trade. America's progress and insistence upon free trade conflicted more and more with the interests and aims of British sugar planters. The latter, in an attempt to cheapen plantation supplies, monopolize the British and American sugar markets, and embarrass their rivals, secured the passage of the well-known Molasses and Sugar Acts discouraging all commerce with the foreign West Indies. It is conceivable that even this legislation might have been tolerable to North America if it had been accompanied with adequate territorial expansion in the tropics. But the peace of Paris revealed the government's intention of maintaining the boundaries of British dominion in the West Indies at substantially the old limits. Finally, the subsequent reforms in colonial administration made clear the determination to enforce all restrictions on colonial commerce. This course led straight to revolution. The interests and aims of American merchants and West India planters were clearly incompatible.

Thus the eighteenth was in general a century of restraint in politics and commerce for the British West Indies. Its history, despite a record of imposing wealth and growth in maritime strength, is wrapt in an atmosphere almost of pathos. For the settlement of the tropics was often a struggle with nature in her most violent moods. The jungle, earthquakes, hurricanes, disease, and an enervating climate seemed at times irresistible foes to the men who came to extract wealth from the soil. That end was achieved but seldom by small proprietors from the lower middle class of England. The production of sugar required above all an enormous outlay of capital, an abundant supply of the most expensive form of labor, slaves, and industrial organization on a large scale. When British capitalists with such equipment entered the field the small farmers could not hope to compete. The Anglo-Saxon society in the West Indies of the mid-seventeenth century gradually decayed. Great numbers died in the struggle with nature, many migrated, and others merged with the negroes and were lost to the white race.

It is impossible within the limits of this volume to describe adequately West India agriculture, slavery, and all the details of plantation economy. With a considerable amount of material already collected, I plan, however, to supplement the present work with a more intimate study of West India plantations and the society resting upon them. This work is based in the main upon the voluminous correspondence from colonial governors to the British Board of Trade. As the greater part of our sources for West India history is still in manuscript in the Public Record Office, the British Museum, and other archives, liberty has been taken to quote freely from contemporary writers. None of the older historians, such as Long, Edwards, or Schomburgk, had proper access to official papers. While many so-called histories are of little worth as records of the past, they remain, however, of permanent value as descriptions of the period in which they were written. This is preëminently true of the works of Edward Long and Bryan

Edwards. Since the material for this volume was gathered, admirable guides to the sources in England have been published by Professor Andrews and Miss Davenport. It is unnecessary, therefore, to enlarge here upon their nature. In the bibliographical note I have listed the principal sources which in the footnotes are indicated only by abbreviations.

With a deep sense of gratitude I acknowledge the painstaking care with which Professor Charles M. Andrews has read my manuscript and proof and offered valuable criticisms. For suggestions from Professor Wilbur C. Abbott also I am very thankful. To Mr. Hubert Hall and other archivists in England I am deeply indebted for access to manuscripts. Professor Herbert L. Osgood's kindly interest in my work from time to time has been a great encouragement. No one who has recently attempted the investigation of colonial history can fail to pay tribute to the careful guidance of Mr. George L. Beer, whose books are models of historical scholarship. Above all, I feel profoundly indebted to two great and beloved teachers, now deceased—Edward Gaylord Bourne and Guy Stevens Callender. To them is owing very largely whatever contribution to early American history this volume may contain. And finally, I owe much to my wife for counsel and an encouraging belief in the value of the records with which this book is concerned.

Frank Wesley Pitman.

New Haven, February 20, 1917.

CONTENTS

APPENDICES

MAP AND STATISTICAL CHARTS

THE DEVELOPMENT OF THE BRITISH WEST INDIES

THE BRITISH WEST INDIES

Colony	How Acquired	Date	Area in Sq. Miles	Chief Town
Barbadoes	Settled	1624–5	166	Bridgetown
Leeward Islands				
Antigua	Settled	1632	108	St. Johns
Barbuda	Settled	1661–2	62	
St. Christopher	Settled	1623	65.5	Basseterre
Nevis	Settled	1628	50	Charlestown
Anguilla	Settled	1650	35	
Montserrat	Settled	1632	32	Plymouth
Dominica	Settled and conquered	1761	291	Roseau
Virgin Islands	Conquered (Tortola)	1672	58	Roadtown (Tortola)
Jamaica and The Caymans	Conquered	1655	4,296	Georgetown (Grand Cayman)
Turks Island	Settled	1678	166	Grand Turk
The Bahamas	Settled (New Providence)	1666	4,466	Nassau
Windward Islands				
St. Lucia	Conquered	1803	233	Castries
St. Vincent	Occupied	1762	147	Kingstown
Grenada	Conquered	1762	133	St. George's
Trinidad	Conquered	1797	1,754	Port of Spain
Tobago	Settled and (1803) conquered	1803	114	Scarborough
British Guiana	Conquered	1803	100,000	Georgetown
British Honduras	Settled and (1798) conquered	1798	7,562	Belize

CHAPTER I

BRITISH WEST INDIA SOCIETY IN THE EIGHTEENTH CENTURY

British colonization in the tropics presents certain characteristics that are in striking contrast to the settlement of New England or the Middle Colonies. The settlement of these Northern Colonies was often of profound religious significance. In the occupation of the West Indies religion played no part. North Americans sought and, in part at least, realized in the new world political and social reforms which did not interest West Indians, who had no particular desire for changes in prevailing English customs. Local patriotism was established early in the life of North America, while in the West Indies devotion to one's island was attained slowly and with difficulty. The Northerners, who came largely from the middle and nonconformist class in England, had imbibed democratic and republican ideas. The West India planters, on the contrary, represented the capitalist class, were often connected with the landed gentry, were Anglicans, and championed the social and political conceptions held by the rural aristocracy of England. So far as they made the West Indies their home, they made it as nearly as possible like the England they loved and to which they ever hoped to return. In New England a free peasantry was drilled in habits that promoted frugality and strength of character. Slavery, on the other hand, was fastened to the West Indies and fostered industrial waste and ways of living that tended to undermine character intellectually and morally.

Elements essential to a healthy and progressive social organism were, therefore, absent in West India society. There is a passage in Seeley which, perhaps better than any other, gives the clue to the above contrast and to the essential character of British colonization in the West Indies. "Thus the old emigration [to North America]," he says, "was a real exodus, that is, it was a religious emigration. Now this makes all the difference. The emigrant who goes out merely to make his fortune may in time forget his native land; but he is not likely to do so; absence endears it to him, distance idealizes it; he desires to return to it with his money, he would gladly be buried in it. There is scarcely more than one thing that can break this spell, and that is religion. Religion indeed may turn emigration into exodus. . . . But if you find a state which is not also in some sense a church, you find a state which is not long for this world." His further observation is especially applicable to the West Indies: "But what is to be found similar to this in our present colonies? They have not sprung out of any religious exodus. Their founders carried no Gods with them. On the contrary they go out into the wilderness of mere materialism, into territories where as yet there is nothing consecrated, nothing ideal. Where can their Gods be but at home? If they in such circumstances can find within them the courage to stand out as state-builders, if they can have the heart to sever themselves from English history, from all traditions and memories of the island where their fathers lived for a thousand years, it will indeed be necessary to think that England is a name which possesses sadly little attractive power."[1] Herein lies the explanation of absenteeism, an evil which was

[1] J. R. Seeley, *The Expansion of England*, London, 1883, pp. 154-155.

widespread in the West Indies and to which the latter part of this chapter is devoted.

The traditional portrait of the West Indies is one of almost oriental magnificence. West Indians themselves, returning to England, probably drew overcolored pictures of the attractiveness of colonial life. Also the luxuries indulged in by many planters in London may have persuaded some that such was their accustomed mode of life. Were life in the tropics so alluring, why were planters so anxious to return to England and colonial officials so loath to take up their resident duties? Have not these glowing descriptions something of the character of advertisements of new communities in all parts of the world? That there were estates where temperance, refinement, gentility, and kindliness reigned, and that landscape gardening was a familiar art, cannot, of course, be denied. The lives and works of the Codringtons, Edward Long, and Bryan Edwards attest all this. But contemporary observers add many touches of sadness to the picture: the terrifying and destructive forces of nature in the earthquake and hurricane, raging fires—sometimes set by malicious bondmen, devastating fevers, the cause and cure of which were ill understood, the degrading bondage of four-fifths of the population, insurrections or the constant fear of them, the waste of the whole system, the frequent recurrence of war with its attendant losses of ships and cargoes, the general absence of the more talented class, and the inadequacy of church, school, or press to awaken society to things of the spirit.

In endeavoring to picture to ourselves West India society in the eighteenth century it will be well to distinguish the following classes: the rich planters, the poor whites, white servants, free negroes and mulattoes, Jews, and slaves. Nearly every account of the islands impresses upon us the widening gulf between rich slave holders and poor whites. The latter were at first in possession of the soil but gradually were bought out and excluded from the chief industry by the capitalist class. Then, even had they devoted themselves to the production of foodstuffs, lumber, and live stock, they could not have competed with North Americans in the supply of these things. Poor whites were more and more shut out of the handicrafts by slaves. Apart from a few cotton hammocks, stockings, and horse nets,[2] there were no manufactures to utilize their labor.[3] It was not until 1781 that a society was founded in Barbadoes to encourage "Hackling, Spinning, Reeling, Weaving, France-knitting &c^{a}, in order to enable them to find usefull and suitable Employment for their numerous, idle, poor people."[4] If, however, it had been possible to start manufactures in the eighteenth century, the home government was op-

[2] Newcastle Papers, Brit. Mus. Add. MSS. 33028, f. 266. Representation of the Board of Trade to Parliament, 1733. Gov. Worsley, in 1724, informed the Board of Trade of the cotton manufactures mentioned.

[3] Representations of the Board of Trade to the Lords and Commons, Jan. 23, and Feb. 1, 1733/4. C.O. 5: 5, ff. 1-24, 55-80, and printed in ff. 81-90.

[4] C.O. 28: 35, Ii 57. In Bridgetown July 5, 1781, was founded "The Society for the Encouragement of Arts, Manufactures, and Commerce." A printed account of its institution and first proceedings (32 pages) was sent to the Board of Trade. Joshua Steele, its president, in a letter to the Board of Trade, Feb. 28, 1782, stated the object of the society to be to encourage the manufacture of point lace, bone lace and edging, fibers of cabbage tree leaves, fibers of silk grass, fibrous materials from plantation stalks, cotton yarn, knitted stockings, and dyestuffs from native products. The government was petitioned to permit the export to Barbadoes of spinning and weaving machinery notwithstanding the laws against it. It was hoped "by that means to furnish the Manufacturers in Great Britain and Ireland with new and valuable Materials; the Quantities of which will probably be found much more than can be wrought up by the Inhabitants of this Island." It was pointed out that in England there was one person to four acres, while in Barbadoes there was one to each acre. Whereas in England three-fifths of the national income was from manufacturing, in Barbadoes two-thirds of the income came from the land and only one-third from the manufacture of sugar and rum. It was thought precarious for so dense a population to rest upon so few industries.

posed to their introduction into the colonies for fear they might diminish the sale of British goods. To some extent the poor whites engaged in coffee and cotton cultivation which did not involve a great outlay of capital. Many migrated to North America; those that remained did not fit into the economic organization and were left idle, shiftless, and inclined to degenerate. Their character and habits are set forth in the descriptive passages that follow.

Among the early accounts of Barbadoes that which is contained in Henry Whistler's *Journal* gives a very vivid description of the people in 1655. Whistler went with the expedition of Penn and Venables that conquered Jamaica. "This Island," he wrote, "is one of the Riches Spotes in the wordell and fully inhabited. But were the pepell sutabell to the Illand it were not to be compared: it is a most riche soile, all wayes Grone and baring frut, and the Chefest commoditie is sugar, and some Indicco, and Cottaine. heare are many plesant frutes, as pine Apeles, and planting, and Buanoes, and orindges, and limes, and Custord apeles, and prickeled pears, and many other frutes: thayer Bred is made of Cassador routes, which is the routes of tres, and vntil it be Ground, and then pressed all the duse out of it, it is poyson; bųt being well ordered it makes good bread: thayer drink is made of patatoe routes, they being boyled thay brus them, and put them in Water, and then straine them, and that they drink: but they must make it to times a day, or els it will be to stall: thes petatoes are the Chefest of thayer foud: specially for thayer servants: heare are routes they call yames which they make yous insted of flower, as for flesh heare is not much, but sume of all sortes: but the best flesh is Porke, and that is far better than our English porke. The genterey heare doth liue far better than ours doue in England: thay haue most of them 100 or 2 or 3 of slaues apes whou they command as they pleas: heare they may say what they haue is thayer oune: and they haue that Libertie of contienc which wee soe long haue in England foght for: But they doue abus it. This Island is inhabited with all sortes: with English, french, Duch, Scotes, Irish, Spaniards thay being Iues: with Ingones and miserabell Negors borne to perpetuall slauery thay and thayer seed: these Negors they doue alow as many wifes as they will haue, some will haue 3 or 4, according as they find thayer bodie abell: our English heare doth think a negor child the first day it is borne to be worth 05^{li}, they cost them noething the bringing up, they goe all ways naked: some planters will have 30 more or less about 4 or 5 years ould: they sele them from one to the other as we doue shepe. This Illand is the Dunghill wharone England doth cast forth its rubidg: Rodgs and hors and such like peopel are those which are generally Broght heare. A rodge in England will hardly make a cheater heare: a Baud brought ouer puts one a demuor comportment, a whore if handsume makes a wife for some rich planter. But in plaine the Illand of it selfe is very delight full and pleasant: it is manured the best of any Illand in the Inges, with many braue houses, and heare is a braue harbor for ships to Rid in. The Illand is but small: but it maintains more soules then any peese of land of the bignis in the wordell. It is but a littell more then 30 mills long and aleuen milles Brod, and it dose ffrayt aboue a hundred sayle of ships a year with Commodites of the growth of the Illand. This Illand may be much improued if they can bring theyer desine of wine mills to perfecktion to grind theyer Shugor, for the mills they now vs destroy so many horses that it begors the planters, a good hors for the mill being worth 50^{li} starling money. The peepel haue a very Generus fashon that if one come to a hous to inquier the way to any plase they will macke him drinke, and if the trafeller does denie to stay to drinke they tacke it very unkindly of him."[5]

In the unsuccessful attack on Santo Domingo made by Penn and Venables in 1655, soldiers were recruited in Barbadoes and St. Christopher. Venables said of them: "Our planters we found most fearful, being only bold to do mischief, not to be commanded as Souldiers nor to be kept in any civil order; being the most prophane debauch'd persons that we ever saw, scorners of Religion, and indeed men kept so loose as not to be kept under discipline, and so cowardly as not to be made to fight; so that had we known what they would have prov'd, we should rather have chosen to have gone ourselves, as we came from England, than have taken in such to our assistance, who, we fear, with some others put upon us in England, have drawn this heavy affliction upon us, dishonour to our Nation and Religion."[6] "Surely a more ungodly army of professed protestants this wicked world cannot afford (and t'were pitty it should), which I conceive to be the inward cause of our misfortune and disgrace."[7]

Governor Willoughby, in 1667, reporting somewhat more favorably, stated that the people had contributed willingly in both person and purse to fortify the island at all landing-places.[8] A year later, however, he declared that Barbadoes was not by two-thirds rendering its former production per acre; the land was worn out and the inhabitants were ready to desert their plantations. The blacks were kept from insurrection through their different tongues and animosities. He feared the creo-

5 Henry Whistler's *Journal*, March, 1654/5, in *Narrative of General Venables*, ed. C. H. Firth, London, 1900, pp. 145-147.

6 *Ibid.*, p. 30.

7 *Ibid.*, p. 136.

8 Gov. William Lord Willoughby to the king, [Sept. 16,] 1667, *Calendar of State Papers, Colonial, 1661-1668*, no. 1576.

lian generation of slaves would eventually overcome their masters. The militia then consisted of six regiments of foot, two of horse, and a life guard, in all about 6000; the forts were few and none too strong, and for artillery no island of half such concernment, said Willoughby, was half so ill furnished. Bridgetown was large and populous but very disorderly built.[9] That same year about three-quarters of the town burned in six hours.[10]

Frontier conditions gave rise to much contention and irreligion. "This country swarms with lawyers," wrote Nicholas Blake in 1669, "the people are much given to law, to their great impoverishment and others' enriching who suck out much of the fat and marrow of the country and all to little purpose; they are for the most part but dablers in the law and very ignorant. . . . Some of the people in most of the parishes come 4 or 5 miles to church, and this great distance is the cause that many people scarce come to church four times in the year, and people become very untoward and ill-behaved and much uncivilized. Indeed the next generation do run the hazard of turning heathen. . . . If things run much longer in the course it now goeth, there will soon be apparent a greater number of fanaticks, noncomformists, etc. than a loyal and truly religious interest." Our alarm is stilled, however, by his remark that the chief church was Presbyterian, and that there were also Independents, Quakers, Jews, and Atheists.[11]

The Church of England in Barbadoes was introduced by its first rector, the Reverend Nicholas Leverton, B.A., of Exeter College, Oxford. But remaining in the island scarcely a year, being disgusted with the conduct of the

9 Same to Lords of the Council, [July 9,] 1668, *ibid.*, no. 1788.

10 *Ibid.*, no. 1816.

11 Nicholas Blake to the king, Feb. 28, 1669/70, *C. S. P. Col., 1699*, Addenda, no. 1113, pp. 592-593.

people, Leverton with some settlers left, in 1625, for Tobago.[12] In 1629, the island was divided into six parishes and a little later into eleven, and after a time a minister was appointed to each parish. The lives of the clergy, according to Willoughby, generally ran counter to their doctrines.[13] In the second half of the seventeenth century ministers were paid a pound of sugar for every acre in the parish.[14] "An Act for the Better Encouragement of the Clergy," passed March 21, 1704, put the church on a sounder basis. A rector was allowed £150 per annum, besides a mansion house and glebe land. This money was to be raised by a tax assessed by the vestry. For the following services his fees were fixed: marriage in the church, 5 shillings, in other places, 20 shillings; funeral in the church yard, 5 shillings, in church, 20 shillings, in the chancel, 50 shillings, and funeral sermon, 40 shillings.[15] Early in the century, the Bishop of London's surrogate in Barbadoes was the Reverend William Walker of St. Peter's parish.[16]

Education in Barbadoes was closely allied with the church. In 1712 or 1713, "The Society for the Propagation of the Gospel in Foreign Parts" sent a chaplain and catechist to its estates in Barbadoes. Christopher Codrington, governor-general of the Leeward Islands, who died in 1710, bequeathed two sugar plantations, "Consett's" and "Codrington's," consisting of 762 acres,

[12] Sir Robert H. Schomburgk, *History of Barbadoes*, London, 1848, p. 92; Edmund Calamy, *Nonconformist's Memorial*, Palmer's edition, I, 371.

[13] Gov. Willoughby to the king, [Sept. 16,] 1667, *C. S. P. Col., 1661-1668*, no. 1576.

[14] Schomburgk, p. 92.

[15] Richard Hall, *Acts of Barbadoes*, London, 1764, no. 104; C.O. 30: 1.

[16] Schomburgk, p. 92. A memorandum of 1713 or 1714 endorsed "Bishop of London about Building Churches in Barbadoes" stated that several persons there were disposed to build new churches or endow old ones provided the Crown would grant them the advowsons of such churches, C.O. 28: 38, ff. 352-354.

three windmills and buildings, 315 negroes, and 100 cattle to the society in trust for the maintenance of a number of professors and scholars, "all of them to be under the vows of poverty, chastity, and obedience; who shall be obliged to study and practice physic and chirurgery, as well as divinity; that by the apparent usefulness of the former to all mankind, they may both endear themselves to the people and have the better opportunities of doing good to men's souls, whilst they are taking care of their bodies." The plantations were then estimated to yield a net income of £2000 a year. As a result of this gift, the institution known as Codrington College was begun in 1716 and the masonry finished in 1721. Colonel William Codrington promised the society timber to repair the old buildings and five hundred guineas to purchase books. Artificers were sent from England for the construction work. But the college remained unfinished for many years because of a debt due the society, which was not settled until 1738. The most flourishing period for the college was from 1740 to 1750, when about eighteen scholars were supported and taught on the endowment, while twenty or thirty outside pupils received instruction. In the latter half of the century the finances of the trust suffered severe depression, but were restored through the efficient management from 1782 to 1795 of John Brathwaite. Codrington College was the only notable school in the British West Indies in the eighteenth century.[17] West Indians, therefore, excepting the few who afforded

[17] Schomburgk, pp. 111-123. In the nineteenth century, Codrington College educated most of the useful and respectable members of Barbadian society and is now affiliated to Durham University. There were a few other schools in Barbadoes, thirty-nine schoolmasters being reported in 1783, C.O. 28: Bundle 42 (unnumbered). Presumably all were licensed by the Bishop of London. See instruction to Gov. Bellhaven in 1721 to allow no school without such license, C.O. 28: 44, no. 11.

tutors, went uneducated; and of those trained in England, a large number did not return to the islands.

While the character of the clergy in general was above serious criticism, there was one glaring exception in Barbadoes, which was typical of an evil more or less prevalent in the West Indies. This was the case of Reverend Thomas Barnard, brother of the canon of Windsor and Headmaster of Eton. Governor Pinfold reported to the Bishop of London in 1762, that Barnard, rector of St. Michael's—the most populous parish and most valuable living in Barbadoes—obtained leave of absence for one year in 1756. "By annual application to the Secretary of State, this was by Royal licenses continued to 1762, six years, during which time Barnard never saw his parish. Last April," continued the governor, "he arrived and after a residence of two months, he petitioned me for leave of one year's absence in order to bring over his Wife and family; which I granted on express condition that it should be the last favour of that sort. Soon after I found he had, during his short stay, made a bargain with his curate to remit to him in England, a certain annual Sum, and the Curate to have the uncertain Surplus—From which it is evident, that instead of Residence, he intends to Apply yearly to the Secretary of State's Office for Royal Licenses of Absence, too easily granted, and as long as his Salary is paid, never see his parish." This, said Pinfold, was a scandalous condition, leaving religious work to poor ignorant curates. "The Rectories in this Island are good, and the Inhabitants in generall, well disposed to be liberal and respectfull to the Clergy: But this Scheme for establishing a perpetual non-residence, this letting out Rectories like a plantation, to a tenant will create a contempt of the clergy and bring a disgrace upon the whole Order." In conclusion, the governor added, however, that the clergy were "men of good Lives & Characters, some eminent, none absolutely deficient in Literature, punctual in discharge of their parochial Duties, & well esteemed and respected by the Gentlemen in the Island.'"[18] It is gratifying to relate that the governor was soon informed that Barnard would have no further leave of absence.[19]

In literature there is little to record in the history of Barbadoes. The first printing press was introduced about 1730, probably by David Harry from Philadelphia. In a few months he sold out to his former master, Samuel Keimer of Philadelphia. Keimer started the first newspaper, the *Barbadoes Gazette,* and printed it until his death in 1738, after which it was continued for many years by his successors. It was at first printed twice a week, but eventually became a weekly. *Caribbeana*[20] published in London in 1741, in two volumes, is mainly a collection of letters and essays by West Indians selected from the *Gazette.* In 1762, George Esmond and Company started a second printing establishment in Barbadoes and began printing *The Barbadoes Mercury,* a weekly which was continued into the nineteenth century.[21]

The appearance of the towns was slovenly and the houses were often dilapidated. This was due in part to the absence of white artisans, and partly to the fact that many planters regarded the West Indies as only a temporary residence which did not warrant unnecessary adornment. Country houses were preferable to those of

18 Gov. Pinfold to the Bishop of London, July 20, 1762, C.O. 28: 50.

19 Egremont to Pinfold, Oct. 2, 1762, *ibid.* The Barnard case became the subject of a representation from the Board of Trade to the Privy Council, March 1, 1764. C.O. 29: 18, pp. 150-151, 156-157. For reference to this and similar cases, see *Acts of the Privy Council, Col.* IV, §§ 541, 632.

20 A recent edition of *Caribbeana* has been edited by Vere Langford Oliver, 2 vols., London, 1909.

21 Schomburgk, p. 124. The first press set up in the French colony of Santo Domingo was in 1762. Peytraud, *L'Esclavage aux Antilles françaises avant 1789,* etc., Paris, 1897, p. 455.

the town, being large and roomy with glazed windows. The approach to them was generally through avenues of shady and ornamental trees. The wealth of some planters was evidenced by their furniture and the quantity and elegance of their plate. There was one principal meal in the day. It began from two to four o'clock and when visitors were present the company rarely broke up till night was well advanced. They used principally a meat diet with very little bread, but entertained handsomely, having good cooks and fine linen, and manifesting a great deal of cleanliness and method. Some gentlemen imported partridges and pheasants alive from England and kept them in mews. Indeed, foreign luxuries of all kinds were indulged in; wines and liquors of every sort were in common use, and drunkenness was prevalent.[22] Slavery here as elsewhere favored a luxurious mode of life among the upper class and this weakened the instinct to accumulate capital. Courts of chancery were continually, perhaps needlessly at times, concerned with the administration of insolvent estates. On plantations whose proprietors resided in England there was often wilful mismanagement and waste. Gambling among all classes, through cards, dice, lotteries, and all sorts of games and bets, was a consuming vice that eluded the laws against it.[23]

One can hardly avoid the conclusion that life in the tropics in the eighteenth century had made serious inroads on the constitution and character of English people in Barbadoes. A traveler at the close of this period recorded that "Debility pervades all ranks. . . . Barbadoes

[22] Labat, *Nouveau Voyage aux Isles de l'Amerique*, etc., 6 vols., Paris, 1722, Chap. 9; Bryan Edwards, *History of the West Indies*, London, 1791-1793, ed. 1801, II, Chap. 1; Daniel McKinnen, *A Tour through the British West Indies in the Years 1802 and 1803*, London, 1804, pp. 14-30.

[23] Hall, *Acts of Barbadoes*, London, 1764, nos. 160 and 186 (1729 and 1744), C.O. 30: 1.

compared with the rest of the West Indies may be esteemed a very healthy island; . . . But from the meagre and sallow appearance of the native yeomanry and citizens, their sunken eyes, relaxed countenances, and languid motions, I felt always on beholding them that the climate was irreconcilable with the constitution of their race. I am afraid also from the mean and disingenuous behavior of some of the inferior white inhabitants of the town, that the climate, and perhaps their association with the blacks, have not a little relaxed in them the strength and integrity of the British moral character.'"[24]

It was nearly a generation after the conquest of Jamaica, in 1655, before that island developed a strong planting community. The continuance of the war with Spain necessitated a military government until 1664 when the first colonial assembly met. The British soldiers who composed the expedition suffered terribly through disease and death and, when given the opportunity, were not disposed to settle down to an agricultural life. Then, absence of civil government deterred any great emigration to the island. Until 1671, the chief interest of Jamaica was buccaneering against the Spanish settlements and commerce. In spite of instructions to the governors to suppress privateering there were from 1500 to 2000 privateers in fourteen or fifteen vessels, who made Jamaica the rendezvous for their depredations and the market for their booty. The war with Holland, in 1665, gave an added stimulus to this activity and for a time secured for it the sanction of the English government as against both the Dutch and Spanish. After the peace of 1667 with Holland, buccaneering under Governor Modyford's sanction continued against Spain, and resulted in the taking by Henry Morgan of Puerto

[24] McKinnen, pp. 30-31.

Principe in Cuba, Porto Bello, Maracaibo, and Panama. But by the treaty of Madrid, in 1670, England secured the Spanish recognition of her American possessions and agreed to suppress hostilities. By the end of 1672, Governor Lynch reported that there were no more British privateers in the West Indies, although a few Englishmen continued their barbarities in French ships. "Privateering," wrote Lynch, "was the sickness of Jamaica, for that and planting a country are absolutely inconsistent." Many Jamaicans, however, instead of settling down to agriculture, traded illegally in slaves and manufactures with the Spanish Main, engaged in the logwood trade at Yucatan and were preyed upon by Spanish coast guards, or turned pirates.[25]

Such was the situation which hindered the economic development of Jamaica in the latter half of the seventeenth century. Whistler's account of the island at the time of its occupation by the English is a worthy appreciation of its resources and a merited condemnation of its inhabitants. "Now I shall give you," he wrote, "the Best account of this Iland that posibell I can. The land is as good as any is in the Indges, and very frutfull if it be planted, but theas pepoel are a very Lasie pepell, for by thayer good willes none will worke, nor take the paines to pland Cassador to make them bred. But nessitie doth move them to it: they doue very few of them tacke care to be rich, for they say that they cannot wont, for Meate they have an abundante, and the hids and talow will buy them clos, and that is all they take care for most of them: hear ar sume small plantations Shouger, But they spend it most in the Illand: hear is sume Cottaine, both silk and other sortes: But the Chefest Commoditees are theas:

[25] G. L. Beer, *The Old Colonial System*, Part I, Vol. II, 47-49; C. H. Haring, *The Buccaneers of the West Indies in the Seventeenth Century*, London, 1909.

Ling a Vittie and fastick woud, and hids and talow, and porke fat tried up and put in gares: and that is not worth a going so far for. The Iland as it is natuorallie [is] the Best in all the Indges: it hath a great deale of Levill ground, and many brave Savanas full of Cattell, and abundat of brave Horses, But they are all wild: and many hoges: and wild foull an abundant: a many parrates: and Muckeas: and plentie of fich: heare are abundant of Alliegators and many larg snakes. This ground will bare anything that they can plant one it: the spaniyard doth say that it will bare all sorts of spices, and Shugor, and Indico, and Cottaine, and tobaco, and very good grapes: but the Ducke of Meden that it did belong to would not sufer them to plant grapes to macke wine, for then he did know they would not care for Spaine. This Illand is Brauelie watered with fresh riufers: and hath 3 braue harbors in the South Side, and one in the North side: But the midellmust in the South Side is one of the Best in the Wordell: in it may ride 500 sayle of ships from 50 fadham water to 8: and you may Corene by the shore with your gones in 5 fadham water; this harbor is land locked, and the trad wind doth blow into the harbor all day and the Land bres out att night: hear are many small Ilands and shoules that lie before the Harbors mouth, But they are plaine to be sene. The worst Ilconuenience of this Harbor is that it is 6 miles from the Toune, But our English doth say that they will remoue and Bild near the water side, for they may Bild such a Toune as that is in a small time, for the houses are but one storie height Becas of the Harrie Cane, for he doth many times com and giue them a visit. This is all I can say of this Iland, for at present it is pore, But it may be made one of the riches spotes in the wordell; the Spaniard doth call it the Garden of the Indges, But this I wil say, the Gardeners have been very bad, for heare is very little more then that which groweth naterallie."[26] Among the images of the Virgin the English found "a blacke Virgin Mary to enveigle the blackes to worship."[27] The principal products of Jamaica were tobacco, cotton, cocoa, hides, woods, and sugar. At this period there were very few sugar mills.[28] West Indians who joined the invading army, and some of whom became Jamaica's first British settlers, were characterized as "raw souldiers, Vagabonds, Robbers, and runagate servants." "Certainly," said the writer, "these islands must be the very scum of scums, and meer dreggs of corruption."[29] After the Restoration, the government undertook to people the island by transporting thither Barbadians of a better character and with some property.[30] Development of Jamaica's immense resources, however, was exceedingly slow. By 1671, the island contained forty-two cocoa walks, nineteen indigo works, fifty-seven sugar and three cotton plantations, besides a number of small plantations.[31]

An event which must have checked the growth of Jamaica for many years was the terrible earthquake that destroyed Port Royal on June 7, 1692. The rector of that place has left what is probably in the main an authentic

26 Henry Whistler's *Journal*, in *Narrative of General Venables*, p. 168.

27 *Ibid.*, p. 130.

28 *Ibid.*, pp. 138-139.

29 *A brief and perfect Journal of The late Proceedings and Success of the English Army in the West Indies*, etc., *by I. S. an Eye-witnesse.* London, 1655, p. 11.

30 An Act of Barbadoes for the better encouragement of such persons as shall now embark for Jamaica, July 22, 1662, *C. S. P. Col., 1661-1668*, no. 336.

31 John Ogilby, *America*, London, 1671, pp. 336-337, including a map with their location and the names of the proprietors. *Cf.* Richard Blome, *Description of Jamaica*, London, 1672, pp. 8-9, who gives 70 sugar works producing 1,700,000 lbs. of sugar, 60 cocoa walks, and over 60 indigo works. Cocoa was then the "most beneficial Commodity" of the island. See also *C. S. P. Col., 1669-1674*, pp. 104, 105.

account of the calamity and, incidentally, a commentary on the society in which he tried "to keep up some show of religion among a most ungodly and debauched people."[32] He stated that the earthquake had "thrown down almost all the houses, churches, sugar-works, mills, and bridges through the whole country. It tore the rocks and mountains, and destroyed some whole plantations and threw them into the sea; but Port Royal had much the greatest share in this terrible judgement of God." Coming down to the shore, he found that the sea "had entirely swallowed up the wharf with all those goodly brick houses upon it, most of them as fine as those in Cheapside, and two intire streets beyond that; I, upon the tops of some houses which lay levelled with the surface of the water, got first into a canoe, and then into a long boat, which put me aboard a ship called the *Storm-Merchant,* where I found the President of the Council safe, who was overjoyed to see me; there I continued that night, but could not sleep for the returns of the earthquake almost every hour, which made all the guns in the ship to jarr and rattle. The next day I went from ship to ship to visit those that were bruised, and a dying, and to pray with them, and likewise to do the last office at the sinking of several corps that came floating from the point, which indeed hath been my sorrowful employment ever since I came aboard this ship with design to come for England, we have nothing but shakings of the earth, and thunder and lightening and foul weather ever since. . . . The day when all this befel us was very clear,

[32] *A Full Account of the late dreadful Earthquake at Port Royall in Jamaica, written in two letters from the minister of that place, from aboard the Grenada in Port Royall Harbour,* London, 1692. This pamphlet is in the British Museum, and a calendar of its contents is in *Royal Historical MSS. Com. Report* XIV, Vol. 4, pp. 266 *et seq.* These letters were also printed in the *Gentleman's Magazine,* Vol. XX, 212-215. The letters are dated June 22 and 28, 1692.

afforded not the suspicion of the least evil; but in the space of three minutes, about half an hour after eleven in the morning, Port Royal, the fairest town of all the English plantations, the best emporium and mart of this part of the world, exceeding in its riches, plentiful of all good things, was shaken and shattered to pieces, and sunk into, and covered for the greatest part, by the sea and will in a short time be wholly eaten up by it; for few of those houses that yet stand are left whole, and every day we hear them fall, and the sea daily encroaches upon it; we guess, that by the falling of the houses and the opening of the earth and the inundation of the waters, there are lost fifteen hundred persons and many of good note." In his second letter the rector stated that "whole streets were swallowed up by the opening earth, and the houses and inhabitants went down together, some of them were driven up again by the sea, which arose in those breaches and wonderfully escaped; some were swallowed up to the neck, and then the earth shut upon them, and squeezed them to death; and in that manner several were left buried with their heads above ground, only some heads the dogs have eaten, others are covered with dust and earth by the people which yet remain in the place to avoid the stench." The country was all broken to pieces and unsettled. At St. Ann's he heard that over one thousand acres of woodland was changed into sea, and carried with it whole plantations. To crown the destruction a French invasion, instigated probably by Jacobites, devastated the southeast portion of Jamaica in 1694 but was driven off by colonial troops, carrying with it considerable plunder.

Soon after the earthquake, the colonial government decided to build a new town, called Kingston, located on the coast a few miles northeast of Port Royal. Here a site of two hundred acres was purchased for £1000 from William Beeston, the lieutenant-governor, and later governor. At first, the place proved unhealthy and many who had survived the earthquake were swept away by disease. Kingston was systematically laid out in 1695, under the direction of the eminent military engineer, Colonel Christian Lilly.[33] It assumed a rectangular shape with the parish church in the center. With the destruction by fire in 1703 of what remained of Port Royal, Kingston became the most important commercial center of Jamaica. Spanish Town, in the interior, remained, however, the seat of government until 1755. Its transfer to Kingston lasted but three years when it was reëstablished until 1872 in Spanish Town, which was always the social center of the planters.[34] In 1714, the assembly created a few new ports of entry on the north shore of Jamaica. The attitude of the English government toward this innovation reveals its policy of discouraging the growth of towns. The customs commissioners had no objection to the act creating new ports "provided the said Act will not encourage the Inhabitants to reside in Townes, and there sett up Manufactures for the Supply of their own Necessities, without Assistance from hence, wch will not only Discourage the Trade carried on from this Kingdom as well by Our own Manufactures, as by the Reexportation of East India Goods and other Forre Goods from hence but will also take off their hands, wch might be employ'd more to the Benefit of this Kingdome, in Planting and raising Sugars and other Commodities of that Island to be ship'd home for the Supply of Our own Forre Marketts from hence, to the prejudice of the Trade and Navigation of this

33 A. E. Aspinall, *West Indian Tales of Old*, London, 1912, pp. 7-8.

34 The transfer of the capital in 1755 gave rise to a long and bitter controversy between the planting and commercial interests. See references to the Board of Trade's representations on the subject in *Acts of the Privy Council, Col.* IV, §236; VI, §§499, 507.

Kingdom.'"[35] Also the more ports the more difficult would it be to collect revenues and enforce the navigation laws. Therefore, the Jamaica act creating outports was not confirmed. Kingston remained the only port of entry until, in 1758, custom-houses were established at Savanna la Mar, Montego Bay, and Port Antonio.[36]

By the opening of the eighteenth century, Jamaica had recovered its state of prosperity. Many rich men acquired immense estates there, but the climate and, possibly, the fear of earthquakes deterred a considerable number from residing upon them. The resident planters took a certain pride in the natural resources of the island and in its position as a practically self-governing colony. But as members of the British empire they were little disposed to contribute toward its defense or extension, so engrossed were they in amassing fortunes. At the outbreak of the war with France and Spain in 1702, the English commander in the Caribbean, Admiral Benbow, wrote of Jamaica: "The Government of this Island now is entirely in the hands of Planters who mind nothing but getting Estates and when so to goe off, having no regard to the King's Interest or Subjects, for at this time we can hardly get fresh Provisions to Support the Sick, the Ships & Soldiers being a great burthen to them as they say & wish they had never come into these parts. The Inhabitants are grown very rich & value themselves for being Judges & Parties in making & executing their own Laws; they doe whatever the desire of Gain leads them to without any regard to the Laws of o^r^ Country.'"[37] Something, no doubt, of the naval officer's impatience with the ways of all civilians influenced the admiral's

[35] Carkess to William Popple, Sec. to Board of Trade, April 3, 1719, C.O. 137: 13, P 37.

[36] Order in Council, June 29, 1758, C.O. 137: 31, Aa 27.

[37] Admiral Benbow to Sec. of State, James Vernon, June 1, 1702, C.O. 137: 45.

estimate of the people. If Jamaicans were not imperially minded, they were, however, outgrowing the reputation for immorality which at one time they probably had merited. "I think the inhabitants," said a traveler in 1711, "not at all deserving the Character I have heard in England of their Lewdness and Debauchery. . . . The Gentry live very handsomly: The Late Governor lived the meanest of any Gentleman in the Island."[38]

In their way of living the planters imitated, as nearly as the climate would permit, the rural gentry of England. Much was written about their pastures of cattle, sheep, and goats; their rabbit warrens, and the huntsmen employed in hunting wild hogs, parrots, and pigeons; their coaches and fine saddle horses; and their sumptuous hospitality. Their diversions were horse-racing, shooting, fishing with angles, nets, and pots, billiards, balls, assemblies, and concerts. Wines, rum, and brandy were in constant use.

The description of Jamaica written by Leslie in 1740 is, perhaps, the most interesting of contemporary accounts. "The Gentlemens Houses," he said, "are generally built low, of one Story, consisting of five or six handsome Apartments, beautifully lined and floored with Mahogony, which looks exceeding gay; they have generally a Piazza to which you ascend by several Steps, and serves for a Screen against the Heat, and is likewise a Way of enjoying any Coolness may be in the Air. In the Towns there are several Houses which are of two Stories, but that Way of Building is disapproved of, because they are seldom known to stand the Shock of an Earthquake, or the Fury of a Storm.[39] The Negroes

[38] Copy of a letter to James Campbell, Port Royal, Aug. 31, 1711, Bodleian Library, Rawlinson MSS., A. 312, f. 22.

[39] Philip Pinnock, for nearly forty years a councilor, chief justice, or member and speaker of the assembly, erected a magnificent dwelling in the parish of St. Andrew. Skilled laborers were imported and the cost to him

have nothing but a Parcel of poor miserable Huts built of Reeds, any of which can scarce contain upwards of two or three.'"[40]

"The common Dress here is none of the most becoming, the Heat makes many Clothes intolerable, and therefore the Men generally wear only Thread Stockings, Linen Drawers, and Vest, a Handkerchief tied around their Head, and a Hat above. Wigs are never used but on *Sundays* or in Court Time, and then Gentlemen appear very gay in Silk Coats, and Vests trimmed with Silver. The Servants wear a coarse *Osnabrig* [coarse German linen] Frock, which buttons at the Neck and Hands, long Trowsers of the same, a speckled Shirt and no Stockings. The Negroes go mostly naked, except those that attend Gentlemen, who take Care to have them dressed in their own Livery, tho' 'tis the utmost Pain to the uneasy Slave. The Laidies are as gay as any in *Europe,* dress as richly, and appear with as good a Grace. Their Morning Habit is a loose Night-gown, carelessly wrapped about them; before Dinner they get out of their Dishabille, and show themselves in all the Advantage of a becoming rich neat Dress. The Servant Maids have

was £25,000 currency. In 1777, he wanted to sell the house for £15,000 currency and desired that his creditors might raffle for it. Stephen Fuller, colonial agent, to Board of Trade, June 5, 1777, C.O. 137: 37, Gg 60. The houses were roofed with northern shingles, which necessitated laws to restrict setting off fire works. An Act of Jamaica, 1734, C.O. 139: 13, no. 69. Sugar canes and houses were very inflammable and many cases of the malicious burning of them occurred. Arson was punishable by death. Act of Jamaica, 1736, C.O. 139: 14, no. 32. Many laws were passed in the West Indies concerning the prevention of fires and arson. See *e.g.*, Hall, *Acts of Barbadoes,* nos. 56, 160.

40 Spanish Town, Port Royal, and Kingston were skirted with negro huts where runaways and rebels were often harbored. The law required that no hut have more than one door. Where four or more huts were built together they should have a seven-foot fence around them with one gate in it. The same act prohibited selling to slaves arms and powder except by certificate. Act of Jamaica, 1744, C.O. 139: 16, no. 19.

generally a Linnen or strip'd Holland Gown, and plain Headclothes. The Negroe Women go many of them quite naked, they don't know what Shame is; and are surprized at an European's Bashfulness, who perhaps turns his Head aside at the Sight. Their Masters give them a Kind of Petticoat, but they don't care to wear it. In the Towns they are obliged to do it, and some of them there go neat enough; but these are the Favourites of young Squires who keep them for a certain Use.

"Learning is here at the lowest Ebb; there is no publick School in the whole Island, neither do they seem fond of the Thing; several large Donations have been made for such Uses but have never taken Effect.[41] The Office of a Teacher is look'd upon as contemptible, and no Gentleman keeps Company with one of that Character; to read, write, and cast up Accounts, is all the Education they desire, and even these are but scurvily taught. A Man of any Parts or Learning would be

[41] Between 1667 and 1736, two hundred and eighteen legacies to churches, the poor, and education were made in Jamaica. But only a few of these appear to have been properly executed. *Journals of the Assembly*, III, 417-420; W. J. Gardner, *History of Jamaica*, London, 1873, ed. 1909, p. 203. Thomas Manning, who died in 1710, left funds for founding a free school with Episcopal masters in the parish of Westmoreland. It was not until 1738 that an act for erecting the school was passed. Act of Jamaica, 1738, C.O. 139: 15, no. 4. John Wolmer, a goldsmith of Kingston who died in 1729, left the residue of his estate to found a free school in Kingston. An act was passed in 1736 to carry this out. C.O. 139: 14, no. 33. In 1740, the assembly provided for building a school in the parish of Vere from several donations that had been given or bequeathed for the poor of that parish. The school might receive as many poor children as it could and, on their reaching the age of twelve or fourteen years, might bind them out as apprentices to artists, tradesmen, or planters. Act of Jamaica, 1740, C.O. 139: 15, no. 31. Peter Beckford, who died in 1730, willed £1000 for a free school in Spanish Town. An act of 1744 vested the fund in a corporate trust. C.O. 139: 16, no. 10. These appear to be about the only educational endowments that were properly administered in our period. Altogether, they offered an elementary education to probably less than fifty boys. There were of course dancing and music masters and some tutors, but most boys who were to be educated were sent to England.

despised and starve. The Gentlemen, whose Fortunes can allow, send their Children to *Great Britain,* where they have the Advantage of a polite generous Education, but others are spoil'd, and make such an inconsiderable figure ever after, that they are the common Butt in every Conversation. . . . 'Tis a Pity, in a Place like this, where the Means could be so easily afforded, something of a publick Nature should not be done for the Advantage of Posterity; but when such a Spirit will appear, is hard to determine. There are indeed several Gentlemen here that are well acquainted with Learning, in some of its most valuable Branches, but these are few; and the Generality seem to have a greater Affection for the modish Vice of Gaming than the *Belles Lettres,* and love a Pack of Cards better than the Bible. To talk of a Homer, or a Virgil, of a Tully, or a Demosthenes, is quite impolite; and it cannot be otherwise, for a Boy till the Age of Seven or Eight diverts himself with the Negroes, acquires their broken Way of talking, their Manner of Behaviour, and all the Vices these unthinking Creatures can teach; then perhaps he goes to School, but young Master must not be whipt; if he learns 'tis well; if not, it can't be helped. After a little Knowledge of reading, he goes to the Dancing School, and commences Beau, learns the common Topicks of Discourse, and visits and rakes with others of his Equals; this is their Method: And how can it be supposed one of such a Turn can entertain any generous Notions, distinguish the Beauties of Virtue, act for the Good of his Country, or appear in any Station of Life so as to deserve Applause? The Laidies read some, dance a great deal, coquet much, dress for Admirers, and at last, for the most Part, run away with the most insignificant of their humble Servants. Their Education consists entirely in acquiring these little Arts. 'Twere a thousand Pities they should not improve their Minds as well as their Bodies, they would appear then the gayest of Creatures."

Of the town life of the planters at Spanish Town, Leslie wrote: "Being an inland Place, its Trade is inconsiderable, but several wealthy Merchants reside there, and the most of [the] Gentlemen of Estates have Houses. They live after a gay Manner; 'tis surprising to see the Number of Coaches and Chariots which are perpetually plying, besides those which belong to private Persons; they have frequent Balls, and lately have got a Playhouse, where they retain a Set of extraordinary good Actors. In short, they live as happily as if they were within the Verge of the *British Court:* And to do them Justice, they seem perfectly polite, and have a Delicacy of Behaviour which is exceedingly taking."[42]

Excessive gaming was a prominent feature of their life. Many dissolute people, overseers, tradesmen, servants, free negroes, slaves, mulattoes, and Indians congregated in gambling houses, and brawls and riots frequently occurred. The favorite games included billiards, lotteries, dice, shovelboard, faro, ace of hearts, passage, hazard, and games with machines or devices of chance. An act of 1744 ordered the closing of the gambling resorts. Backgammon, skittles, ninepins, and bowls, however, were permitted. Also persons having estates that yielded £100 a year or who had £1000 in clear property were allowed to gamble privately in their own houses in all games but hazard and passage, provided no one lost more than £20 at any one time.[43] Evidently the vice had the sanction of the whole community and its suppression was not seriously looked for. Efforts were made to popu-

[42] Charles Leslie, *History of Jamaica*, London, 1740, pp. 28, 30, 31, 35-39. *Cf.* the anonymous *Description of Jamaica* in Duke of Wharton's *Poetical Works*, London, 1727, II, 16; Edward Long, *History of Jamaica*, London, 1774, II, 231 *et seq.;* Edwards, *West Indies*, II, Chap. 1.

[43] Act of Jamaica, 1744, C.O. 139: 16, no. 7.

larize a watering resort at the Bath in St. Thomas in the East in imitation of Tunbridge Wells and Bath in England, but its proximity to the rebel Maroon negro settlements hindered its growth.[44]

An act to encourage good and able ministers to come to Jamaica was passed in 1707,[45] but there is little to record in the religious history of the island. In the eighteenth century there were nineteen parishes. Two of the churches in 1740 had organs, one in Spanish Town and one in St. Andrews. The others were small, plain houses scarcely to be recognized as churches, and the ministers were said to have troubled them little, their doors being seldom opened.[46] Though always under his jurisdiction, the authority of the Bishop of London was not recognized by act until 1748. There were some Quakers and possibly a few other dissenters, but not enough to maintain a church.[47] Moravian missionaries to the slaves arrived in 1754 and preached at the Foster and Barham estates. From the five missions founded both white and black exhorters went forth to convert the slaves. But proprietors and overseers raised serious opposition to their preaching, and by the close of the century less than a thousand negroes had been baptized.[48]

The color line in Jamaica, as elsewhere in the tropics, was carefully drawn. Marriage between whites and blacks was prohibited.[49] Mulattoes born of slave mothers followed the status of the mother. Many slaves, however, were freed and educated, sometimes in England. An act

44 Act of Jamaica, 1731, C.O. 139: 12, no. 36.

45 *Ibid.*, 1707, C.O. 139: 9, parchment 106.

46 Leslie, *Jamaica*, p. 30.

47 Gardner, *Jamaica*, p. 199.

48 J. H. Buchner, *The Moravians in Jamaica, History of the Mission . . . 1754-1854*, London, 1854.

49 Either by custom or law. An act of Montserrat inflicted a penalty of £100 on any minister who married a white person to a negro. C.O. 391: 70, p. 51 (Feb. 16, 1762).

of 1733 provided that no one who was not three degrees removed in lineal descent from the negro ancestor exclusive should be allowed to vote and no one should be deemed a negro after the third generation as aforesaid.[50] Nevertheless, many mulattoes who could not avail themselves of this law were granted the rights and privileges of free-born whites by special acts. Such acts usually recited that the persons were children of white planters, had been educated and brought up in the Church of England, and had come into the possession of considerable property.[51] The white blood of the West Indies was being assimilated into the black far more rapidly than was the case in the plantation colonies of North America. A Jamaican lumber merchant traveling in the Carolinas observed that intercourse with negro women was sanctioned and advertised less than in the West Indies. In Charleston he met only one man who acknowledged keeping a mulatto mistress and he was "pointed at."[52] While the increasing number of mulattoes in the West Indies points to the corrupting influence of slavery under a tropical sun, it must be remembered that illegitimacy is more apparent in a society of mixed blood than in a wholly white community. The sanction, however, which was given to illegitimacy in the tropics must indicate a breaking down of the English moral constitution and was so recognized by the best people of the community.[53]

From an early period, the Jews formed a numerous but unwelcome portion of the population of Jamaica. In 1736, they owned about fifteen settlements producing

[50] An act of Jamaica to secure the freedom of elections, etc., C.O. 139: 13, no. 58.

[51] *E.g.*, many such acts for Jamaica during 1738-1741 are in C.O. 139: 15.

[52] Two letters by G. Moulton dated Newbern, N. C., Dec. 20, 1772, and Charleston, S. C., Jan. 23, 1773, Brit. Mus. Add. MSS. 22677, ff. 69-76.

[53] See, *e.g.*, the opinions of Long, *Jamaica*, II, 330-338, and Edwards, *West Indies*, II, 25-27.

sugar and other commodities besides lands and houses in Kingston, Spanish Town, Port Royal, and other places. Two synagogues were erected, one in Spanish Town and the other in Kingston. The shopkeepers were for the most part Jews. They were in communication with Jews in England who, it was said, "are almost the only Persons that send any dry, fine Goods to Jamaica, at their own Risque, and on their own Account, and the Jews in Jamaica, the only Persons almost that have any large Quantitys of all kind of Goods lying in their Ware houses or Shops, for the Supply of the Inhabitants of the Island, and for making proper Sortements of Goods for the Spaniards, when there is either a Want or Call for them."[54]

The prosperity of the Jews, due to their greater industry and temperance, excited the jealousy of the Christians. Accordingly, they were excluded from all public offices, were ineligible for military service, and were subject to special taxes. Also the act by which assistance was given to immigrants did not extend to Jews, nonjurors, or Quakers. From these unjust burdens the Jews petitioned the king for relief.[55] The government appeared reasonable and instructed Governor Edward Trelawny to withhold his assent from any act imposing additional duties on the Jews.[56] The assembly was so insistent that the governor at first assented against his

[54] Wm. Wood to Sec. Popple, Feb. 18, 1735/6. Wood enclosed a petition from Benjamin Bravo, a Jew who had resided in Jamaica ten or eleven years, had returned to England, and was seeking relief for the persecuted Jews in Jamaica. C.O. 137: 22, V 11. See also Long, *Jamaica,* I, 573.

[55] Petition to the king, Feb., 1735/6, C.O. 137: 22, V 10. Many Christian merchants also signed the petition. William Wood, who had been in Jamaica, sympathized with the Jews and thought their treatment was detrimental to the increase of inhabitants. *Ibid.,* V 11.

[56] Privy Council to Gov. Trelawny, March 22, 1738/9, C.O. 137: 23, W 7. Instructions repeated to Knowles in 1752. See *Acts of the Privy Council, Col.,* III, §366; IV, §35.

instructions to a special act in 1739.[57] After a hard fight with Trelawny the assembly finally gave way to the principle of equal taxation, not, however, without issuing a scathing manifesto against the Jews.[58]

The feature of West India society that accentuated all the evils of a community resting on slavery was the general absence from their estates of the highest class of planters. In the domains of government, militia, and industry, the absentees left the islands impoverished for want of talent and leadership in public life, and of humane, conscientious, and responsible direction in private industry. The absence of proprietors did much to promote that careless, cruel, and extravagant management of plantations that became so frequent a subject of comment toward the middle of the eighteenth century. Out of such a condition, moreover, slave insurrections frequently arose, for life under agents and overseers was a far more burdensome and hopeless experience than it was under responsible and often merciful masters. There gradually grew up in England a West India aristocracy of great wealth and political power. This bound the

57 Trelawny to Board of Trade, May 10, 1739, C.O. 137: 23, W 10.

58 *Journals of the Assembly*, III, 570-572. An example of the hatred that was prevalent toward Jews may be seen in the following passage: "The Jews make up a very considerable Part of the Inhabitants, and have a larger Share of Trade than the Christians in proportion to their Number. . . . I do declare that these were the worst Set of Rogues that ever I knew, in the whole Course of my Life; a Set of meer low-life Theives, (as bad as the Negroes themselves, who are all naturally Theives) the meaner Part of whom held a strict Correspondence with all theiving Negroes, from whom they received the stolen Goods. . . . Some of them acquired great Estates." *Memoirs of the Life and Travels of Dr. James Houstoun*, London, 1747, p. 277. In Barbadoes also, the Jews were reported to "have almost entirely engross'd the business of Shop-keeping from the Christians, and . . . Subsist upon an unlawful commerce with Our Slaves, whom they encourage to commit Thefts and Robberies, and are the common receivers of all Stolen Goods." James Dottin, president of council to Board of Trade, April 2, 1741, C.O. 28: 25, Aa 106; and to Newcastle, C.O. 28: 45, f. 265.

sugar colonies closely to the mother country and secured for them a peculiarly powerful lobby in the imperial parliament. Thus an extraordinarily favorable position in the economic organization of the empire was by legislation conferred upon them. On two occasions, indeed, in 1733 and again in 1764, parliament was persuaded by the "planting interest" to sacrifice the prosperity of New England—a policy that hastened the dissolution of the empire itself.

In Barbadoes, as early as 1669, the evil results of the non-residence of many planters occupied the attention of the colonial assembly. An act of that year stated that "several of the most eminent plantations of this island do little or no parochial duties, by reason the representative owners of such plantations have removed themselves to England, or elsewhere, and in their stead and place have left their particular attorneys, agents, and overseers, to manage their estates for them in their absence, whereby the country hath not only a far less choice of able men to officiate in the highest places of trust, but the burthen likewise of inferior offices doth thereby lie more heavily on the poorer sort of people." It was therefore provided that attorneys, agents, or overseers of absentees' estates of over sixty acres should be eligible for election as surveyors of highways and constables.[59] This meant the initiation of a generally inferior class of men into the public service of the island.

No attempt, however, was made by Barbadoes to discourage the return to England of planters, as was later the case in Antigua and Jamaica. The practice was looked upon as perfectly natural. Littleton, later agent for Barbadoes, writing in 1689, said: "By a kind of magnetic force England draws to it all that is good in the plantations. It is the center to which all things tend.

[59] Hall, *Acts of Barbadoes*, London, 1764, no. 49; C.O. 30: 1.

Nothing but England can we relish or fancy: our hearts are here, wherever our bodies be. If we get a little money, we remit it to England. They that are able breed up their children in England. All that we can rap and rend is brought to England.''[60] In 1698, Davenant estimated that the West Indies had sent back annually to England about three hundred of their offspring, with the advantage, he added, that the fathers went out poor and the children came home rich.[61] A pamphleteer of 1701 said that some of the planters who came to England bought lands, others sent their children to be educated, so that their annual expenditure there amounted to nearly fifty thousand pounds.[62] By 1706, the number of absentee proprietors from Barbadoes in England was more than twenty, and they had already formed the habit of holding meetings to memorialize the government on affairs of the island.[63] Colonial governors usually returned from Barbadoes with large fortunes.[64] Governor Henry Worsley, for example, who served from 1722 till 1731, received annually £2000 from the British government, £6000 from Barbadoes, and about £2000 from fees in the island, making a total of about £10,000 sterling money.[65] Also many

60 *Groans of the Plantations*, London, 1689, p. 34.

61 Charles Davenant, *Two Discourses on the Public Revenues and on the Trade of England*, London, 1698, II, 96.

62 *Some Considerations humbly offered to both Houses of Parliament, Concerning the Sugar Colonies, and Chiefly the Island of Barbadoes*, London, 1701, p. 4 (Brit. Mus. 1391, b. 1).

63 Memorial from Bromley and other absentees to the Board of Trade, Nov. 7, 1706, C.O. 28: 9, N 50. *Cf.* C.O. 29: 10, f. 179.

64 H. St. John to Robt. Harley, Nov. 5, 1706. St. John had just heard from George Granville of the death of his brother, Sir Bevil Granville who was governor of Barbadoes from 1703 till 1706. ''This misfortune touched George to the quick. I hear, by the bye, that his brother has brought from the Barbadoes enough to make him easy, though not in any degree such a fortune as governors usually amass.'' *Royal Hist. MSS. Com. Report on MSS. of Marquis of Bath*, 1904, I, 121.

65 Memorandum of Income of Gov. Worsley, 1722, C.O. 28: 39, ff. 59-60.

colonial office holders, preferring to reside in England, farmed out their offices to persons of small fortunes who extorted exhorbitant fees from the colonists. Even deputies sometimes sublet positions. In this way port fees came to be higher than in many other parts of the empire and constituted an unnatural burden upon commerce.[66]

In Jamaica the evils of absenteeism were recited in an act of 1700, which tried to compel patentees of offices, at least, to reside in the island.[67] Governor Hamilton, in 1715, hesitated to appoint two able gentlemen to offices on account of "The many Inconveniences that frequently happen by the non Residence of Pattentees." At the same time he lamented the "want of public spirit and due regard to future Advantage and posterity, so necessary for the good of the whole, which in some measure I conceive may be Attributed to the General Inclination of the Inhabitants, Natives as well as others, sooner or later to go home, as their fraise is, most people pleasing themselves with that thought, their present Interest is Cheefly Consider'd the better to enable the prosecution of that design."[68]

Education in England, so frequently the reason for absence from the West Indies, often bred in young men a dislike for plantation affairs and weaned them from the land of their birth. "I wish you would consider," said Governor Lawes to the legislature of Jamaica, "of making some proper Provisions to Educate our Youth at Home; which will beget in them a natural and stronger Affection for the Place they are born and bred in, and a greater Love for their native Country then their Fathers

66 Address of assembly of Barbadoes to Gov. Worsley, Oct. 5, 1727, C.O. 28: 44, f. 278.

67 Act of Jamaica, 1700, C.O. 139: 9, f. 51. The act was repealed May 2, 1700.

68 Gov. Hamilton to Board of Trade, Aug. 30, 1715, C.O. 137: 11, O 10.

ever shewed; and, perhaps, become better Benefactors to the Island then any we can yet boast of.''[69] Again he declared to the same body: ''If you will not make a Public Provision to Educate Your Children here, in Religion, Vertue, and useful Learning, nor for the Regulating the Markets, so that Poor Free People, and White Tradesmen may live among Us, the number of Planters will never multiply; though you Gentlemen may have many Plantations without Masters to inhabit them; the Fatal Consequences whereof may easily be foreseen, and I hope timely prevented, by the Prudence of this Assembly.''[70] Long estimated, in 1774, that three-fourths of the planters' children were sent to England for their education, and that probably not two out of three of these returned to the island. One reason for this, however, was that the rule of primogeniture was followed to preserve estates intact. Annuities or fortunes generally sufficient to support the younger children in England were left them. There was, therefore, no inducement for them to return to the West Indies.[71]

A great lack of councilors was complained of in 1729,[72] and in the following year Governor Hunter wrote: ''Mr. Law is in England for his health, so that I am frequently at a loss for a quorum.[73] A little later it was said of the

[69] Gov. Nicholas Lawes' speech to the council and assembly, Oct. 20, 1719, C.O. 137: 13, P 65.

[70] June 16, 1720, *ibid.*, P 86. For evidence of the wealth of Jamaica planters, most of whom resided in England, see *Abstracts of Jamaica Wills, 1625-1792*, where 312 wills are abstracted, Brit. Mus. Add. MSS. 34, 181. Among the names of absentees or merchants of Jamaica residing in England in 1716 are the following: Melworth, Sir Justus Beck, Morris, Colonel Thompson, Colonel Edlyn, Major Ayscough, Moore, Forrard, Rigby, Gregory, Broderick, Cockburne, Colonel Oldfield, Bowles, Compere, Major Hemings, Parrot, Mitchell, and Wood. C.O. 137: 22, f. 171.

[71] Long, *Jamaica*, I, 438.

[72] Memorial of Charles Delafaye to Board of Trade, received May 23, 1729, C.O. 137: 18, S 57.

[73] Gov. Hunter to Board of Trade, March 12, 1729/30, *ibid.*, S 82.

sugar colonies that "none of the Gentlemen there seek to settle their Posterity in those excessive hot Climates."[74] "Whenever any Person has made his Fortune," wrote another in 1740, "he seldom fails to transport his Family and Effects to England."[75]

Between the planters of Jamaica who resided in England and those who remained in the island there gradually developed in the eighteenth century a diversity of interests which culminated in a sharp controversy in the years from 1746 to 1752. This arose from the attempt of the Jamaica government to force absentees to return to the island by subjecting their estates to additional taxation. Thus the so-called Deficiency Act of 1718, compelling proprietors to maintain a certain number of white people in proportion to their slaves and live stock, or to pay a fine for each deficiency, required the estate of an absentee to maintain a considerably greater number of white persons than the estate of equal size owned by a resident.[76] Similar acts were passed annually and the additional fine, or tax as it came to be regarded, upon absentees amounted to from twenty to fifty per cent.[77] In 1746, the Jamaican planters residing in England presented to the government a petition praying for relief from the additional taxes, amounting to fifty per cent, levied on them by these annual acts. As a precedent for such action they cited a law of Antigua of similar character which had recently been disallowed.[78] As early as

[74] *Answers to All Objections made to the Bill for Supporting the Sugar Colonies*, [1731], p. 1 (in C.O. 5: 4, ff. 270-271).

[75] *Importance of Jamaica to Great Britain*, [1740], p. 56.

[76] West to Board of Trade, July 8, 1719, C.O. 137: 13, P 53.

[77] Same to same, July 16, 1724, *ibid.*, 14, Q 103. Residents should maintain 1 white person for every 30 slaves, and another for every 150 horses; non-residents should maintain 1 white person for every 20 slaves, and another for every 100 horses. Act of 1723.

[78] Petition to the king, Jan. 15, 1746/7. Referred by Committee of

1719 Richard West, the standing counsel to the Board of Trade, had submitted to the Board of Trade with reference to such taxation the question as to "how far this may be prejudiciall to persons residing in Great Britain and inconsistent with that equality which ought to be observ'd in the levying of Publick Taxes."[79] In the absence of any complaint, however, from the planters, the board took no action. But in 1746 the West India interest in England was well organized and the requests of its lobby were not to be denied.

The views of the residents of Jamaica were ably presented to the Board of Trade in a defense of their policy written by Governor Edward Trelawny. "I think the absentees," said he, "have misrepresented this matter; the Deficiency Bill is not a money Bill; 'tis true it raises money, but it raises it only in Consequence of the Penalties inflicted on those that do not comply with it. It were better for the Island that it was entirely complied with & not a farthing rais'd by it. As the intent of the Bill is, that a sufficient Number of white Men in proportion to our Slaves should be kept up, (an intent which ought never to be out of our thoughts), I would ask any of these absentees whether, suppose a Plantation of 320 Negroes, he doth not in his conscience think that the Master with 10 Servants is more likely to keep these Negroes quiet than 16 Servants without the Master. There is no comparison, such is the nature of Slaves, the presence of the Master is all in all. So that the absentees are not hardly treated, they are favourably treated; the addition[l] burthen that is impos'd upon them is not adequate to the inconvenience, I may say danger they bring upon the Country; & I truly think that there is hardly any evil

Council to Board of Trade, Feb. 17, 1746/7, C.O. 137: 24, W 125. The Antiguan case is discussed on pp. 38-39, note 82.

[79] July 8, 1719, C.O. 137: 13, P 53.

we labour under, but what is owing ultimately to the small number of Inhabitants; it makes the duties of Magistracies & all Commissions, Civil & Military, very hard on the Residents, & there is no Choice. If Jamaica was like England, it would be hard indeed to lay an additional burthen on a Man on acco^{t} of his absence, when it doth not signifie to the Community whether he is absent or present on his Estate, but the case is not so here.'"[80]

Three months later, Trelawny put the matter even more concretely. Spaniards had recently made depredations on the southeastern coast for the purpose of stealing negroes. In view of this the custos of the parishes of St. Thomas in the East and St. David wrote to Trelawny: "I join with you in opinion that a couple of small Crafts fitted out of these two Parishes, would be of great Service, & it's what I have thought on myself for some time past, & would readily contribute more than my quota, if such a thing could be effected; but there is a great obstacle in the way, which is, that most of the great Estates in these two Parishes, have no Proprietors in the Island & of course cannot regularly bear their proportions; another is, the great scarcity of white People." "By this one instance," said the governor, "your Ldps will judge how much all Publick-works, howe necessary soever, which are not provided for by Law . . . must suffer, by Gent^{men} of Fortune in the Island living themselves out of it; & I hope, it will appear the Island has more reason to complain of their absence, than they of the Deficiency Bill as it now stands." Hanover parish, where, he believed, all the proprietors but one, Colin Campbell, were in residence, though a much exposed parish, was entirely secured.[81]

The organization of the absentees worked quietly but

80 Gov. Trelawny to Board of Trade, Oct. 14, 1747, C.O. 137: 24, W 141.

81 *Ibid.*, Jan. 19, 1747, C.O. 137: 25, X 1.

very effectively. Their legal agent was Ferdinand John Paris, while the residents of Jamaica were represented by John Sharpe. May 1, 1747, the Board of Trade recommended an instruction against the Jamaica law. June 1, Sharpe petitioned to be heard for Jamaica, but was not given a hearing until June 25, 1748. But it was to no avail, for on June 20, 1748, an order in Council approved an instruction to Governor Trelawny directing him not to pass any act for taxing absentees in greater proportion than residents without a suspending clause. In November 1749, an address to the king from the assembly of Jamaica arrived. This set forth the fear the island was under on account of the great number of slaves in proportion to the whites, and prayed that it might be permitted to bring the former pressure to bear upon absentees to force them to return. In April 1750, the islanders were given another hearing by the Board of Trade. A final consideration of the matter was made by the board on May 5, 1752, both Paris and Sharpe being present, when it was decided to leave the absentees undisturbed in their triumph over the resident planters.[82]

[82] *Acts of the Privy Council, Col.*, IV, 40; C.O. 137: 25, X 17, 41, 44; C.O. 391: 59: May 1, 1747, June 25, 1748, April, 1750, May 5, 1752. On the last date, Sharpe called attention to a slave rebellion that had recently broken out on the estate of an absentee. The money from the additional tax on absentees was to go toward maintaining the royal regiment in the island.

Antigua passed acts levying double taxes on estates of absentees in 1740, 1741, and 1742. The absentees petitioned the king for relief in 1742. Some of them had resided in Great Britain from their birth to care for their affairs and estates there. The petitioners were W. Codrington, Samuel Martin, John Duer, Thomas Kerby, S. Bethell, Edward Otto Bayer, Walter Lydserfe, W. Smith, William Farquier, and John Johnson, C.O. 152: 24, Y 52. Gov. Mathew of the Leeward Islands wrote the Board of Trade that the island had instructed its agent to defend its policy, and he stated that similar acts were passed in Antigua June 1, 1706, May 24, 1707, June 24, 1711, and March 30, 1714, to which no objections were made. Also a similar law had been passed in St. Christopher April 15, 1743, *ibid.*, 24, Y 61. The additional tax in Antigua was 8 shillings per slave on estates of

For that degraded state of society in the West Indies of the eighteenth century, characterized as it was by cruelty, slave rebellions, or the constant fear of them, industrial waste, political and social corruption, and destitute of practically all the finer values of life,—where shall the responsibility be placed? Must not a large share rest upon that West India aristocracy in England, which, grasping its profits, did little or nothing to ameliorate the society that created them?

West Indians living in London were often conspicuous for their wealth, social position, and political power.[83]

absentees, to be used toward supporting a militia company from the Virgin Islands. The opposition of the governor's council was overcome by the assembly. Mathew to Board of Trade, June 13, 1744, *ibid.*, 24, Y 71, 73. The standing counsel, Francis Fane, recommended a disallowance of the additional tax on the ground of no precedent and violation of personal liberty. "I think," said he, "a Man may live where he pleases within his Majesty's Dominions." Also he thought absentees could recover from the treasurer of Antigua money already paid in. Letter to the Board of Trade, Nov. 26, 1744, *ibid.*, 24, Y 73. The policy of the resident planters was defended in London by Sharpe and Douglass, colonial agents for Antigua and St. Christopher respectively. Fane to Board of Trade, Feb. 27, 1744, *ibid.*, 25, Y 75. James George Douglass, in defending the double tax acts of St. Christopher (passed July 12, and Sept. 5, 1744), stated that absentees of that island were "very numerous, one-half of the property of the whole Island being owned by Absentees." To Board of Trade, received March 19, 1745/6, *ibid.*, 25, Y 76, 164. The government finally disposed of the matter by an order in Council, March 5, 1744/5, approving an instruction to the governor of Antigua not to pass any act whereby the lands, negroes, or other property of any absentee, who is an English subject, should be higher taxed than that of the rest of the British subjects resident in Antigua, without a suspending clause. *Ibid.*, 25, Y 133. This at first caused great discontent in Antigua. Gov. Mathew to Board of Trade, July 6, 1745, *ibid.*, 25, Y. 139. A similar order in Council, July 23, 1746, prevented the passage of additional taxes on absentees in St. Christopher or any of the Leeward Islands. *Ibid.*, 25, Y 164. On September 19, 1746, Governor Mathew was able to report of Antigua that "The Assembly here are come, sometime since, to a Right way of thinking as to omitting an Additional Tax on Absentees." To Board of Trade, *ibid.*, 25.

83 Lord John Sackville, writing to his brother Lord George, June 5, 1754, speaks of a certain gentleman who "lives in Cavendish Square, and is very

It was revealed in 1740 that absentee planters of Jamaica were in the habit of holding meetings at taverns and coffee houses, supposedly in the interest of the island, and charged the expenses of such meetings to the colony.[84] After the middle of the century, these men frequently knew nothing by personal experience of the islands, having inherited their property there from the former generation. A writer in 1764 said of them: "The gentlemen of the West Indies, who have large estates there, and reside in England, have never acquired those estates by their own industry. I believe this will be readily granted me, by those who know them. Their estates have been raised by the hardship, sweat, and toil of their forefathers, among few capable competitors, in the infancy of colonies."[85] Comparing them with the Virginians, Adam Smith said: "Our tobacco colonies send us home no such wealthy planters as we see frequently arrive from our sugar islands."[86] A traveler at the end of the century, while impressed with the wealth of Jamaica, was surprised at the absence of most all its owners. "Considering this amount of property, and the many eminent and wealthy families of Jamaica to be met with in every circle of fashion where the English language prevails, it is naturally a matter of some surprise to a stranger, who visits the island, to find it almost totally deserted by its principal inhabitants. In one of the northern and richest districts, it is said that of eighty proprietors not three are to be found at this time on the spot. Their plantations continue under the management of agents, who, to im-

rich, most of his estate lying in Jamaica." *Royal Hist. MSS. Com. Report on MSS. of Mrs. Stopford-Sackville*, 1904, I, 40.

84 This appeared from the accounts of Lovel Stanhope, the colonial agent. Minutes of Jamaica Council, Dec. 18, 1760, C.O. 137: 32, Bb. 13, 14.

85 Anon., *Some Observations . . . of our New West-India Colonies*, London, 1764, p. 22.

86 *Wealth of Nations*, Book I, Ch. xi, Pt. 1, ed. Canon, I, 159.

prove their circumstances, submit to a temporary banishment from their homes; whilst all the wealth of the soil is transported and consumed in remote countries.'"[87] To recur to a simile of Seeley, this indeed was a wilderness of materialism, where little was consecrated or ideal. The gods of its people were in the home land, and they found little courage to stand out as state-builders. To them the dearest of all names, traditions, and memories was England.

[87] Daniel McKinnen, *Tour through the British West Indies in the Years 1802 and 1803*, London, 1804, p. 108.

CHAPTER II

WHITE LABOR IN THE SUGAR ISLANDS

The most striking feature, perhaps, of the British sugar trade in the eighteenth century was the inadequacy of supply to meet the effective demand of the empire. One of the contributing causes of this phenomenon was the deficiency of the labor supply. In the first half of the seventeenth century there was a prevalent impression that England was over-populated, and opinion favored emigration to the new world as a cure for this condition.[1] From about the time of the Restoration, however, the belief as to population was just reversed, and, indeed, until near the early nineteenth century, it was generally supposed that England was under-populated. Roger Coke wrote in 1670: "Ireland and our Plantations, do in proportion to England more exhaust it of men, than the West Indies do Spain."[2] "Ireland and our Plantations Rob us of all the growing Youth and Industry of the Nation, whereby it becomes weak and feeble, and the strength as well as Trade, becomes decayed and diminished."[3] Nevertheless a few writers distinguished between the effects of emigration to colonies producing the same products as England and the result of migration to the tropics. Thus Josiah Child, writing in 1668, observed that "The People that evacuate from us to Barbadoes,

[1] G. L. Beer, *Political Science Quarterly*, March, 1908, XXIII, 83 *et seq.*, and authorities there cited.

[2] Roger Coke, *A Discourse of Trade*, London, 1670, pp. 12-13.

[3] *Ibid.*, 46.

and the other West India Plantations, . . . do commonly work one English-man to ten or eight Blacks; and if we keep the Trade of our said Plantations intirely to England, England would have no less Inhabitants, but rather an encrease of People by such Evacuation, because that one Englishman, with the Blacks that work with him, accounting what they eat, use and wear, would make employment for Four Men in England . . . ; whereas peradventure of ten men that issue from us to New England and Ireland, what we send to or receive from them, doth not employ one man in England.'"[4] The agent for Barbadoes, in 1688, wrote: " 'Tis strange we should be thought to diminish the People of England, when we do so much increase the Employments—England is a well or spring inexhausted, which hath never the less water in it, for having some drawn from it.'"[5] Davenant in the same year estimated that the population of England since 1600 had increased 900,000, which, he thought, could not have been if the plantations were such drains of the people.[6] Cary, in 1695, voiced the popular impression in telling us that many thoughtful men were doubtful of the advantage of colonies to England because "Inhabitants being the Wealth of a Nation, by how much they are lessened, by so much we are poorer than when we first began to settle our Foreign Colonies." But Cary himself believed that productive colonies, instead of decreasing population, really increased its supporting power.[7] The population of the kingdom at that time he estimated at eight millions.[8]

[4] Josiah Child, *A New Discourse of Trade,* London, 1668, ed. London, 1718, pp. 215-216.

[5] Littleton, *Groans of the Plantations,* London, 1689, pp. 29-30.

[6] Charles Davenant, *Collected Works,* ed. Chas. Whitworth, London, 1771, II, 96.

[7] John Cary, *Essay on Trade,* London, 1695, pp. 65-66.

[8] *Ibid.,* 163.

During the eighteenth century the belief continued that the population was too small and care should be taken to preserve it. But the advantages of colonies were more widely recognized, and writers were less and less disposed to see in them a cause of weakness to the homeland. To some extent this change of opinion must, of course, have benefited the sugar colonies. A writer on commerce in 1721 estimated the population of England at seven millions.[9] He added, what must have been the feeling of the landed class, that all the inhabitants should be confined to England "since if any Numbers of them should leave the Kingdom, this could not chuse but make a great abatement in the 42 Millions that are now yearly paid to the Rents of our Lands, and the Labour of our People."[10] A writer in 1763 stated that at that period many manufactories in England were idle for the want of hands, which he thought was only partly due to the demand for soldiers.[11] Campbell, writing in the same year, said that "The old objection, which from an appearance of truth had some degree of weight before this subject was fully understood, that people going to our plantations weakened the mother country, is now, from our better acquaintance with the subject, incontestably obviated."[12] The wants of the colonists diversified the industries of England, which enabled her to support a larger population.[13] The older view, however, was not entirely abandoned. Pulteney remarked in the House of Commons, April 29, 1772, "It has always been, and I hope now will be, the care of the legislature to invite

9 Charles King, *British Merchant*, London, 1721, I, 165.

10 *Ibid.*, 170.

11 Anon., *Propositions for Improving the Manufactures*, etc., London, 1763, p. 20.

12 John Campbell, *Candid and Impartial Considerations on . . . the Sugar Trade*, London, 1763, p. 28.

13 *Ibid.*, 29.

foreigners, in order to supply the waste of people produced by emigrations to our widely extended colonies.'"[14]

From these passages it is apparent that the labor market of England was by no means in a favorable condition to satisfy the requirements of the West India sugar planters. In the first decade of sugar culture in Barbadoes, Ligon[15] showed clearly that white indentured servants from England worked side by side in the fields with the negroes. He further added that they were subjected to harder labor and treatment than the slaves, inasmuch as their five-year term of service was so much shorter. He omitted entirely any consideration of their adaptability for tropical field labor but stated the number of negroes to be over twice that of the Christians. During the Civil War and under the Commonwealth and Protectorate, Barbadoes and Jamaica received a supply of such servants in the form of royalist adherents, Irish rebels, and prisoners, and the practice of transporting prisoners increased under Charles II and James II.[16] A visitor to Barbadoes, in 1655, wrote that "the custom of all merchants trading thither is to bring as many men and women as they can. No sooner doth a ship come to an anchor than presently the Islanders go aboard her, enquiring what servants they can buy, if they are above

14 *Parliamentary History*, XVII, 484.

15 Richard Ligon, *True and exact history of the island of Barbadoes*, etc., London, 1657, pp. 44-45. Ligon's estimate of the population of Barbadoes, in 1647-1650, of about 50,000 whites and over 100,000 blacks appears to be a gross exaggeration.

16 *Royal Hist. MSS. Com. Reports*, VI, 106 b, 108 a; VII, 117 a; VIII, i, 500 a, 572 a, 514 b; IX, ii, 126 a, b, 154 a; X, Pt. 5, pp. 85, 25, 35, 47, 73, 91, 94; XII, Pt. 7, p. 314, mentioning 860 followers of Monmouth sold as servants to Barbadoes; *C. S. P. Dom. 1649-1650*, p. 384; *1651-1652*, p. 432; *1652-1653*, p. 504; *cf.* 232, 546 ("passengers" for Barbadoes); *1654*, p. 305 (Pirates to be transported); p. 333; *1655*, pp. 62; 108; 329; *1659-1660*, p. 504 (Pirates from Cork gaol); *1660-1661*, pp. 28, 58, 87, 221, 238, 320; *1661-1662*, pp. 16, 196, 353, 461-462, 502; *1663-1664*, pp. 175 a, 229, 230, 537; *1664-1665*, p. 491.

seventeen years of age they serve but four years, according to the law of the Island, but if under 17, then left to the discretion of the merchant, as he can agree with the planter; these servants planteth, weedeth, and manureth their ground all by hand, in which lieth their estates.'"[17]

Kidnapping and forcible transportation to the sugar islands was no infrequent occurrence in the seventeenth century.[18] This practice appears to have been most widely carried on at Bristol prior to the Revolution of 1688. The Lord Keeper Guilford, who visited Bristol in 1680, found the mayor and some of the justices and alderman in the habit of supplying the West India merchants with petty criminals at so much per head. Guilford censured the mayor and aldermen and prosecutions against them pended till the Revolution, after which the abuse seems to have disappeared.[19] Josiah Child's characterization of such white servants as the colonies obtained appears to be accurate: "Virginia and Barbadoes," he said, "were first peopled by a sort of loose vagrante People, vicious and destitute of means to live at home (being either unfit for Labour, or such as could find none to employ themselves about, or had so misbehaved themselves by Whoreing, Theiving, or other Debauchery, that none would set them on work) which Merchants and Masters of Ships by their Agents (or Spirits, as they were called) gathered up about the Streets of London, and other places, cloathed and transported to be employed upon Plantations; and these, I

[17] Francis Barrington to Sir John Barrington, June 6, 1655, *Hist. MSS. Com. Rep.*, VII, 572 a.

[18] *C. S. P. Dom., 1652-1653*, p. 542; *1660-1661*, pp. 391-392; *1664-1665*, p. 164 (the case of three Hertford Quakers); *1664-1665*, p. 140; *Hist. MSS. Com. Rep.*, VII, 372 a (a case in 1682).

[19] *The Lives of the Right Hon. Francis North, Baron Guilford, Lord Keeper of the Great Seal*, etc., by Hon. Roger North, 2 vols., London, 1740, 1742. New ed., 3 vols., London, 1826, II, 24.

say, were such as, had there been no English Foreign Plantation in the World, could probably never have lived at home to do service for their Country, but must have come to be hanged or starved, or dyed untimely of some of those miserable Diseases, that proceed from want and vice; or else have sold themselves for Soldiers, to be knocked on the Head, or starved, in the Quarrels of our Neighbours, . . . or else, if they could, by begging or otherwise, arrive to the Stock of 2s. 6d. to waft them over to Holland, become servants to the Dutch, who refuse none.'"[20]

While the planters themselves looked upon such white men as hopeless for their purposes, the belief survived for a generation, at least, that white men would be used in the cultivation of sugar. A Jamaica planter, in 1665, wrote home that the island would take all the negroes of the African Company if they could be had on eighteen months' credit. But why, he added, should not his Majesty send out one family from each parish, not your convict gaol-birds or riotous persons, rotten before they are sent forth, and at best idle and only fit for the mines, but such well-disposed Englishmen as would come, not as indentured servants, but for meat, drink, and wages, as in England until they could make provision for themselves? These were the conditions he had authorized his son to offer.[21]

No supply of whites, however, sufficiently numerous or efficient was forthcoming. By 1665, the ratio of negroes to whites in Barbadoes was reported at six to one.[22] In

[20] Josiah Child, *New Discourse of Trade,* 1668, ed. London, 1718, pp. 183-184.

[21] John Style to Sec. Lord Arlington, July 24, 1665. *C. S. P. Am. & W. I., 1661-1668,* no. 1023. A letter from Barbadoes, in 1667, asked for an open trade to Guinea, and also a larger importation of white servants, especially Scotch. Gov. Willoughby to king, July 1667, *ibid.,* no. 1539.

[22] Thomas Newton's petition to Chas. II, May 15, 1665. *Ibid.,* no. 989.

1668, Child estimated it at eight or ten negroes to one white servant.[23] In 1712, there were 3537 white men to 41,970 negro slaves, a ratio of twelve to one, the total of white men, women and children being 12,528.[24] Jamaica, after 1670, showed a similarly increasing ratio. The same was true of the Leeward Islands.[25] Two works dealing at length with the cultivation of sugar in 1689 and 1690 show that the planters believed the industry at that time depended absolutely upon slave labor.[26] Indeed, from this time through the eighteenth century the West India planters employed in the production of sugar practically no white men, except as overseers.

An increasing need of white people was no less keenly felt, however. The two leading reasons for their immigration were, (1) that they might serve as a military protection against invasion and insurrection, and (2) that by raising garden truck and live stock on the lands less favorable for sugar they might render slaveholders less dependent on outside sources of supply. So small was the population of white men able to bear arms that,

[23] Josiah Child, *New Discourse on Trade*, p. 190.

[24] Lowther to Board of Trade, Aug. 16, 1712, C.O. 28: 14, T 15.

[25] Estimated Population of

Jamaica:

	White	Black
1658	4500	1400
1660	4500	4000
1665	7700	9598
1670	8000	8000
1673	7764	9504
1722	7100	80000

Long, *Jamaica*, I, 376; *Jamaica Handbook*, 1899, p. 34.

Antigua:

	White	Black
1711	2854	11838

C.O. 152: 10, O 15.

St. Christopher:

	White men	Negro slaves
1717	799	7973
1720	2800	7321

of which 808—men
C.O. 152: 13, Q 46.

Nevis:

	White	Black
1720	1343	5689

(364 men)
C.O. 152: 13, Q 46.

Montserrat:

	White men	Black
1719	286	4192

C.O. 152: 13, Q 40.

[26] *Groans of the Plantations*, London, 1689, pp. 6 *et seq.*; Dalby Thomas, *Historical Account of the Rise and Growth of the West India Colonies*, etc., London, 1690, in *Harl. Misc.*, II, 357 *et seq.*

on several occasions, slaves were armed for military service. At a council of war in Jamaica, in 1712, it was resolved that one negro or mulatto slave to every four white men in each regiment be armed, besides those that attended their masters, and that the number of pioneers and baggage negroes be left to the discretion of the respective colonels or commanding officers of each regiment.[27] Again, in 1719, it was agreed to send an officer to all the plantations in St. Elizabeth and Westmoreland parishes "to Inroll the Negros and to take care they are furnished with proper Arms and understand the use of them."[28] At Antigua, in 1720, negroes were commonly employed on the forts and fortifications.[29] An act of Jamaica, in 1739, provided for the purchase for the Crown of negroes from each parish to be employed for building and repairing barracks and cutting roads.[30] The same island, in 1741, ordered a levy of slaves for military service.[31] For the same purpose at Jamaica, in 1748, Governor Trelawny proposed that the Crown purchase five or six hundred slaves which he believed the island would maintain.[32] In 1759, negroes were enlisted in the four principal Leeward Islands and used in the capture of Guadeloupe.[33] In 1712, Governor Hamilton of Jamaica wrote that the militia on the island amounted to only about 2700 infantry and horse, and that there was a probability of a negro insurrection unless some of the regular forces were allowed to remain after the peace.[34]

27 Copy of Proceedings of a council of war, Sept. 1, 1712, C.O. 137: 10, N 30.

28 Minutes of council of war, June 29, 1719, C.O. 137: 13, P 77.

29 Act of Dec. 19, 1720, C.O. 152: 13, Q 73.

30 Acts, Jamaica, C.O. 139: 15, no. 21.

31 *Ibid.*, no. 46.

32 Trelawny to Board of Trade, Mar. 12, 1747/8, C.O. 137: 25, X 3.

33 Gov. George Thomas to Board of Trade, July 13, 1759, C.O. 152: 29, Cc 43.

34 Hamilton to Board of Trade, Oct. 10, 1712, C.O. 137: 10, N 19.

On November 14, 1715, he wrote: "Should these companies be recalled or broke, I shall not think my self safe where I am from the Negroes."[35] Accordingly, it was decided that the governor of Jamaica might retain a detachment of the royal army indefinitely, or until there were enough white people to defend the colony against foreign enemies and negro slaves.[36]

The principal sugar islands sought by legislative experiments to remedy this deficiency of white population. In 1703, Jamaica passed "An Act to Encourage the Importation of White Men." By this measure every planter was obliged to keep one white man for his first ten negroes young and old, and two for the first twenty, and one for every twenty thereafter; in like manner for the first sixty head of cattle, horses, etc., one white man, and one for every one hundred after the first sixty; the receiver general was to pay the passages of the men and they were to be placed with the planters according to their deficiencies or pay for each deficiency a fine supposed to be equivalent to the cost of a servant's maintenance. For each servant so placed the planter was to reimburse the receiver general, who was to apply that money towards paying the passages of others, thus creating a perpetual fund. Though £500 was appropriated to start this fund, the Jamaica council reported in 1716, that they "Don't find it was ever apply'd," and they computed that, if fines on the existing deficiencies were collected, at least two thousand servants might be provided for. By the same act every boat, wherry, or canoe was obliged to have a white man or forfeit 40s. per month, which, in 1716, it was estimated, would have provided for at least 200 servants more.[37]

35 Hamilton to Stanhope, Nov. 14, 1715, C.O. 137: 11, O 2.

36 Instructions to Gov. Pitt, 1716, Arts. 4-7, C.O. 137: 12, O 106.

37 Acts of Jamaica assembly; Memorial from Jamaica council to Board of Trade, Mar. 13, 1715/6, C.O. 137: 11, O 47.

A supplementary act of 1712 obliged vessels trading about the island, hackney coaches, and wains to employ white people, but this measure, like the preceding, went unenforced.[38] In neither of these acts was there any provision to encourage the introduction of women.

The Deficiency Law of 1720 radically modified the old act and became the model for similar annual bills during the eighteenth century. By this law,[39] every person had to keep a white man for every thirty slaves, young or old, he or she possessed, and also one white servant for every hundred and fifty horses, mares, mules, asses, or neat cattle, and so in proportion for a greater or lesser number. In case a planter was deficient in the number of white men he had to pay 7s. 6d. per week. To encourage the keeping of supernumerary white servants on plantations the master of such supernumerary servants should receive out of the money arising by virtue of this act the sum of 7s. 6d. for every supernumerary white servant in proportion to the aforesaid number of slaves and cattle. All masters of wherries and canoes that had not one white man plying in them should pay 7s. 6d. per week. All vessels trading about the island for freight should have not over one-fourth part of the crew slaves under penalty of £20. Masters or mistresses of families not actually residing on the island were to be allowed

[38] Acts of Jamaica assembly, C.O. 137: 11, O 47. Gov. Hamilton wrote the Board of Trade, April 26, 1715: "I am sorry to find that instead of our increasing in People, There is hardly one parish in the Island that is not weaker in Men fitt to bear Arms, then before the Peace, The true Cause of which proceeds in great measure from the Late Assembly's Letting fall the Deficiency Act, . . . and their not making any other necessary provision for the Incouragement of Familys to come, reside and settle with them." C.O. 137: 10, N 139. It should be said, however, that their attempted repeal of the Act of 1703 was disallowed. Gov. Nicholas Lawes to Board of Trade, Sept. 1, 1718, C.O. 137: 13, P 19.

[39] Acts of Jamaica assembly; Lawes to Board of Trade, Feb. 2, 1719, C.O. 137: 13, P 72.

only twenty-four negroes, young and old, to a deficiency. The total yearly fine for a deficiency under the act of 1703 was £13 10s. This was so much below the actual cost of maintaining a servant that the majority of planters preferred paying the fine to keeping the man, "By which practice the End and design of the Acts which was to people the Island were defeated."[40] This act, therefore, raised the fine to £19 10s. computed to be adequate to the subsistence of a man. But Lawes tells us that the demand created by this act at once raised the price of white people. The increased penalty for absentees was thought reasonable inasmuch as the master's presence on an estate generally tended to restrain the slaves from insurrection.[41] Similar acts, or renewals of this measure, were passed annually. From the revenue thus arising it was customary to provide for the two companies of the royal army stationed in the island.[42] In 1730, by an alteration in the Deficiency Law, white women, white boys and girls over fourteen years of age, servants, might stand for deficiencies, and the fine for one deficiency was increased to £6 10s. per quarter, or £26 per year.[43] It was believed that the planters would rather choose to keep their complement of white people than to pay so high a tax.[44] By the act of 1735, the tax on absentees, except children at school, men over sixty years, or mentally unsound persons, was increased by requiring them to provide one white person to every twenty slaves or 100 horses, mules, asses, and neat cattle. Also tavernkeepers, retailers of

[40] Lawes to Board of Trade, Feb. 2, 1719, C.O. 137: 13, P 72.

[41] C.O. 137: 14, Q 56; Q 101; Q 50; *ibid.*, 17, S 21; S 42; *ibid.*, 18, S 58; S 73.

[42] Gov. Robert Hunter to Board of Trade, May 10, 1730, C.O. 137: 28, S 102.

[43] Hunter to Board of Trade, May 10, 1730, C.O. 137: 18, S 102; *ibid.*, 137: 19, S 120; Att'y Gen'l's approval, Mar. 22, 1731/2.

[44] *Ibid.*, Apr. 21, 1731, C.O. 137: 19, S 128.

rum, sugar, or rum punch, or keepers of billiard tables in or within six miles of any town should respectively keep one white man such as the commander of the militia should approve as able to bear arms under penalty of £6 10s. a quarter, or £26 a year.[45]

The Deficiency Act, originally for the purpose of increasing the white population, had, by 1736, been transformed completely into a revenue measure, and during the remainder of the century it preserved essentially this character.[46] In 1736, it was estimated there were in Jamaica 90,000 negroes and 3000 whites able to bear arms, most of the latter being hired or indentured servants of which only 1000 were kept by planters, and would be so kept anyway. A deficiency, therefore, of 2000, if fully paid for at the rate of £26, would far exceed the amounts actually received.[47] In England, in 1749, the act was regarded as a money bill, though Governor Trelawny, at that time, still had hopes that it might

[45] Act passed May 3, 1735, C.O. 137: 22, V 22: Deputy Receiver General's Accounts.

[46] The revenue produced by the deficiency tax was in the following years as follows:

Year	Amount	Year	Amount
1721	£ 150	1745	£14,598
1722	450	1746	15,363
1723	2,640: 13: 5½	1747	16,029
1725	7,829: 4: 11½	1748	14,851
1726	8,423: 15: 0½	1749	9,606
1727	——	1750	9,303
1728	9,718: 18: 7¼	1751	6,310.
1729	10,126: 19: 11¾		
1730	8,455: 7: 1¾		
1731	6,012: 8: 10¼		
1732	4,624		
1733	4,124		
1734	3,330		

Deputy Receiver General's Accounts, C.O. 137: 16, R 61; *ibid.*, 22, V 22; *ibid.*, 27, X 202.

[47] John Gregory, president of Jamaica council, to Board of Trade, Apr. 10, 1736, C.O. 137: 22, V 22.

answer its original purpose.[48] The average annual revenue produced from 1750 to 1754 was £7500,[49] and it was then classed among the revenue bills.[50]

As a means of increasing the labor supply the deficiency laws were clearly of no use whatever, and, as a military expedient, their utility is very doubtful. The indentured servants were described as "Artificers and Labourers that come from Europe, that . . . soon grow lazy and Indolent"; they were "Runagadoes and a loose Sort of People."[51] They were mostly Irish Catholics, and Governor Robert Hunter thought that, as they had no property and were Romanists, they might prove of disservice rather than of use in a war with France or Spain.[52] He wrote of them as a danger to the island "from pouring in upon us in such Sholes, as they have done of late years," and that they were "a lazy useless sort of people who come cheap and serve for deficiencies, and their hearts are not with us," and he thought their introduction should be restrained.[53] The cost of hiring and victualing indentured servants, in 1736, was computed at £30 a year per man, and the cost of bringing them from England was at least £15 sterling each. Besides, there was the risk of their dying soon after landing or deserting their masters, which they frequently did, and returning to England.[54] None of the revenue arising from the Deficiency Act ever appears to have been spent in the purchase of servants; up to 1737, none were imported but what appear to have sold for more

[48] Trelawny's address to council and assembly of Jamaica, Oct. 3, 1749, C.O. 137: 25, X 46.

[49] Gov. Knowles to Board of Trade, Jan. 12, 1754, C.O. 137: 27, X 200.

[50] C.O. 137: 27, X 198.

[51] Lawes to Board of Trade, Dec. 6, 1719, C.O. 137: 13, P 62.

[52] Hunter to Board of Trade, Dec. 24, 1730, C.O. 137: 19, S 124.

[53] *Ibid.*, Nov. 13, 1731, C.O. 137: 19, S 146.

[54] John Gregory to Board of Trade, Apr. 10, 1736, C.O. 137: 22, V 22.

than the receiver general was impowered to give.[55] Governor Trelawny, writing, in 1741, of the effects of the acts, said: "The Deficiency Laws have not only raised mony for the public purposes but likewise caused a greater demand for white Servants: who tho' they are generally idle bad Men, yet several of them have after serving their time proved industrious & are at present Owners of Good Plantations." Further he added: "The Deficiency Laws have obliged the Planters to employ more white People than they wanted, & occasioned servants to turn Planters."[56]

With convict labor, Jamaica also experimented without success. Six or more shiploads of Scottish Jacobites were sent to the West Indies in 1716.[57] Jamaica, in the same year, made a bid for its share of such prisoners.[58] An act of parliament of 1717 provided that the English courts might contract for the transportation of their convicts whose labor should be sold in the colonies for seven or fourteen years according to the offense. Also persons convicted of piracy were subject to transportation.[59] The governor of Jamaica thus described the sort of people brought in by this act: "Several People have been lately sent over out of the Gaols in England, upon the Encouragement of An Act of Parliament pass'd the last Sessions, . . . Those people have been so farr from altering their Evil Courses and way of living and becoming an Advantage to Us, that the greatest part of them are gone and have Induced others to go with them a

[55] The Jamaica council's reasons for passing the Deficiency Bill of 1737, Feb. 19, 1736/7, C.O. 137: 22, V 47.

[56] Trelawny to Board of Trade, Nov. 21, 1741, C.O. 137: 23, W 47; W 49.

[57] *Hist. MSS. Com. Rep.* "Stuart Pap.," II, 453: Tho. Dahnahoy to the Duke of Mar, Sept, 9, 20, 1716; Andrew Ramsay to Walkingshaw of Barrowfield, Dec. 10, 1716. *Ibid.*, III, 304-305.

[58] Memorial of Jamaica council, Mar. 13, 1715/6, C.O. 137: 11, O 47.

[59] 4 Geo. I, c. xi.

Pyrating and have Inveglied and Encouraged Severall Negroes to desert from their Masters and go to the Spaniards in Cuba, the few that remains proves a wicked Lazy and Indolent people, so that I could heartily wish this Country might be troubled with no more of them.'"[60]

An act of 1723 laid an import tax of ten pounds per head on all convicts brought into Jamaica, which, however, was opposed in England as not being in conformity with 4 George I. c. xi.[61] Finally, Jamaica levied an impost on convicts of £100 per head, which, wrote Hunter, "seem'd to the body of the people here absolutely Necessary to prevent such from coming from the Northward, for none are Imported directly from England hither, and our Trading people in our Towns who used to Sleep with their doors open heretofore are since the arrival of some of these Convicts here oblig'd to keep Watches on their Counting and Store houses," several robberies having recently been committed.[62] Hunter was instructed, in 1731, however, not to pass for the future any act laying a duty on convicts.[63] The impost was at once repealed. From the experience of Jamaica, it is evident that prison labor was of no assistance to the West India sugar planters.

Experiments with such white labor as was available convinced the planters that it was more expensive than slave labor. From the lists we have of plantations with the number of white servants and negroes, it is evident that sugar could be produced with a less number of white men than in the proportion of one to thirty slaves.[64]

[60] Lawes to Board of Trade, Sept. 1, 1718, C.O. 137: 13, P 19.

[61] Richard West to Board of Trade, undated, C.O. 137: 14, Q 103.

[62] Hunter to Board of Trade, Nov. 13, 1731, C.O. 137: 19, S 146.

[63] Order in Council, Dec. 9, 1731, C.O. 137: 20, S 168.

[64] See the table of whites and blacks on each estate in St. Andrew's Parish, Jamaica, in 1753, C.O. 137: 28, Y 46 (6 folios). The same is shown in an elaborate table for Montserrat in 1730, C.O. 152: 18. Moseley, who

In the Leeward Islands, legislation to encourage the importation of white servants was passed in 1716.[65] The military situation here was even more critical than in Jamaica. Antigua had, in 1722, a militia of 1400, St. Christopher 1100, Nevis 250, and Montserrat not over 200.[66] All the most wealthy and best educated of St. Christopher resided in England; the rest were mainly Irish Catholics disaffected to English government.[67] In Antigua, the principal island in the group, the white servant act was not only ineffectual but, in 1724, there was reported an actual decrease in the number of white people.[68] The total white population of the Leeward Islands from 1724 to 1734 is reported to have decreased from 12,420 to 10,262.[69] Governor Mathew, writing in 1734, said that the servant acts had never been obeyed nor the penalties insisted upon. "Besides," wrote Mathew, "these Laws hitherto procure Such unwilling, Worthless, Idle Vagabonds, as from whom but Little Service can be hop'd for, on Military Emergencys. Most of these Serve for a Term of Years, without wages, poorly Cladd, hard fedd, a worse State here than a Common Soldier, if Possible. Are these the Men that are to Die in Our Defence? My Lords, we must have a recruit

was familiar with sugar culture, said that the production of sugar was simple enough to be carried on entirely by slaves; the planters preferred paying the deficiency tax in his time, as white servants were expensive and a less number than the legal proportion was sufficient for the purpose of making sugar. Benjamin Moseley, *Essay on Coffee* (ed. 1785), pp. x-xi.

[65] Gov. Walter Hamilton to Board of Trade, July 15, 1717, C.O. 152: 12, P 45, 46.

[66] Gov. John Hart to Board of Trade, July 11, 1722, C.O. 152: 14, R 43.

[67] Hart to Board of Trade, Jan. 20, 1722/3, C.O. 152: 14, R 57.

[68] Address of the lieut. gen. and council to Gov. Hart, Nov. 2, 1722, C.O. 152: 14, R 48; address of assembly to Gov. Hart, Feb. 6, 1723/4, *ibid.*, R 49; address of Gov. Hart to council and assembly of Antigua, Mar. 21, 1724/5, C.O. 152: 15, R 129.

[69] Representation of Board of Trade, Jan. 14, 1734, Brit. Mus. Add. MSS., no. 35, 907, f. 64.

of a better Sort, or better none at all''; he suggested that more regular troops be sent to the islands.[70] In the following years reports from the Leeward Islands indicate that the white population either remained stationary or actually declined. This may be seen from the following table.[71]

Antigua			St. Christopher			Nevis			Montserrat		
Year	White	Black	Year	White	Black	Year	White	Black	Year	White	B'lk
1723	5200	19800	1720	2800	7321	1720	1343	5689	1724	1000	4400
1729	4088		1724	4000	11500	1724	1100	6000	1729	1053	5855
1744	4955	27892	1729	3697	14663	1729	1296	5648	1744	1528	5945
1753	3261		1734	3881	17335	1744	1166	6511	1755	1430	8853
1756	3435	31428	1744	3222	19174	1755	1118	8380			
			1755	2713	21891						

Long stated, in 1774, that the demand for indentured British servants was then almost discontinued.[72]

Not only was agriculture left to negro labor, but, in an increasing degree, all trades and handicrafts were carried on by slaves, and the belief was prevalent that their presence in these vocations retarded the immigration of respectable whites. Many persons in Jamaica, some of them widows and orphans, derived their whole income from hiring out their slaves as carpenters, masons, etc. The Jamaica council, in 1716, thought some restrictions should be imposed on bringing up slaves to trades.[73] The variety of occupations for which slaves were employed is indicated in the newspaper advertise-

[70] Gov. William Mathew to Board of Trade, Aug. 31, 1734, C.O. 152: 20, V 46.

[71] C.O. 152: 13, Q 46; *ibid.*, 152: 14, R 101; *ibid.*, 152: 18, T 57, 58; *ibid.*, 152: 20, Y 46; *ibid.*, 152: 25, Y 154; *ibid.*, 152: 28, Bb 77, Bb 81.

[72] Long, *Jamaica*, I, 511.

[73] Memorial of Jamaica council to Board of Trade, Mar. 13, 1715/6, C.O. 137: 11, O 47.

ments of sales and runaways.[74] Lawes thought it a great evil that owners of slaves permitted them to ramble throughout the island provided they paid their master weekly or monthly sums agreed on; this liberty not only afforded them opportunity of plotting insurrections, but encouraged thieving to enable them to pay their owners.[75] Richard Harris declared, in 1732, that even if white people went to Jamaica there was no employment for them.[76] Gregory, in 1737, urged the assembly to legislate against this "evil" but "They didn't relish the idea."[77]

The same situation prevailed in Antigua. In a report of the justices on the great negro conspiracy of 1736, they regarded the admission of slaves into the "Occupations truly proper only for Freemen" as an underlying cause for insurrection. They recommended that slaves be disabled from being tradesmen, overseers, drivers, distillers, shopkeepers, hawkers, peddlers, sailors, fiddlers for gain, or from keeping horses, or working out for themselves.[78] Tomboy, a leading conspirator, was a master mason, and was permitted by his master, Thomas Hanson, to take negro apprentices and to make what profit he could of his own and their labor. He paid his master only a monthly consideration and devoted his surplus earnings to the interests of the conspiracy.[79] Of the participants, six were gibbeted, five broken on the

[74] See, *e.g.*, *Jamaica Weekly Courant*, XLVII, Apr. 15, 1719, C.O. 137: 13, P 50.

[75] Lawes' speech to Jamaica council and assembly, Oct. 20, 1719, C.O. 137: 13, P 65; Lawes to Board of Trade, Mar. 31, 1720, C.O. 137: 13, P 84.

[76] Richard Harris to Board of Trade, July 26, 1732, C.O. 137: 20, S 163. Harris was an English slave trader whom the Board of Trade frequently interrogated on West India affairs.

[77] John Gregory, president of the Jamaica council, to Board of Trade, July 22, 1737, C.O. 137: 22, V 51.

[78] Report of John Vernon and others, Antigua, Dec. 30, 1736. Copy in Gov. Mathew to Board of Trade, Jan. 17, 1736/7, C.O. 152: 22, W 94.

[79] *Ibid.*

wheel, and seventy-seven burned alive. There were three waiting men, thirteen carpenters, eight coopers, one coppersmith, one sugar boiler, two masons, one butcher, twenty-six drivers, three coachmen, one head field negro, one millwright, three fishermen, one wheelwright, carpenter, and mason, one "obiaman" or medicine man, and three fiddlers. Thirty-six other negroes were banished.[80] As a matter of fact, laws were made in most of the islands to prevent slaves hawking and peddling from town to town, but, obviously, they could not be strictly enforced.[81]

[80] Mathew to Board of Trade, C.O. 152: 23, X 7. One Hercules who was executed was "an Excellent Tradesman and all most the Support of the poor family that own'd him." Extract from an unnamed correspondent's letter to Wavel Smith, secretary of the Leeward Islands, Nov. 10, 1736, C.O. 152: 23, X 32.

[81] See, *e.g.*, Antigua Act of 1739, "An Act to prevent all persons, Negroes, and other Slaves from hawking & Retailing Wares and Merchandizes thro' the several plantations, Towns, or parishes of this Island." *Acts passed in the Island of Antigua, 1734-1764,* 2 vols., London, 1734-1764.

CHAPTER III

THE SLAVE TRADE

The culture of sugar on a profitable scale required a larger, more highly organized, and more continuous labor force than any other agricultural pursuit in the eighteenth century. In the early stages of the industry, the planters made use of such white labor as was available, but to the second generation it was apparent that only African slaves could fill the necessary conditions. The Portuguese sugar growers in Brazil had pointed the way to such a solution of the labor problem. Ligon tells us that Barbadoes, in 1647-1650, had Indian slaves from the Spanish Islands and the Main. The women made good house servants, and the men excellent hunters, but he does not say they were used as field hands.[1] Many Indian slaves in the eighteenth century were shipped from Charleston to the Sugar Islands.[2] There was also an extensive trade in Indians from the Mosquito Coast to the French and British West Indies.[3] Indians, however, if at all, were made very slight use of by the British sugar planters. Barbadoes had no Indian slaves in 1747, though the French were then reported to have had many.[4] Bryan Edwards stated that they were unfitted for hard toil on the plantations.[5]

1 Richard Ligon, *History of Barbados*, London, 1657, p. 54.

2 Channing, *History of the United States*, II, 390; Logan, *History of Upper South Carolina*, I, 182; *cf.* Schaper, *Sectionalism in South Carolina*, p. 292.

3 *Parliamentary History*, XIX, 62.

4 Gov. Thomas Robinson to Board of Trade, C.O. 28: 27, Bb 57.

5 Edwards, *West Indies*, II, 45.

The necessary conditions prevailed in the sugar islands to make negro slavery a successful solution of the labor problem. Holing the field with a hoe, manuring the soil, planting cane slips, weeding by hand or hoe, and cutting the canes with a bill, were tasks easily learned by the Africans.[6] They were operations also which permitted concentrating the laborers into large gangs of twenty or more, driven by a white overseer or an experienced negro. The processes of crushing, boiling, clarifying, and claying, demanded, however, an alertness and exercise of judgment which required the constant supervision of the proprietor or white overseer. There was no need for a variety of crops, or even rotation of crops, for an abundant supply of provisions, horses, and cattle could be had from the Northern British Colonies. Sugar also could afford the expense of slave cultivation, for the profits from it, as Adam Smith pointed out, were greater than those from any other cultivation either in Europe or America.[7] In the eighteenth century, there can be little doubt that slavery was more profitable to the West India planter than free labor, and it is very probable that the production of sugar would not have taken place as soon as it did if slavery had not existed to furnish a sufficiently large and continuous body of labor.[8]

The interdependence of the slave trade and the sugar industry was universally accepted by the English prior to about 1780. A petition of West India merchants and planters to the House of Commons, in 1739, stated that

[6] A competent French writer said that under good masters the slaves succeeded perfectly in agriculture and carpentry. Auberteuil, *Essai sur l'état des colonies*, etc., Paris, 1776, I, 130-146. In sugar culture, one slave to about 2 acres was required; in cotton, one to from 5 to 10 acres; in Indian corn, one to 30 or 40 acres. *Cf.* Hammond, *Cotton Industry*, p. 47.

[7] Adam Smith, *Wealth of Nations*, ed. *World's Classics*, I, 365.

[8] John Stuart Mill, *Political Economy*, Bk. II, Ch. v, Sec. 2, ed. 1871, I, 308.

the slave trade to the coast of Africa "entirely depends" on the sugar trade.[9] Hippisley, in 1764, expressed the prevailing opinion when he wrote that "The impossibility of doing without slaves in the West-Indies will always prevent this traffick being dropped. The necessity, the absolute necessity, then, of carrying it on, must, since there is no other, be its excuse."[10] In England, slavery was also favored as expedient politically. "Negro labour will keep them in due Subserviency to the Interest of their *Mother Country;* for while our Plantations," said Postlethwayt, "depend only on Planting by Negroes our Colonies can never become independent of these Kingdoms."[11] The governor of Barbadoes was able to report, in 1766, with great satisfaction of that island's ready obedience to the Stamp Duty as in marked contrast to the "outrageous" conduct of the Northern Colonies.[12] It is at least an interesting consideration that the decline in production of cane sugar in the British West Indies in the nineteenth century was intimately connected with the limitation and final abolition of compulsory labor. The export of sugar from Jamaica, in 1805, was 150,352 hogsheads, in 1834 this had fallen to 88,770 hogsheads, and in 1865 to an annual average of 30,000 hogsheads. There is little doubt that this decline was owing to the abolition of the slave trade and slavery

9 Petition of West India merchants and planters to House of Commons. Presented by John Sharpe to Board of Trade, Mar. 14, 1738/9, C.O. 152: 23, X 52; Commons Journal XXIII, 284. Many expressions of the same belief might be cited.

10 John Hippisley, *Essays*, London, 1764, pp. 17-18. *Cf.* the following typical statement, made ten years later, that the slave trade "supplies the Islands in the West Indies with such slaves as are absolutely necessary to the Cultivation of the Lands there, and without whose labour the Lands would be of no value." Memorial of Soc. of Bristol Merchants to Board of Trade, Nov. 17, 1774. B. T. 6 (Africa): 2, B 33.

11 Malachi Postlethwayt, *The African Trade, the great Pillar,* etc., London, 1745, pp. 13-14.

12 Charles Pinfold to Board of Trade, Feb. 21, 1766, C.O. 28: 32, Ff 68.

itself. The freed negroes, in 1838, under advice of fanatical Baptist missionaries, refused to work for the wage tariff offered. A general strike took place. The young canes were choked with weeds and rotted in the ground. Estate after estate went out of cultivation. Others were sold at a merely nominal price but, from the lack of continuous labor, remained useless.[13] Under a wage system with only intermittent labor and the possibility of strikes, the cane sugar industry became a most precarious undertaking. When we realize that cane juice ferments within twenty-four hours the advantage of compulsory, or coolie, labor over such a régime is obvious.[14] Thus the revival of the original labor problem in the middle of the last century, coming in conjunction with the development of quick means of transportation, served as a powerful impulse to substitute for the cane of the tropics the beet sugar of the temperate zone.

The Royal African Company's monopoly of the slave trade greatly aggravated the West India planters toward the end of the seventeenth century. It was complained, in 1689, that negroes which formerly sold for £7 had been increased to £20 in price by the African Company and that the supply of negroes was so inadequate that the planters had to "scramble" for them.[15] Cary said that the company demanded ready payment from the planters while interlopers gave long credit. Cary also declared the supply of slaves to be very deficient and voiced the general opposition to the company's monopoly.[16] In 1698, the slave trade was thrown open to independent traders, though the African Company re-

[13] William Reed, *History of Sugar*, London, 1866, pp. 60 *et seq.* *London Morning Post*, Nov. 22, 1865, contains an interesting article on the decline of Jamaica.

[14] *Cf.* Reed, p. 53.

[15] *Groans of the Plantations*, London, 1689, p. 6.

[16] Cary, *Essay on Trade*, 1695, p. 77.

tained its ports on the coast and taxed each trader for their maintenance. These duties expired in 1712, and from then till 1730, the company maintained its establishments at its own expense. From 1730 to 1750, it received for this purpose subsidies from the government. In 1750, the old company was replaced by the last African company, which was also subsidized, that it might continue the English occupation of a line of military posts on the West Coast of Africa which was expected to control the trade of the hinterland.

In the period of exclusive trade, during the nine years from 1680 to 1688 inclusive, the Royal African Company delivered into all the British colonies 46,396 slaves or 5155 per annum. In the ten years immediately after the opening of the trade, 1698-1708, the total number of slaves imported by separate traders into Jamaica, Barbadoes, and Antigua alone, amounted to 86,908—probably 90,000, since the slaves in seven ships to Antigua were not listed,—an average of 9000 per year. The number brought into the same islands during this decade by the African Company was 17,760, or 1776 per annum.[17] The importance of the African Company as a direct importer of slaves steadily declined, and became, in fact, insignificant during the remainder of the century. Governor Hamilton of Jamaica wrote, in 1712, that the Board of Trade "will observe that there is no Proportion between the Import of Negroes by the Company and Separate Traders; the latter so much exceeding."[18] By 1750, the separate traders had ceased buying slaves at the factories of the Company.[19]

[17] *An Account of the Numbers of Negroes delivered into Barbadoes, Jamaica, and Antigua, taken from the Accounts sent by the Governors of those islands to the Commissioners of Trade*, Brit. Mus. Pamphlet: 8223. e. 4, f. 33.

[18] Hamilton to Board of Trade, Mar. 8, 1711/2, C.O. 137: 10, N 11.

[19] Capt. Thomas Pye to the Admiralty, Mar. 18, 1749/50, C.O. 388: 44,

The independent traders resided mainly at London, Bristol, and Liverpool. Their export of English manufactures was a matter of increasing profit to England. The general growth of the trade is indicated in the following table:[20]

British Ships Cleared from England for Africa:		
Year	Number	Tons
1713	11	2,000
1714	20	2,105
1715	23	2,791
1723	55	4,200
1730	111	10,416
1737	109	10,560
1748	89	9,706
1749	83	8,178
1750	82	7,635
1763	163	18,939
1764	163	17,798
1765	135	14,924
1787	137	22,263

	Imports from Africa	Exports to Africa
1698	£ 2,496	£ 70,588
1699	19,226	96,295
1700	26,889	155,793
Average		
1726–7–8	32,441	157,905
1753–4–5	32,096	227,996
1773–4–5	64,085	764,935
1784–5–6	95,123	666,606

In the first half of the century it was the general practice for the slavers to return from the West Indies to England with sugar. But as a cargo of sugar was of much less value than a load of slaves, the balance was paid in specie or bills of exchange.[21] But in the latter half of the eighteenth century, the planters preferred to ship by vessels which came annually from Great Britain, especially built to carry hogsheads of sugar in larger quantities and more carefully. Therefore the Guinea ships often went home in ballast or else offered advanced prices for sugar rather than return dead freighted.[22]

Cc 73. The African Company exported from the Gold Coast, in 1758, 2203 slaves.

[20] B. T. 6 (Africa): 7.

[21] See, *e.g.*, C.O. 142: 14, B. T. Naval Office Lists, Jamaica (1709-22), where this is the universal practice.

[22] Examinations of Charles Spooner and Stephen Fuller before a committee of the Privy Council, Mar. 1, 1788, B. T. 6: 9, f. 550; 10, f. 102.

London, in 1725, sent out to Africa 87 vessels carrying from 100 to 600 slaves each, 26,440 in all. Bristol, the same year, cleared 63 vessels, each carrying 150 to 450 slaves, or 16,950 in all.[23] Liverpool, in 1743, had 55 vessels in the trade, each carrying 100 to 450 slaves, in all 15,400.[24] In 1749, the same port sent out 70 ships for 23,200 negroes.[25] Bristol, in 1749, had decreased its African trade to 47 ships and 16,640 slaves. From 1750 to 1776 inclusive, Bristol sent out 588 vessels to Africa. Liverpool in the same period sent out 1868 vessels.[26] London had greatly declined: the ships cleared for Africa in 1747 were by one account only 3; in 1748, 7; in 1749, 8; 1750, 6; 1751, 7; 1752, 14; 1753, 13.[27] In 1771, Liverpool employed 107 ships for 29,250 slaves, London 58 for 8136, Bristol 23 for 8810, Lancaster 4 for 950,—a total of 190 English ships in the trade, exporting from Africa 47,146 slaves.[28] By 1804, Liverpool possessed six-sevenths of the whole English trade to Africa.[29]

The goods sent from England to purchase the slaves consisted of woolens, firearms from Birmingham, Sheffield and other places, powder, bullets, iron bars, brass pans, malt spirits, tallow, tobacco pipes, Manchester goods, glass beads, certain linens, ironmongery and cutlery, toys, and certain East India goods.[30] In the early eighteenth century, a large proportion of the negroes was purchased by the separate traders at the

23 Bristol merchants to Board of Trade, Mar. 18, 1725, C.O. 388: 25, S 38; London merchants to Board of Trade, *ibid.*, S 74.

24 C.O. 388: 43, Bb 89.

25 C.O. 388: 45, Dd 5.

26 B. T. 6 (Africa): 3, B 91; B 92.

27 B. T. 6 (Africa): 7.

28 Long, *Jamaica*, I, 492, n. a; Edwards, *West Indies*, II, 65.

29 Young, *West India Commonplace Book*, p. 9.

30 Long, *Jamaica*, I, 491.

establishments, or factories, of the African Company, but before 1750 they had given up trading with the company.[31]

It was usual to take in twice as many males as females, for women were in less demand on the plantations, and breeding was not encouraged. The men were from sixteen to thirty years of age. This depletion of the male Africans reacted on native institutions by favoring polygamy, which prevented an exhaustion of the population.[32] The men averaged four feet six inches in height. Boys and girls were taken in smaller numbers. A little over two slaves per ton was the average burden of a ship. A vessel of 120 tons would carry thirty white sailors.[33] A surgeon with medicines was on board each slave ship. The men were chained hand and foot during the whole voyage; the women were not chained and recovered from their despondency sooner than the men. In good weather they were brought up on deck from 8 a.m. till 4 p.m. Their food consisted of rice, yams, and beans,—no animal food.[34] Some authorities stated the mortality of the slaves en route at five per cent; others put it as high as one-twelfth of the number. The mortality of white sailors also was greater in the African than in any other trade.[35]

[31] Capt. Thomas Pye to the Admiralty, Mar. 18, 1749/50, C.O. 388: 44, Cc 73. Pye had just visited the coast.

[32] John Hippisley, *Essays*, p. 6.

[33] Examination of John Newton, a slaver about 1748, Feb. 1788, B. T. 6 (Africa): 9, f. 9.

[34] Examination of Capt. Hall, a slaver about 1772, Feb. 22, 1788, B. T. 6 (Africa): 9, f. 72.

[35] In six voyages, between 1775 and 1786, John Mathews, a Liverpool slaver, carried 2576 slaves, of whom 110 died, and of his 211 white sailors, 18 died and 2 were drowned. Examination before the P. C. committee, Mar. 4, 1788, B. T. 6 (Africa): 9, f. 343. Robert Norris, in five voyages, between 1769-1777, lost 78 slaves out of 2175, and 10 sailors out of 160. Exam. before the P. C. committee, Mar. 4, 1788, B. T. 6 (Africa): 9, f. 218.

The negroes from the Gold Coast and Whydah, who were fed on Indian corn, were the best and showed the least mortality. Those from the Windward Coast, whose food was rice, were the next best in health; and those from the Bight of Guinea, whose diet was yams were subject to the greatest mortality.[36] No systematic accounts of the total exports of slaves to the British colonies were kept in the eighteenth century. From the broken lists it is possible, however, to form estimates which are approximately true. Between 1680 and 1786, the total import of slaves into all the British American

According to the muster rolls of 88 Liverpool slave ships that returned to Liverpool in the year 1786 and up to Sept. 1787, of the original crew of 3170, 1100 were discharged or deserted in Africa or the West Indies, 642 died, 1428 returned to Liverpool. Of 24 Bristol slavers having 910 sailors, 239 were discharged or deserted, 216 died, and 455 returned. Exam. of Tho. Clarkson before the Board of Trade, July 27, 1788, B. T. 6 (Africa): 11. Nine typical voyages from Liverpool, reported to the Board of Trade, in 1788, furnish the following details of the voyage between Africa and the West Indies:

VESSEL	OWNER	TONS	SLAVES CARRIED				Total	MORTALITY				Total	Passage Days
			M.	W.	B.	G.		M.	W.	B.	G.		
Brooks	Jas. Brooks & Co.	297	351	127	90	41	609	10	5	3	1	19	49
Bud	Th. Foxcroft & Co.	297	69	30	56	47	200	2	4	1	2	9	49
Golden Age	Th. Hinde & Co.	377	—	—	—	—	624	—	—	—	—	41	40
Betty	John Copland	260	—	—	—	—	450	—	—	—	—	5	42
Kitty	" "	137	—	—	—	—	282	—	—	—	—	18	42
Venus	Tarletons & Blackhouse	146	87	80	56	77	300	3	7	-	3	13	43
Brothers	Baker & Dawson	325	353	257	25	15	650	17	16	4	3	40	50
Rose	Jos. Ward & Co.	147	124	79	83	54	340	7	9	5	4	25	35
Jane	Wm. Boats & Co.	242	—	—	—	—	528	—	—	—	—	41	44
							3983					211	

M, W, B, G, indicate Men, Women, Boys, and Girls respectively. The vessels carried about two slaves per ton. The profit per slave was from £2 to £3.

36 Examination of John Mathews, Liverpool slaver, before a committee of the Privy Council, Mar. 4, 1788, B. T. 6 (Africa): 9, f. 343.

colonies has been estimated at 2,130,000 or an annual average of 20,095.[37] The high water mark was reached, it would appear, in 1768, when 104,100 slaves were estimated to have been exported from Africa,[38] of whom the English took off 53,100, the North Americans 6,300, the French 23,500, the Dutch, 11,300, the Portuguese, 8,700, the Danes, 1,200, making a total of 104,100. From 1735 to 1763, the average annual English export from the Gold Coast was 13,000 slaves. In 1771, the British export of Gold Coast negroes declined to 6000.[39]

Barbadoes, by 1724, was reported to have contained no ungranted or uncultivated land.[40] Again, in 1729, it was stated that Barbadoes ''has many Years been all clear'd, and the necessary Works fix't''[41] The demand for slaves here, therefore, was occasioned not by the presence of virgin soil, but by the exhaustion of its land. In 1714, that island required the ''utmost art, Industry and Manure, and that too annually,'' or it would yield no crop.[42] In 1717, 30 acres, which in the French Islands could be worked with 30 to 40 negroes and a few horses and cattle, in Barbadoes required 150 negroes, 50 to 60

[37] Edwards, *West Indies*, II, 65.

[38] Brit. Mus. Add. MSS. 18960, f. 37. In 1788, Robt. Norris, a Liverpool trader, estimated the total export at 74,200.

English exported . . .	38,000
French	20,000
Dutch	4,000
Danes	2,000
Portuguese	10,000
	74,200

B. T. 6 (Africa): 9, ff. 137 *et seq.*

[39] *Parl. Hist.*, XIX, 302.

[40] Worsley to Board of Trade, Oct. 18, 1724, C.O. 28: 18, W 51.

[41] Francis Freelove to William Wood, July 1, 1729, C.O. 28: 21, Y 10.

[42] William Sharpe, president and commander-in-chief of Barbadoes, to council and assembly, May 4, 1714, C.O. 28: 14, T 44.

head of cattle, and a dozen horses.[43] Worsley stated, in 1723, that one negro in Martinique would cultivate the same amount of ground as well as two in Barbadoes.[44] In 1724, some planters looked to St. Lucia, Dominica, and St. Vincent for new lands to provide for the "worn out soyles of Barbados and the Leeward Islands."[45] In 1731, though the planters attributed the higher price of English sugar to the impoverished lands, yet, it was said, they were not "willing" to move to places where twice the quantity of sugar might have been made by the same labor as there.[46] In 1734, the Board of Trade represented to the House of Lords that the worn-out state of Barbadoes necessitated many more hands than in the fresh lands of Hispaniola and other islands.[47]

The annual importations of negroes into Barbadoes can be estimated, at least approximately, from the treasurer's accounts of that island.[48] According to these accounts, and the naval office lists, the number of slaves imported into Barbadoes was as follows:[49]

[43] Agents for Barbadoes to Board of Trade, rec'd Sept. 25, 1717, C.O. 28: 15, T 127.

[44] Worsley to Board of Trade, Sept. 24, 1723, C.O. 28: 17, V 119.

[45] Thomas Ekines to Hans Sloane, London, May 2, 1724, Brit. Mus. Sloane MSS. 4047, f. 170.

[46] *Importance of the British Plantations*, 1731, p. 113; *A Short Answer to an elaborate Pamphlet*, 1731, p. 4.

[47] Representation of Board of Trade to House of Lords, Jan. 14, 1734/5, C.O. 5: 5, f. 114.

[48] By an act of Nov. 18, 1705, Barbadoes laid an import duty of 5s. per head on new negroes, payable by the importer. This act was continued annually as one of the regular revenue bills. No export duty was laid. Gov. Henry Grenville to Board of Trade, Dec. 14, 1752, C.O. 28: 30, Dd 21; Gov. Charles Pinfold to Board of Trade, June 1, 1762, *ibid.*, 28: 32, Ff 25.

[49] This table has been compiled from C.O. 28: 18, W 90, 91; 24, Aa 28-29, 35, 41, 54; 25, Aa 71, 87, 89; 27, Bb 56; 29, Cc 12, 13, 21, 41, 89, 136; 30, Dd 12, 54, 61, 76, 103; 31, Ee 10, 26, 34; 32, Ff 4, 16, 29, 36, 47, 58, 59, 86, 101.

Year	Total Number	Year	Total Number
1708	1227	1733	1532
1709	683	1734	953
1710	1170	1735	1375
1711	1075	1736–1746	wanting
1712	3031	1747	1802
1713	6467	1748	1182
1714	5107	1749	1044
1715	5546	1750	3100
1716	4084	1751	4908
1717	5534	1752	3761
1718	7126	1753	3815
1719	3705	1754	6139
1720	1192	1755	4244
1721	1253	1756	2579
1722	818	1757	2778
1723	2064	1758	2601
1724	3251	1759	1177
1725	2670	1760	3388
1726	3914	1761	642
1727	3007	1762	3069
1728	4141	1763	4092
1729	4661	1764	5104
1730	2908	1765	3228
1731	3832	1766	4061
1732	3081		

The number imported into Barbadoes by the African Company from 1708-1725, inclusive, was 4226, against 49,777 by the private traders, *i.e.*, the company imported only about one-twelfth of the slaves.

From this list, which is approximately correct for long periods, it appears that the average annual import of slaves into Barbadoes in the period 1708-1715 inclusive was 3038, 1715-1725 inclusive was 3170, 1726-1735 inclusive was 2940, 1746-1756 inclusive was 3177, 1757-1767 inclusive was 3014. It is possible that some slaves were reëxported, but in the absence of complete lists, it has not been possible to determine how many. In 1748-1750 inclusive, however, out of 5326 imported, according to the treasurer's accounts, we are told the island kept about 5100 for its own use.[50] From the Naval Office lists of slaves imported into the Leeward Islands, it would appear that very few came into those islands from Barbadoes. Thus in the years 1721-1727 inclusive, only one ship, with eighty slaves, entered at Antigua from Barbadoes;[51] from 1721-1730 inclusive, four vessels entered at St. Christopher from Barbadoes with 216 slaves;[52] from 1721-1726 inclusive, no slaves were entered at Nevis from Barbadoes;[53] from 1721-1729, none were entered at Montserrat from Barbadoes.[54] In 1734, we are told that Barbadoes had not sent any negroes to the Spanish since the Asiento Contract of 1713.[55] It seems probable that the import of slaves into Barbadoes from 1708 to 1767, as indicated by the above figures, remained constant at about 3000 a year. This represents the annual increment necessary to sustain the industries of that island. From 1712 to 1762, the importation, at this rate, would have amounted to 150,000 negroes, while, during that same period, the slave population increased from 41,970[56] to only a little over 70,000.[57]

In 1730, Barbadoes was the only British sugar island sufficiently stocked with slaves. The other colonies, at that time, were said to have been not above half supplied.[58] During the Spanish and French Wars, 1739-

50 Computation presented to Board of Trade by Mr. Hardman, Jan. 11, 1750/1, C.O. 388: 45, Dd 3.

51 C.O. 152: 15, R 190-191.

52 C.O. 152: 19, T 150.

53 C.O. 152: 15, R 204.

54 C.O. 152: 15, R 205; *ibid.*, 152: 18, T 67.

55 James Dottin, president of Barbadoes council, to Board of Trade, C.O. 28: 24, Aa 35.

56 Lowther to Board of Trade, Aug. 16, 1712, C.O. 28: 14, T 15.

57 Gov. Charles Pinfold to Board of Trade, June 1, 1762, C.O. 28: 32, Ff 25.

58 Charles Dunbar, ''State of the British Sugar Colonies,'' in William Mathew to Board of Trade, Oct. 3, 1730, C.O. 152: 19, T 142.

1748, the supply lessened, but, excepting those years, Barbadoes appears to have been well supplied with black labor.

In the Leeward Islands complaints of an insufficient supply of slaves were more frequent. The demand here was determined partly by the presence of uncultivated sugar lands, but, in a larger degree, by the need of an increasing force of negroes on the impoverished lands. In 1719, Hamilton, governor of this group, complained of the decay of the negro trade in recent years and the increasing price of slaves which was "a great Hindrance to the Improvement of the Sugar Plantations."[59] To supply their deficiency all these islands clandestinely imported slaves from the Dutch at St. Eustatius, paying for them sometimes in specie but more often in sugar—thus evading the king's customs (of four and one-half per cent *ad valorem*) in the West Indies and in England and permitting the Dutch to become carriers of a British enumerated commodity.[60]

Dunbar, the surveyor of customs, stated that the Dutch brought two or three thousand slaves each year to St. Eustatius which, from the suitableness of Dutch cargoes on the African Coast, were sold to the English and French at lower rates than the English charged. He knew of no law to prevent such imports in English vessels.[61] Richard Harris, the English slaver, stated that Dutch goods for the African trade, such as powder, French brandy, pipes, and East India goods, were usually

[59] Hamilton to Board of Trade, July 20, 1719, C.O. 152: 12, P 207.

[60] Hamilton to Board of Trade, Antigua, May 15, 1717, C.O. 152: 12, P 2; same to same, Sept. 16, 1718, *ibid.*, P 140. Hamilton did not know whether or not to treat the slaves as foreign produce.

[61] Charles Dunbar to Commissioners of Customs, Sept. 25, 1718 (extracts), C.O. 152: 12, P 213. The Commissioners of the Customs, in their letter of transmission to Board of Trade, Sept. 11, 1719, stated that they were aware of no law against such trade.

cheaper in Holland than in England. The African Company, for several years past, had not imported a fifth part of the goods needed in the slave trade. Recently, the English traders had been granted liberty to take in Dutch goods in Holland, which would enable the English, Harris thought, to sell as low as the Dutch.[62] The African Company also complained of the St. Eustatius trade with the English planters, but admitted that the Dutch bought their slaves cheaper and sold them lower.[63] The customs collector seized several slaves that were landed in St. Christopher, and brought them to trial as Dutch merchandise illegally imported, but the vice-admiralty judge always acquitted the traders.[64] Between 1721 and 1723, Governor Hart stated that at least one thousand slaves were thus imported into St. Christopher.[65] In 1727, Hart wrote, however, that a full stop had been put to this trade at St. Christopher, for the island had been well stocked by English traders who were then selling lower than the Dutch.[66] In 1731, interlopers from the Leeward Islands were still trading for slaves at St. Eustatius where they could frequently be had twenty per cent cheaper than from the English.[67] After this time, little is heard of this practice.

The Leeward Islanders themselves carried on a direct trade to Africa for slaves.[68] These they purchased chiefly with rum, which sold on the African Coast for five

62 Richard Harris to Board of Trade, London, Sept. 21, 1719, C.O. 152: 13, Q 5.

63 Representation of the Royal African Co., to Board of Trade, Oct. 2, 1719, C.O. 388: 21, Q 19.

64 John Helden, collector of customs, to Royal African Co., London, Sept. 20, 1719, C.O. 388: 21, Q 19. Helden had observed that this trade began within the last four years.

65 Gov. Hart to Board of Trade, Feb. 15, 1726/7, C.O. 152: 15, R 188.

66 *Ibid.*

67 *Importance of the British Plantations*, 1731, pp. 32-33.

68 Hamilton to Board of Trade, July 20, 1719, C.O. 152: 12, P 207.

shillings sterling per gallon, though the first cost was not one shilling.[69] A few slaves from Madagascar, along with East India goods obtained from pirates of the Indian Ocean in exchange for ammunition, were brought in by West India interlopers.[70] New slaves imported in 1726-1727, brought from £17 to £27.[71] The number of slaves delivered into Antigua, from December 25, 1720, to December 25, 1729, was 12,278 or 1362 per annum.[72] In 1750, Antigua's annual average import was estimated at 2000 slaves.[73] St. Christopher, from December 25, 1721, to December 25, 1730, imported 10,358 or 1035 per annum.[74] In 1750, the annual import for this island was estimated at 2000 slaves.[75] Montserrat imported, from 1721 to 1729 inclusive, 3210 slaves or about 357 per annum;[76] in 1750, its annual import was computed at 500.[77] Nevis, from December 25, 1721, to December 25, 1726, imported 1267 negroes, or 253 per year;[78] in 1750, the annual import was reckoned at 500.[79] An estimate of the total import of slaves into the Leeward Islands, from 1720 to 1755, would give, at 1500 per annum for Antigua, 37,500; at 1500 per annum for St. Christopher, 37,500; at

69 Hart to Board of Trade, July 12, 1724, C.O. 152: 14, R 101.

70 Order in Council, Oct. 2, 1721, forbidding the plantations to trade to the East Indies, C.O. 388: 23, R 53.

71 William Mathew to Board of Trade, Nov. 1, 1727, C.O. 152: 16, S 27; List of Imports, Mar. 25-Nov. 9, 1727, *ibid.*, 152: 16, S 70.

72 Naval Officer's and Collector's Accounts, C.O. 152: 15, R 190-191; *ibid.*, 152: 18, T 78.

73 Hardman to Board of Trade, rec'd Jan. 11, 1750/1, C.O. 388: 45, Dd 3.

74 Naval Officer's and Collector's Accounts, C.O. 152: 15, R 191; *ibid.*, 152: 19, T 150.

75 Hardman, C.O. 388: 45, Dd 3.

76 Naval Officer's and Collector's Accounts, C.O. 152: 15, R 205; *ibid.*, 152: 18, T 67.

77 Hardman, C.O. 388: 45, Dd 3.

78 Collector's Accounts, C.O. 152: 15, R 204.

79 Hardman, C.O. 388: 45, Dd 3.

400 per annum for Nevis, 11,000; at 350 per annum for Montserrat, 8750; making a total of 93,750. The slave population in those islands during the same period showed only the following increase:[80]

	1720	*1755*	*Increase*
Antigua	19,186	(1756)31,428	12,242
St. Christopher	7,321	21,891	14,570
Nevis	5,689	8,380	2,691
Montserrat (1719)	4,192	8,853	4,661
			34,164

The correspondence from these islands leaves the impression that they were generally in need of more slaves than they could get. In 1730, Charles Dunbar, surveyor general of the customs, reported that none of the sugar islands, except Barbadoes, were above half supplied with negroes.[81] Mathew, governor of the Leeward group, wrote of Antigua in 1734, that the slave importation "is fallen to almost nothing."[82] He then stated that out of a total area in Antigua of 72,000 acres, about 50,000 was manurable cane land and added: "But want of Negroes occasion'd these Acres to be not wholly improv'd," for by the last return there were under canes only 24,408 acres. He added that this observation applied to St. Christopher, Nevis, and Montserrat as well as Antigua.[83] In St. Christopher, "This want of Negroes," continued Mathew, "will arise rather from

[80] C.O. 152: 13, Q 40, 46; *ibid.*, 152: 28, Bb 75, 76, 77, 81. Returns from governors to Board of Trade.

[81] Charles Dunbar, "State of the British Colonies," in William Mathew to Board of Trade, Oct. 3, 1730, C.O. 152: 19, T 142.

[82] William Mathew, "State of the Leeward Islands," Aug. 31, 1734, C.O. 152: 20, V 46.

[83] *Ibid.*, f. 1.

the Lands growing poorer, Consequently more to be Tended and Manur'd, than from want of Strength to put more of it, than is at present, in Culture." Of Montserrat he wrote: "As to its Produce it is Stinted, for want of Negroes, and Suffers in every Point as Antigua Does"; the island had only 6176 negroes, which was an increase of but 321 in the last four years. Again, in 1745, Mathew stated that the four Leeward Islands were wholly cultivated except what is barren land or pasture, "but with more Negroes [they] might make more of each Species."[84] In 1750, the planters of Antigua framed a memorial on the difficulties they suffered from the want of such negroes as were fit for hard labor. The strongest slaves were those from Whydah and the Gold Coast, but in recent years very few of them had been imported. The memorial states that "The loss of this most valuable Trade we conceive is owing partly to the rivalship of foreign Nations, but principally to the desertion of the Bristol and Liverpool Merchants who chuse rather to trade to Calabar, Angola, and the Bite; because Negroes of those countries being much less valuable, are purchased at cheaper rate; and yet are imposed upon us for want of better, at the price of the best sort, to the great injury of the Planters, who are deeply involved in debt, by the mortality and failure of that kind of Negroes, and must be inevitably ruined for want of Caramantee, Fantee and Poppa Negroes, unless that Trade be restored again to Britain.[85] St. Christopher also was getting sickly negroes that soon died, whereas the French got the

[84] William Mathew to Board of Trade, July 6, 1745, C.O. 152: 25, Y 139.

[85] Memorial of the Governor, Council and Assembly of Antigua to Board of Trade, rec'd Jan. 23, 1750/1, C.O. 152: 27, Aa 4. In substance the same is the Memorial of Lieut. Gov. Gilbert Fleming and the Council and Assembly of St. Christopher to Board of Trade, Dec. 20, 1750, *ibid.*, 152: 27, Aa 16.

best laborers.[86] Though completely cultivated, in 1756, Antigua still needed an increase in its slaves owing to "the Necessity the Planters are under to manure their Lands, which are getting impoverished by long Culture."[87]

In Jamaica it was possible to determine more accurately, from collectors' accounts, the number of slaves imported. In 1719, that island laid, as a revenue measure, a duty on the import of negroes, and in 1707 an export tax on slaves was also passed; these continued to be regular sources of revenue. From the entry lists at Jamaica, there were imported, from 1700 to 1786, 610,000 negroes.[88] Between 1702 and 1775 inclusive, the total import of slaves was 497,736 and the total export 137,114,—the annual average being an import of 6726 slaves and an export of 1853. The annual average increment for Jamaica was, therefore, 4873 slaves. Those imported in the first thirty-four years exceeded those imported in the last thirty-four by 38,751, and those exported in the latter period were less than half those exported in the former period.[89]

The demand for slaves in Jamaica was keener than in any of the other British sugar islands. This was primarily due to the presence of a vast area of uncultivated sugar land. The deficiency of supply which generally prevailed was attributed in large measure to the Asiento of 1713 with Spain by which the South Sea Company agreed to furnish the Spanish Americans with 4800

[86] Lieut. Gov. Fleming's address to council and assembly of St. Christopher, Dec. 1750, C.O. 152: 27, Aa 6; Aa 13-14: the reply.

[87] Gov. George Thomas to Board of Trade, Feb. 20, 1756, C.O. 152: 28, Bb 75.

[88] Edwards, *West Indies*, II, 64. Edwards took these figures from the Naval Office Lists at Jamaica.

[89] Macpherson, *Annals of Commerce*, III, 575. For an account of the annual import and export of slaves at Jamaica, and the number of ships engaged, for each year from 1702 to 1775 inclusive, see Appendix, p. 391.

slaves per annum. Jamaica was found to be best situated and provisioned as a depôt from which the company could carry on this traffic. The negro ships from Africa put in here to refresh their cargoes, sort them, receive orders from the Spanish markets, and purchase other slaves if necessary from independent traders, or even from the planters. The ready payments in specie, paid by the Spanish, made that market very attractive. Hence in Jamaica, where long credit and payment in sugar was asked, very few slaves were sold by the Company.[90]

With opportunities at hand for the employment of more labor than they could get, the planters looked on this export of negroes as a great grievance. Governor Lawes wrote, in 1717, that Jamaica was seventy times bigger than Barbadoes and could support twenty times more inhabitants. He thought there was room for one hundred thousand families more, who might produce sugar and other commodities to the value of a thousand pounds each a year. But negroes were the support of planting and they were from £25 to £40 per head, and little or no credit was given to new settlers. "The Asiento," declared Lawes, "carries all the able, stout, and Young Negroes, or Such as they call peic'd India to the Spaniards and Sell[s] none to the planters but old, Sickly, and decrepid, or what are called Refuse, if a Choice Negro is Sold to a planter, he must give as much or more than the Spaniards, & that in ready Money. . . . But this Asiento . . . deprives the planters of the Best, and only Sells them the worst of Negros."[91] The planters resorted, in 1716, to a duty of 20s. per head in

[90] Thus between Oct. 8, 1716, and Aug. 3, 1717, the Company imported into Jamaica 1576 slaves; of these 1248 were, without landing, sent to Cartagena, Porto Bello, Vera Cruz, and Cuba, and 349 were sold in the island. South Sea Factor's Account, C.O. 137: 12, O 179.

[91] Sir Nicholas Lawes, Memorial to Board of Trade, rec'd Nov. 11, 1717, C.O. 137: 12, O 178.

addition to the 10s. import tax,[92] on all slaves reëxported from Jamaica. From this time, the slave duties, renewed annually, became regular revenue measures. The incidence of the export duty it was believed would fall entirely on the Spaniards, to whom the slaves were reëxported. It appears at first to have had in it a certain idea of protection to the slaveholders, for Lawes observed: "But if the Case be so that the Company cannot afford to pay that Duty out of the profitts of the Asiento; then it may be supposed the Negroes imployed on our plantations are of Greater Advantage to this Kingdome, than selling them to the Spaniards."[93]

The South Sea factors viewed this measure as a great hardship and persistently opposed it.[94] In 1718, the slave export duty act of that year was disallowed,[95] and Governor Lawes was instructed to withhold his assent from any act laying a duty on slaves brought in for refreshment only and not landed.[96] Jamaica continued to tax all negroes landed or reëxported.[97] The independent traders also complained, for it had been their common practice to call at Jamaica, and to trade so long as the

92 Virginia's import duty on negroes was a similar revenue measure whose disallowance in the interests of the traders was secured in 1724. Journal of the Board of Trade, Jan. 17, 1723/4, C.O. 391: 33, f. 15. This case was used as a precedent against Jamaica, Harris to Board of Trade, Mar. 20, 1724/5, and Nov. 13, 1727, C.O. 137: 16, R 8, 10.

93 Sir Nicholas Lawes, Memorial to Board of Trade, rec'd Nov. 11, 1717, C.O. 137: 12, O 178. In the same document Lawes said that the duty was based on a similar act laid some years earlier [1693] for raising money.

94 Daniel Wescomb, secretary of the South Sea Co., to Board of Trade, Nov. 15, 1717, C.O. 137: 12, O 179.

95 Order in Council, Jan. 9, 1717/8, C.O. 137: 12, O 190, 193.

96 Additional instruction to Gov. Lawes, Jan. 16, 1717/8, C.O. 137: 14, Q 26.

97 Wescomb to Board of Trade, Mar. 2, 1721/2, C.O. 137: 14, Q 25. Between 1721-1725, the South Sea Company's factors paid, on slaves reexported in 67 ships, £9086, and in the latter year still owed £1090 on 13 shiploads. Account by Charles Lloyd, Rec. Gen'l, C.O. 137: 16, R 58.

market was active, and then sail with the remainder of their slaves to Virginia, the Carolinas, and other places.[98] The act of 1725 revived the duties on slaves brought for refreshment only,[99] but forbade their being landed unless they were sold. In his defense of the bill, the Duke of Portland, then governor, said the South Sea Company "us'd to land all their Negroes brought in here for refreshment, which infected the Inhabitants, wth all their Malignant Fevers, Small Pox and other dangerous distempers."[100] A memorial from the planters at this time condemned the company not only for preventing their getting slaves to open new lands but for depriving them of even the necessary increment to maintain the old plantations.[101] In 1727, both the company and private traders

98 Richard Harris to Board of Trade, Mar. 20, 1724/5, C.O. 137: 16, R 8. Harris said that this practice of private traders tended much to the advantage of England and should not be obstructed. Naturally, he thought the company could pay the duty by shifting it onto the Spanish. The standing counsel was of the same opinion. Richard West to Board of Trade, C.O. 137: 14, Q 103. Petition of South Sea Co. to H. M. against the Act of 1723, rec'd Dec. 10, 1724, C.O. 137: 16, R 3.

99 Wescomb to Board of Trade, Mar. 3, 1724/5, C.O. 137: 16, R 4.

100 Portland to Board of Trade, Feb. 8, 1724/5, C.O. 137: 16, R 13.

101 The Memorial from Jamaica Merchants and Planters to Board of Trade, Mar. 11, 1724/5, states: "And it is Certain, that since the South Sea Company hath settled its factories in the said Island, and brought up Negroes there, in order to transport them to the Spanish Colonies, that the price of Negroes is considerably risen in the Island, By which means it is become difficult and expensive for the Inhabitants to procure a Sufficient number of Negroes to Cultivate their plantations, they being not able to give so high a price for them as the South Sea Company who buy up the best Negroes and carry them to the Spaniards, so that the Inhabitants are forced to take up with such Negroes as they can get to Cultivate their Sugar Plantations that are already Setled, and prevents more being sett on foott which certainly obstructs the further Strengthening of the Island." C.O. 137: 16, R 5. *Cf.* also J. Ayscough, president of Jamaica council to Board of Trade, Oct. 2, 1726: "The best sort are exported by the South Sea Company to make good the Asiento Contract." Others, of the middle and lower sort, were reëxported by the private traders, but very few to the Northern Colonies. C.O. 137: 16, R 74.

renewed their opposition to the duties.[102] Relief was obtained for the company, at least, by an instruction against any such duties for the future on slaves brought in for refreshment only.[103] Richard Harris petitioned in vain in behalf of the private traders.[104]

The South Sea Company succeeded in retaining the favor of the Court, but by those who voiced the planting interest it was heartily censured for being an interest which was dwarfing the more productive industries of the empire.[105] It was pointed out by a Jamaican, in 1728, that since the company had established the factory at Havana, the Spanish had settled over thirty new sugar works whose product the English ships were carrying to European markets. An export duty on slaves sent to foreigners and the removal of the import duty was advocated, therefore, as a reasonable means of cheapening labor to the English planter. The South Sea Company offered for sale in Jamaica less than one-half its slaves, and those were its worst; the reason being that the charge of transporting poor negroes to the Spanish was as great as that for the best and the latter would bring double the price. "By those *Means* the *Spaniards* are now furnish'd with the Choicest Negroes which are to be had, and our own Plantations with the worst; whereas they formerly bought the good and bad together."[106] Not one-fifth of Jamaica's sugar lands were then settled and without a better slave supply no improvements ever could be made.[107]

102 Wescomb to Board of Trade, Mar. 14, 1726/7, C.O. 137: 16, R 70.

103 Order in Council, Nov. 2, 1727, C.O. 137: 17, S 9.

104 Harris to Board of Trade, Nov. 13, 1727, C.O. 137: 17, S 10.

105 In 1721, it was said there could not have been a better contrivance to so weaken Jamaica and that it would soon fall into the hands of the French. Charles King, *British Merchant*, 1721, III, 257.

106 *Observations on the Asiento Trade*, 1728, p. 21.

107 *Ibid.*, pp. 21-23, 32. "The *Plantations absolutely depend* on Negroes,

The Board of Trade, in 1729, instructed the governors to recommend substitutes for the import duties on negroes, inasmuch as such taxes in some measure raise the price of labor in the plantations.[108] But the Jamaica act of 1731 increased the import duty from 10s. to 15s., and the export duty to 30s. per slave. The violent opposition which this bill[109] met led the government to instruct the governor to withhold his assent from any duties of import or export on slaves for the future.[110] But the colony, and its governors also, ignored instructions and disallowances[111] and continued the duties annually.[112]

It is improbable, however, that these export duties had much effect in diminishing the export of negroes to Spanish America; the marked reduction after 1739 is to be attributed rather to the war with Spain and the lapse of the Asiento. Trelawny, in 1749, stated that then only

and without them must be thrown up, or be of inconsiderable value.'' *Ibid.*, p. 14.

108 Board of Trade Jour., Oct. 14, 1729, C.O. 391: 38, f. 261.

109 Petition of Jamaica merchants to H. M., referred to Board of Trade, July 7, 1731, C.O. 137: 19, S 129; petition of Bristol merchants, referred July 7, 1731, *ibid.*, S 130; petition of Liverpool merchants, referred July 7, 1731, *ibid.*, S 131. Gov. Hunter to Board of Trade, Nov. 13, 1731, said the colony was greatly distressed for revenue, which occasioned the act, *ibid.*, S 145. An elaborate defense by the council and assembly is given in a representation to H. M., Feb. 10, 1731/2, C.O. 137: 20, S 155.

110 Order in Council, Dec. 9, 1731, approves the instruction which was dated Dec. 10, C.O. 137: 20, S 168.

111 Gov. Hunter to Board of Trade, Feb. 19, 1731/2, reports his assent to the Act of 1732, C.O. 137: 20, S 156. Order in Council, Oct. 13, 1732, renewed the instructions against the duties, *ibid.*, 137: 20, S 190. A number of petitions for and against these acts are given in C.O. 137: 20, S 159, 160, 161, 162, 163, 164, 165, 180; 21, T 34, 36, 38.

112 In 1735, a new instruction for Governor Cunningham was authorized by which ''reasonable'' duties on negroes purchased in the island might be laid. Order of the Committee of the Council, May 10, 1735, C.O. 137: 21, T 41; order in Council, July 9, 1735, approving, *ibid.*, 137: 22, V 2. Gov. Trelawny to Board of Trade, June 8, 1749, stated that since this instruction the duties had remained the same each year, 10s. import and 20s. export. *Ibid.*, 137: 25, X 37.

one-fourth of the negroes brought in were reëxported.[113] The revenue produced by these duties averaged, in 1754, £10,000 per year.[114] Prior to 1739, between a third and a half of the slaves brought to Jamaica were reëxported to Spanish America. Contemporary observers regarded this deflection of labor as a primary cause of the comparatively dormant state of the sugar industry in Jamaica. It is conceivable, however, that without the incentive which the Asiento supplied to the slave trade, Jamaica would not have been even so well supplied as it was with African labor.

Another influence generally believed to be detrimental to the upkeep of slavery in the British West Indies was the ever increasing demand for negroes in the French islands. The French government had sought to encourage their introduction by its decree of September 27, 1720, which accorded an exemption of half the duties in France on sugars bought with a cargo of negroes.[115] The Dutch, who formerly sold their negroes to the English, now found a better market in the French who came to St. Eustatius in sloops loaded with sugar from Martinique and Guadeloupe.[116] But by 1726, the English traders had begun to rival the Dutch in supplying the French. The trading depôt was the neutral island of St. Lucia. "The British Ships," wrote Governor Hart, "go into a Harbour there, called Petit Carnage, Where they sell

113 Trelawny to Board of Trade, June 8, 1749, C.O. 137: 25, X 37.

114 Gov. Charles Knowles to Board of Trade, Jan. 12, 1754, C.O. 137: 27, X 200. The amounts for 1745-54 were:

1745	£ 9,514	1749	£ 6,376	1753	£10,000*
1746	7,283	1750	5,741	1754	11,150
1747	12,994	1751	5,366		
1748	17,628	1752	12,299		

*(appropriated —total not yet paid in.)

Rec. Gen'l's Acc., C.O. 137: 27, X 202.

115 Boizard et Tardieu, *Histoire de la législation des sucres, 1664-1891*, Paris, 1891, p. 6.

116 Gov. Hart to Board of Trade, Feb. 15, 1726/7, C.O. 152: 15, R 188.

their Slaves for money or sugar: This place having the advantage of St. Eustatia being within a few hours Sail of Martinique. If they sell for money they generally resort to these Islands to purchase Sugars for Great Britain: But if they dispose for Sugars, then they carry them for Ireland directly.''[117]

In a few years, the French ventured even to Barbadoes where a surplus supply of slaves could be bought at cheaper rates. In 1742, after the Spanish market was closed, Governor Robinson gave the following account of this trade: ''The Market that most eagerly offered was with France, as the French could not only carry down the Slaves they purchased to Markets where we had no access, but as they themselves had the Settlement of the Neutral Islands in View, & wanted Slaves to answer those Ends.

''This I take to be the obvious & most natural Cause of the Progress of this Trade, rather than of the Trade itself; for as I observ'd before, in a less & more private Degree, it had for many years passt been carried on:—I wish I could spare the Hint, that It had not heretofore received some Countenance from the Ships of War that had been Station'd in this Road . . . The large Prices those Dealers paid for Slaves gave great Encouragement to this Communication.'' Finally ''They ventured up to Barbados, & the Trade that was formerly carried on below, is now, I fear, conducted in the Houses of our Merchants here.''[118] Coppers and stills for their sugar

[117] *Ibid.*

[118] Gov. Thomas Robinson to Board of Trade, Nov. 27, 1742, C.O. 28: 26, Bb 1. This trade brought, in return for slaves, Spanish pistareens of silver, which passed at 1s. 3d. (=1s. sterling) though their specie value was believed to be a third less, soap, candles, wine, gold and silver brocades, laces, silks, calico, dimities, and cambrics. All this trade was, of course, in violation of the navigation acts, but the seizures in the last six years were insignificant, the governor's share of which did not annually exceed seven pounds.

works were also supplied to the French prior to the Austrian Succession War. The balance in favor of Barbadoes from this trade was thought to have been ten to twenty thousand pounds per annum, paid in specie.[119] A proposal, in 1757, to permit British ships to supply the French with slaves in return for indigo was rejected by Pitt as impolitic.[120] The only foreign trade of Barbadoes, in 1762, was reported to be in slaves and provisions to the Dutch at the Berbice and Demerara rivers on the Main in exchange for timber for sugar mills.[121]

The competition of the French on the African Coast also tended to raise the price of slaves. This attracted especial attention after 1739, when French traders on the Gold Coast, and particularly at Anamaboe, startled the English by their increase, and carried off, it was thought, ten times as many slaves as the British.[122] Captain Pye, who sailed along the African Coast in 1749, reported to the Admiralty that English, Dutch, and Danish traders were all complaining against the French, who had everywhere raised the price of negroes and were getting more and more control of the trade. At the African Company's forts on the Gold Coast he learned that the English company's best customers for slaves were the French on whom the former had come absolutely to depend for

119 Robinson to Board of Trade, Feb. 20, 1746/7, C.O. 28: 27, Bb 57.

120 William Pitt to Board of Trade, Sept. 30, 1757, concerning a petition of Hutchinson Mure to H. M., C.O. 388: 47, Ff 68; B. T. Jour., Feb. 2, 1758, C.O. 391: 65.

121 Gov. Charles Pinfold to Board of Trade, June 1, 1762, C.O. 28: 32, Ff 25.

122 Charles Hayes to Thomas Hill, secretary of the Board of Trade, Dec. 31, 1744, C.O. 388: 43, Bb 100. Hayes was an officer of the African Company on the Coast. He connected the amazing growth of the French sugar industry with their superior supply of slaves. Robinson stated, in 1742, that a ship just arrived from Guinea brought 123 slaves which the master was five months in procuring, and that he left forty French vessels on the Coast, trading for Negroes. Thomas Robinson to Board of Trade, Barbadoes, Aug. 29, 1742, C.O. 28: 25, Aa 112.

supplies, as English private traders had refused trading with them, and he was asked to leave the French traders undisturbed.[123] The managers of the African Company stated that it had been usual for them to supply the French with slaves, and they justified this trade by saying that it prevented the French trading directly with the natives. Moreover, Rhode Island and Bristol merchants no longer bought slaves from the company's factors.[124]

The reëstablished African Company, in 1750, forbade all trade with foreigners, excepting with the Portuguese in tobacco and gold. Such a restriction, remarked one of its agents, would "render that Service not worth any Man's Acceptance who means honestly to comply therewith."[125] The company's governor wrote home that his two chief agents, Husbands and Boteler, were determined to trade with the French; it was no use regulating against it; if the English refused to trade, the Dutch would take it up; the French offered the best prices and so controlled the market.[126] Postlethwayt, in 1756, expressed alarm at the French getting so many of the best slaves from Anamaboe (Gold Coast) and Whydah, and connected this fact with the extraordinary growth of their sugar plantations.[127] It continued the regular practice for the French

123 Capt. Thomas Pye to Admiralty, Mar. 18, 1749/50, C.O. 388: 44, Cc 73.

124 Richard Stockwell, governor of Cape Coast Castle, William Husbands, and Thomas Boteler, agents of the Company, to Capt. Pye, dated Cape Coast Castle, Jan. 18, 1749/50, C.O. 388: 44, Cc 75.

125 Thomas Boteler, agent at Cape Coast Castle, to Thomas Melville, president of the council at Cape Coast Castle, July 3, 1751, C.O. 388: 45, Dd 54.

126 Melville to the African Committee, Cape Coast Castle, July 23, 1751, C.O. 388: 45, Dd 56. The committee, *i.e.*, the governing body of the last African Company, adhered, however, to their instructions against trade with the French. *Ibid.*, Dd 57, 58.

127 Postlethwayt, *Short State of the Progress of the French Trade*, etc., London, 1756, pp. 83-86.

to buy their slaves of the English on the coast and in the West Indies.[128]

It is evident that the rise in price of slaves was not wholly due to the need of slaves on the new sugar lands of Jamaica, but was influenced also by a similar need in the French sugar islands. In 1726, new slaves sold in the Leeward Islands at £18 sterling,[129] Jamaica prices are given for 1739 at £35 currency, or about £24 sterling; for 1739-1741 at £47 currency or about £35 sterling; and for 1741-1745 at £50 currency or about £37 sterling.[130] In 1749, an able Anamaboe slave, purchased on the Gold Coast for £9 or £10, sold in the British West Indies for £29 or £30 sterling. In 1776, a negro of the same description, costing £27 18s. on the coast, sold in Jamaica at near £50 sterling.[131] In 1788, the average price given for Gold Coast negroes was £40 sterling; creole slaves were then bringing £50.[132] Jamaica, especially after the discontinuance of the Asiento trade, in 1739, suffered from an insufficient supply. "The markets have every year grown worse supplied; so that, at present," said Long in 1774, "the planters are unable to procure, at any terms, the number they require; besides . . . these are sold in course at most extravagant prices."[133]

For the expansion of the British sugar industry, in the eighteenth century, Jamaica contained the only large

128 Examination of Charles Spooner, planter, before Board of Trade, Mar. 1, 1788, B. T. 6 (Africa): 9, ff. 545-546. "The French have all along purchased Negroes from Foreigners. At present they buy them from the English." A number of the planters and traders examined in 1788 testified to the large supplies of slaves furnished the French in time of peace.

129 Gov. William Mathew to Board of Trade, Nov. 1, 1727, C.O. 152: 16, S 27.

130 *Importance of the Sugar Colonies to Great Britain*, 1745, pp. 7-8.

131 (Mr. Temple Luttrell), *Parl. Hist.*, XIX, 302.

132 Examination of Charles Spooner before Board of Trade, Mar. 1, 1788, B. T. 6 (Africa): 9, ff. 191-192.

133 Long, *Jamaica*, I, 506.

area suitable for the purpose. English public opinion showed great impatience with the slow progress of sugar culture in that island; it was thought that the planters were pursuing the methods of monopolists. There was some truth in this criticism. But in view of the evidence above presented, it can hardly be denied that difficulty in obtaining a sufficient stock of the right kind of slaves was also an important reason for Jamaica's slow development.

CHAPTER IV

THE EXCLUSION OF SMALL PROPRIETORS FROM BARBADOES AND THE LEEWARD ISLANDS

The competition for the best sugar lands in the British West Indies between the wealthy planter with fifty or more slaves and the small farmer with from six to ten slaves was clearly defined in the early years of the eighteenth century. Sugar culture was so hazardous an undertaking that the contingencies which might force the poor settler, with his land and slaves already mortgaged to local factors, to sell out to his stronger neighbor were numerous and of frequent occurrence. Hurricanes, droughts, and fires in the canes were dreaded calamities. The frequency with which they are reported and the accounts of them indicate that every few years hundreds of weaker settlers were forced into bankruptcy and their estates amalgamated with those of the rich. Great planters, through their connection with important mercantile houses in England, were usually able to save themselves during such crises. The restricted area of sugar land, coupled with a labor system most profitable only where extensive agriculture of the one crop type was possible, naturally made this struggle for land in the West Indies a keener competition than probably ever occurred for the cotton belt, a century later, in the United States.

In Barbadoes, as early as 1668, it was noticed that small planters, under pressure of taxes and high prices of provisions, were migrating to the French islands where the cost of living was lower, thus tending to leave the English islands inhabited only by masters of great plantations and their blacks. It was the military danger arising from this process which at that time caused alarm.[1] Emigration to the Northern Colonies assumed large proportions in the first half of the eighteenth century. "Many hundred Familys have gone from Barbadoes to Carolina and Pensilvania," wrote William Gordon in 1720. Most of them were debtors, who, in North America, however, became "Some of the Toppingest Inhabitants" and paid their old debts.[2] A depression in the prices of sugar and rum, in 1730 and the years immediately following, seems to have entailed great loss upon the poorer sort of people. "There are at present great numbers of Plantations abandoned in Barbadoes," declared a pamphlet in 1731.[3] Another writer of the same year stated that "Many of our planters are necessitated to forsake their antient and well-built estates in this island, and shelter themselves in Pensilvania and Virginia, and other northern colonies; many of which instances might easily be given."[4] The lowest prices for sugar were reached in 1733. The following year, the colonial agents complained of the defenseless condition of the island, due to the decrease of white settlers, which, they stated, "On the best information we can receive is owing to the great decay the Sugar Trade hath of late fallen into and to the weak and defenceless condition in which this island is at present . . . many of the Inhabitants from the great decay in their Sugar Trade having been necessitated to retire to North America and no new Setlers having come either in their place or to supply

1 Account of the English Sugar Plantations (Jan. 23), 1668, *C. S. P. Col., 1661-1668*, no. 1679.

2 William Gordon to Board of Trade, July 14, 1720, C.O. 28: 17, V 36.

3 *Answers to all Objections to the Bill*, etc., 1731, p. 2.

4 *Present State of the British Sugar Colonies*, pp. 10-11.

the great decrease in their White Inhabitants by the death of the old Setlers.'"[5] Lord Howe, the governor, reported that the whites were migrating not only to the Northern Colonies but also to the foreign islands.[6] English, Irish, and North American merchants, he added, "have Suffer'd very much of late Years in their Fortunes by the running away of the Inhabitants of this Island, who are indebted to them, with all their Slaves and Effects to South Carolina where they have been protected by a law of that Province from any Arrests or Suits for Such Debts as they have contracted.'"[7] It seems likely that, had the prices remained as low as they were at that time, Barbadoes would have been largely abandoned for the newer lands in Jamaica and the foreign colonies. Without an encouragement in trade, said one of the planters in 1735, "it will not be possible for the Planters, who are greatly in debt, long to Subsist, but as many of them have ruin'd their Plantations by running off the Island with their Negroes, & Land without Stock, is of little Value, so it is much to be feared others will follow their Example, & in a few years this Island will be possessed only by a small Number and not have Strength enough to manure above a quarter of it. Our produce of late Years has scarce Sold for more than it cost & if no profit ensues from great Labour and

[5] John Sharpe, Peter Le Heup, and George Teissier to Board of Trade, May 31, 1734, C.O. 28: 24, Aa 10. See prices of sugar below, pp. 134, 186.

[6] Lord Howe to Board of Trade, Nov. 7, 1734: "I believe you will find their numbers lessen'd, and daily decrease by the Inhabitants running off this Island, Some to our Northern Colonys and I am afraid many to Foreign Governments; I must here observe their Leaving this place is by some imputed, and I believe it may be an inducement to a few of them, to the weak and defenceless Condition in which this Island is at present and the great Danger there is of it's being taken in Case of War." C.O. 28: 24, Aa 28.

[7] *Ibid.* It should be noted that the South Carolina laws protected the debtors only for periods of five years or more. Channing, II, 417; Cooper, *Statutes of South Carolina*, II, 124.

Industry there will be a temptation to give it over and let the Land lye fallow.'"[8] So prevalent had the breaking up and abandoning of plantations become, in 1737, that the distress of Barbadoes was a common topic of conversation for merchants at the London coffee houses. "A great part of the Lands," wrote John Ashley, "lye unmanured for want of Negroes and Cattle, and the buildings unrepaired for Want of Spare Money, whereby some plantations do not make, by a large proportion, what they might otherwise Do.'"[9] To ease the condition of the insolvent planter, Ashley, who for several years was customs collector at Barbadoes, proposed[10] that his estate be worked under trustees appointed by the creditors and the profits applied toward the planter's debts. It does not appear, however, that the scheme was ever carried out.[11]

Some of the large planters also, at this time, left Barbadoes with their slaves and capital in search of fresh lands. "Many of the best planters," wrote Governor Robinson[12] in 1744, "were purchasing in the Dutch Settlements. This, I take the Liberty to assure Your Lordships, is a Truth most remarkably observed by my self, within these Twelvemonths; many of our Inhabitants

8 James Dottin, president of the council of Barbadoes to Board of Trade, Aug. 2, 1735, C.O. 28: 24, Aa 32.

9 John Ashley, "The Settlement and Insolvency Bill," 1737, f. 17. This manuscript of 25 folios describing the situation of the indebted planters, with a draft insolvency bill to alleviate their condition, was received and read by the Board of Trade, July 6, 1737. Of these debtors, said Ashley, "many hundreds are forced away and Settled amongst Foreigners." C.O. 28: 25, Aa 60.

10 *Ibid., passim.*

11 An insolvency act passed Jan. 21, 1740, was intended to relieve the debtors, prevent their deserting the island and carrying away their effects to the prejudice of their creditors. James Dottin to Board of Trade, Apr. 2, 1741, C.O. 28: 25, Aa 106.

12 Gov. Thomas Robinson to Board of Trade, May 10, 1744, C.O. 28: 26, Bb 11.

having purchased, and others Removed to Isaqueba: the Consequence I apprehend will, in time, be very fatal to the Charribbee Isles; . . . because

"(1) The Gentlemen who purchase in Isaqueba, in order to settle their Sugar-Works, of which the Dutch are at present perfect Strangers, strip this Island of many Usefull Workmen, who, no doubt, where they are more wanted will find a great encouragement to stay; and, in Consequence, others will be incited to follow them: By this means, our Militia will be broke in its Strength, and the planters here deprived of their further Service as Workmen.

"(2) The Dutch will have an in-sight, not only in improving their Sugar works, but in Boyling, Curing, and Improving the Sugar themselves, as well as in all parts of Distilling, and meliorating their Rum, Melasses & Sugar. And as their Land is now New, and very fertile, and the produce, in consequence Larger & cultivated at less Expence, they will be enabled to supply and under sell us at Foreign Markets.

"(3) The great opening it will give to Frauds in the Revenue, where sugars, and spirits thus produced in the Dutch Settlements, may, while in the Hands of the English, who reside there, be Collusively Shipped to the Northern, or other Colonys.

"(4) The precarious Title the English may have in his New-purchase: For tho' with his Land, he purchases, in imagination, such a Number of blacks as he thinks equal to the Culture, yet, such blacks are not in Fact Slaves, as in the English Plant[s] but Indians by Birth, and as it were Villians Reguardant to the Soil, so that when the Dutch are sufficiently improved in the knowledge of the Sugar-Trade &c tho' they may not immediately disturb the English Proprietor in the possession of his Land, yet the Indian by which that Land is cultivated, may be encouraged to revolt; and by this indirect means, the Land will become equally useless to the English purchaser, and he must in Consequence soon relinquish it, or part with it back to the Dutch, on their own Terms." Three years later, Robinson again spoke of the danger to the English sugar islands of this moving away of the planters, and thought a stop should be put to it.[13] The white population of Barbadoes, from 1740 to 1748, showed a decline of 2551, "The Reason of which Decrease," wrote Governor Grenville, "is, That several of the Inhabitants have Quitted the Island with their Familys." The decrease of blacks, in the same period, appears to have been 3313.[14] The census accounts, such as they are in the eighteenth century,[15] indicate that the population—white and black—reached its maximum about 1740, after which the population remained stationary or actually decreased. In 1762, the governor reported that, according to the parish registers of births and burials, "the Inhabitants are within ten Years decreas'd 442—in which are not computed those who have left the Island."[16] The population, which in this year was 18,419 whites and something over 70,000 blacks,[17] had still further decreased by 1786 to 16,187 whites and 62,115 blacks and 1,838 free people of color.[18]

In the production and shipping of Barbadoes there was

13 Gov. Thomas Robinson to Board of Trade, Feb. 20, 1746, C.O. 28: 27, Bb 57.

14 Gov. Henry Grenville to Board of Trade, Feb. 8, 1748, C.O. 28: 29, Cc 28.

15 The deputy secretary's remark appended to the census of 1757 states: "Tis Allow'd on all Hands That the above is a very imperfect List & fall(s) at Least ⅓d short, of the real Number of Inhabitants wch from the Nature & Disposition of the people cannot more exactly be come at." Richard Husbands, "Census for 1757," C.O. 28: 31, Ee 8.

16 Gov. Charles Pinfold to Board of Trade, June 1, 1762, C.O. 28: 32, Ff 25.

17 *Ibid.*

18 Edwards, *West Indies*, I, 351.

a corresponding decline during these years. In 1736, Barbadoes produced 22,769 hogsheads of sugar, from 1740-1748 the annual average was 13,948 hogsheads, and in the years 1784, 1785, and 1786 the annual average had decreased to 9554 hogsheads.[19] The shipping to Barbadoes showed a general decline in tonnage from 1735 to 1748, then a revival followed by a further decline. The domestic shipping also shows a general decrease from 1735 to 1760.[20]

[19] Edwards, *West Indies*, I, 352. See also Chart I, facing p. 98.

[20]

	Number of Vessels trading to Barbadoes, exclusive of those belonging thereto.			Number of Vessels belonging to Barbadoes.		
Year	Number	Tons	Men Aboard	Number	Tons	Men Aboard
1735	392	16,814	2523	56	1790	314
1736	282	16,180	2316	49	1760	301
1737	270	16,114	2292	58	1782	309
1738	306	17,112	2614	51	1801	321
1739	290	16,586	2574	46	1701	302
1740	253	15,292	2474	51	1775	299
1741	361	21,380	3356	36	1340	237
1742	376	22,832	3358	31	1265	211
1743	300	17,178	2622	37	1540	216
1744	180	10,116	1511	43	1637	237
1745	218	12,423	1822	29	1283	169
1746	265	17,623	2559	24	1302	163
1747	214	14,982	2198	26	1590	229
1748	221	15,390	2184	31	1586	249

Number of All Vessels entered at Barbadoes.

	Number	Tons—Annual Average			
1751	511	21,010	(1745-9)	36	1622
1752	574	34,491½	(1749-53)	58	2477
1753	608			40	1466
1754	650			52	1603
1755	604	65,206	(1753-56)	45	1444
1756	446			35	1411
1757	292			33	2011
1758	230	20,740	(1756-59)	39	1626
1759	302			34	1132
1760	246	21,460	(1759-60)	36	978

The above accounts to 1749 were enclosed with Gov. Henry Grenville's letter

These facts make clear the general progress of the sugar industry in Barbadoes. The island was the oldest of the British sugar colonies, all its land had been brought under cultivation by slave labor at the opening of the eighteenth century. A general exhaustion of the soil had been reached as early, at least, as 1689, at which time the artificial fertilization of the ground was universally practiced.[21] Such intensive agriculture necessitated larger forces of slaves which, in turn, made the planters more dependent on outside sources of supply.[22] At the same time, more cheaply produced sugar from the fresh lands in Jamaica and elsewhere still further decreased the profits of Barbadian planters. Under these circumstances, the smaller planters were gradually forced out of the island. This process appeared to be at its climax from 1730 to 1740. Some of the large planters themselves, in the following decade left the island and purchased fresh lands in South America. But an increasing demand for sugar and a decided rise in its price saved Barbadoes from anything like general abandonment. The large planters continued to find it profitable to produce sugar on the old lands even under a most expensive system of intensive agriculture.

In the Leeward Islands the same tendency of the small planters to give way to large scale production took place. The four principal islands in this group were

to Board of Trade, Feb. 8, 1748, C.O. 28: 29, Cc 28; those for 1751-1760 were with Gov. Charles Pinfold's letter to Board of Trade, June 1, 1762, C.O. 28: 32, Ff 25.

21 *Groans of the Plantations*, 1689, *passim.*

22 William Sharpe, president of the council and commander-in-chief of Barbadoes, in an address to the council and assembly, May 4, 1714, comparing the younger sugar colonies with Barbadoes, said: "Their Fresh fertile Soil yields them, without Manure many Cropps from one planting; our's requires the utmost Art, Industry, and Manure, and that too annually. If this Deficiency be not supply'd, they will undersell, and consequently drive us from the Market." C.O. 28: 14, T 44.

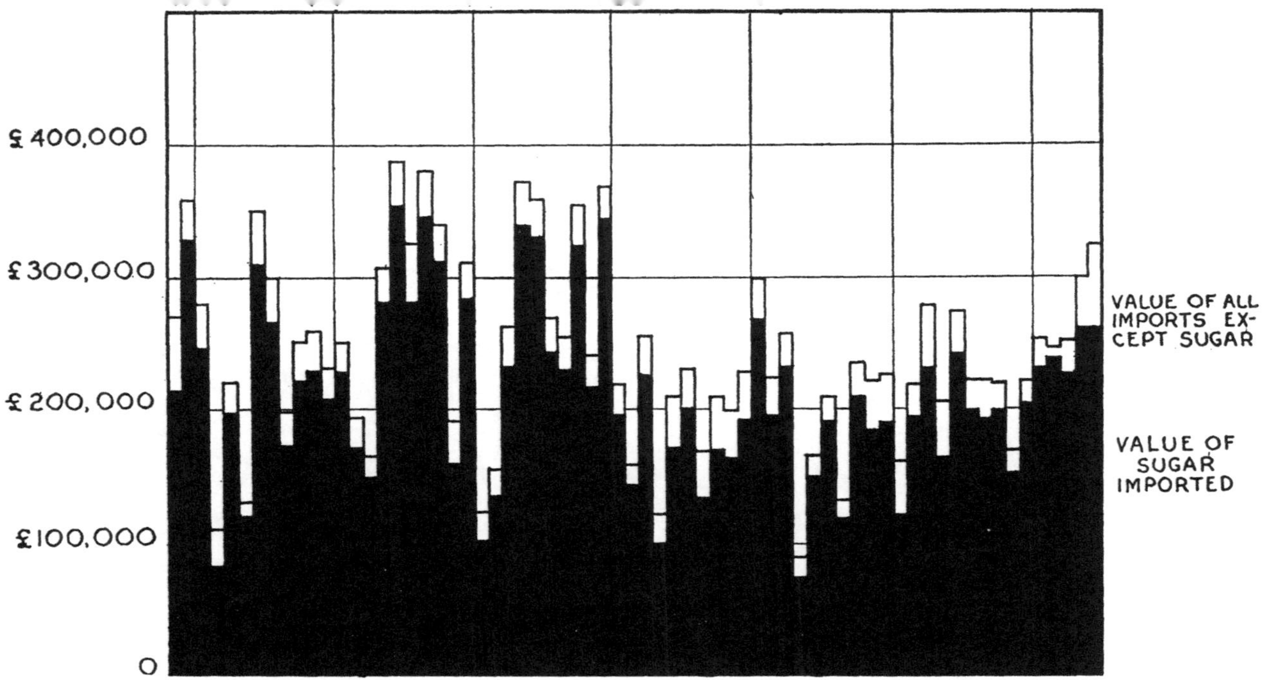

CHART I. VALUE OF THE YEARLY IMPORTS INTO ENGLAND FROM BARBADOES

The shaded portion of the above chart represents the total value of imports of sugar into England from Barbadoes; the upper area which is not shaded represents the total value of all other imports into England from Barbadoes. The diagram is based on tables compiled from Custom House Accounts, Ledgers of Imports and Exports. Statistics for Antigua, Montserrat, Nevis, and St. Christopher show similar characteristics.

brought under sugar culture during the latter half of the seventeenth century. No unpatented land remained in any of them in 1724.[23] A certain area of virgin soil remained in all these islands and also pasture and provision grounds. But in the thirty years following, all the virgin land as well as a great part of the ground for pasture and provisions, had been brought under canes. "Almost every Inche of the Leeward Islands is cultivated," wrote Governor Thomas in 1754, and "from thence the Price of fresh Provisions is double what it is at Jamaica, where they have great Tracts of Pasture Land."[24] The artificial fertilization to which the planters were driven, required, as compared with fresh land, a very much larger amount of labor per acre. To keep up production, overseers more and more turned provision lands into cane fields, and for their food supplies depended on outside sources. Poorer planters, however, who tried to follow the methods of wealthy neighbors found themselves running deeply in debt for slaves and provisions. When a failure of crops or a great depression in prices came, the small proprietors were either undone completely or sold out at a great sacrifice to the large planters. The methods of the large planters in the Leeward Islands were thus described, in 1717, by a resident of St. Christopher: "According to the present method of the Possessors (the richest Persons having large Quantities of Ground) they plant as long as the Grounds will bear without Husbandry, then leave them in

23 Gov. John Hart to Board of Trade, July 12, 1724, C.O. 152: 14, R 101. The amount of "manurable land" was then given as follows:

Antigua	55,000 acres
St. Christopher . .	22,000 acres
Nevis	11,000 acres
Montserrat . . .	9,000 acres

24 Gov. George Thomas to Board of Trade, May 22, 1754, C.O. 152: 28, Bb 16.

a poor Condition, and proceed to take up Fresh Grounds, & by being in Favour with the Governours, they frequently got grants of what People of less Substance had made fertile, which in process of Time would not only have made the whole Island unprofitable to the Crown, but depopulate it also by being unfruitfull & barren, and the Tempers of the Suffering People have been so sour'd at this Usage, that above a hundred of them, have already left the Island. . . . There are few Families can manage above 200 Acres, Tho' some at present enjoy above thrice that Quantity; And from thirty Years Experience I have observed; that Grants of large Quantities of Lands to Favorite Families have been the greatest Detriments to the Leeward Islands.''[25]

The dispersion of the poor inhabitants of the Leeward Islands among the other settlements was deemed by the government, at this time, a great danger to the islands from a military standpoint,[26] and called out strong protests and demands for reform. Thus Governor Hart, in 1724, told the Antiguans that "The real cause why there are so few White People is, that the wealthy Inhabitants of this Island have Ingross'd such vast Tracts of Land, that there is not Room for a Number of poorer Inhabitants to invite them to Settle amongst You.''[27] A drought in Antigua and Nevis, in 1726, lasted eight months and occasioned the loss of many cattle and negroes. "And many of the midling and poorer Planters,'' wrote Governor Hart, "will be utterly undone. For they are not only

[25] "Philo Patria'' to Board of Trade, Aug. 23, 1717, C.O. 152: 12, P 42.

[26] Stanhope to Sec. Popple, Treas. Chambers, Nov. 27, 1717, C.O. 152: 12, P 55.

[27] Gov. John Hart's address to the Antiguan council and assembly, Dec. 5, 1724, C.O. 152: 15, R 130. In his address of Jan. 15, 1724/5, Hart said: "The Bulk of the Lands are now in a few hands, and Cultivated by Negroes.'' C.O. 152: 15, R 132.

disappointed of their Crops for two Years But are oblig'd to buy Provisions for themselves and Negroes, upon Credit from the Merchants.'"[28] The development of great plantations in Antigua, at the expense of small estates, was thus described, in 1734, by the colonial agent who had been in Antigua: "The decrease of white men in the Said Island I apprehend to be owing to Several Causes. Epidemicale Distempers have destroyed Numbers, Dry Weather, Want of Provisions, And Inability to pay their Taxes have obliged others to go off. Land has been at so high a price from the Smallness of the Quantity in the said Island that the Settlers of Ten or Twenty Acres who formerly rais'd Only provisions have been tempted to sell their Possessions to the Sugar Planters & have thereupon quitted the Island; but I must observe to their Lordships that this Alteration tho' it may have Occasion'd the loss of some Inhabitants, has been in General beneficial to the Trade Navigation & revenue of Great Britain, All the improvable land being by that Means employ'd in the raising of Sugar, & provisions coming to them now, Chiefly from his Majesties other Dominions in English Bottoms. But notwithstanding this alteration, there are very few persons in the Said Island at present possest of above, or even so Much as 300 Acres of Land fit for Sugar; and without Such a quantity or Something near it, no Planter Can be enabled to bear the great Expence of the Buildings & Utensils Necessary for making of Sugar, Especially Considering the Low price that Comodity has sunk to, for Several years past.'"[29] Many of these poor families in the Leeward Islands moved to the Danish island of Santa Cruz; Governor Mathew was much annoyed by this "Spirit of

[28] Gov. John Hart to Board of Trade, May 20, 1726, C.O. 152: 15, R 166.
[29] Yeamans to Board of Trade, May 27, 1734, C.O. 152: 20, V 29.

turning Renegado.'"[30] The migration of whites from Nevis, which attracted special notice after 1730, was attributed to the great fall in prices of sugar during that decade.[31]

The French half of St. Christopher, by the peace of Utrecht, was ceded to the English. This cession put at the disposal of the Crown about 20,000 acres or more of sugar land. Before this area was finally distributed, it was allotted in provisional grants[32] to English planters, and so remained until 1726. During the interval, various schemes for its settlement were proposed. These model plans sought to remedy two dangers to which the island was exposed by the large scale production of sugar: first, military insecurity in case of insurrection or invasion, occasioned by the exclusion of the poor whites; and second, undue dependence on the Northern Colonies and other places for food, lumber, horses, and cattle.

In Governor Douglas's plan, the worst lands and those along the seashore were to be divided into ten acre plots and given gratis to poor white families. Such land was to be held by a kind of feudal tenure which required the holder either to serve in the militia or furnish an armed man. It was expected that these plots would be suitable for raising stock and provisions. Thus the island would be skirted by a fringe of armed white people to defend it against invasion or an insurrection of the slaves. The interior was to be sold in plantations of not more than two hundred acres to one family. Each planter should

30 Gov. William Mathew to Board of Trade, Mar. 19, 1733/4, C.O. 152: 20, V 35; same to same, May 31, 1736, C.O. 152: 22, W 47.

31 Thomas Butler, agent for Nevis, to Board of Trade, rec'd June 6, 1734, C.O. 152: 20, V 30.

32 Gov. Walter Hamilton to Board of Trade, Apr. 10, 1716, C.O. 152: 11, O 131; same to same, Apr. 12, 1721, with list of the holders, C.O. 152: 13, Q 68, 69. Official notice of sale and nullification of all provisional grants, Aug. 6, 1726, C.O. 152: 15, R 176.

keep one white servant for every forty acres the first year, and one for every twenty acres after three years.[33] Such land should be put under canes within a limited time or be subject to such a duty as the Crown would probably derive if they were planted. This scheme was strongly advocated by the succeeding governor, Walter Hamilton.[34] "For I must Observe to your Lordships," wrote Hamilton, in 1716, "what has weakened these Islands most has chiefly been Occasioned by rich mens buying out the Poor of their Little Settlements by this means they have been in time drove off the Islands."[35] For us the chief interest of these schemes lies in the supposed evils which they point out in the existing agricultural system. They received no support either from the planters or the government, and were soon dropped.

The estimation in which these fresh lands in St. Christopher was held is clearly shown by the great number and amounts of the bids made for them. The principle of sale by auction was decided upon in 1715.[36]

[33] Gov. Walter Douglas to Board of Trade, July 14, 1713, C.O. 152: 10, O 26. A similar scheme was suggested in an unsigned memorandum of 1714. This provided 4000 acres in 10 acre plots for the poor whites, and 14,000 acres for sugar plantations:

5	plantations	of	300	acres	each
8	plantations	of	250	acres	each
15	plantations	of	200	acres	each
20	plantations	of	150	acres	each
25	plantations	of	100	acres	each
30	plantations	of	50	acres	each
20	plantations	of	25	acres	each
			14,000	acres	

Also 2000 acres were to be divided between the governor, lieut. governor, the church, and the schools. C.O. 152: 10, O 78.

[34] Gov. Hamilton to Board of Trade, Apr. 12, 1715, *ibid.*, O 87.

[35] Hamilton to Board of Trade, Apr. 10, 1716, C.O. 152: 11, O 129. *Cf.* same to same, Mar. 15, 1717, C.O. 152: 12, P 106.

[36] Order in Council, Apr. 15, 1715, C.O. 152: 11, O 152. The treasury was authorized to appoint commissioners for the sale by this order.

Persons desirous of purchasing the whole or parts of the French cession were to deliver, on or before August 24, 1717, their sealed proposals to the Board of Trade.[37] A large number of bids were made, some by groups of capitalists for the entire area, some by West Indians, but absentee planters in England were the principal and highest bidders.[38] One company of promoters proposed to purchase 10,000 acres for £16,000, one-third payable on accession, and two-thirds within a year. They promised that 7000 acres would be cultivated within two years. This land would produce, they were assured, 3000 lbs. of sugar per acre; this would mean, in customs to the government, £5 per acre or £35,000 for the 7000 acres.[39] Many of those then occupying the land on provisional grants, offered through their London merchants £5 per acre for estates ranging from 100 to 300 acres.[40] This was thought a reasonable price for the lands.[41] Several of the large bidders proposed to give six to ten acre plots to the poor whites, and to maintain a white servant for every 20, 30, or 40 acres for the sake of defense.[42] A number of bids were made for 200 acre estates at £6 per acre or as much more as any should offer.[43] For four plantations in "Capisterre" (Cabeca terre) of 2000 acres as much as £8 per acre was offered.[44]

37 Announcement by treasury commissioners, 1717, C.O. 152: 12, P 18; advertisements in the *Daily Courant* (London), *ibid.*, P 21.

38 A large collection of these proposals are filed away in C.O. 152: 12.

39 Stanhope to Board of Trade, June 22, 1716, enclosing the scheme, C.O. 152: 11, O 143; Charles Douglas to Board of Trade, July 6, 1716, *ibid.*, O 145: same proposal.

40 Proposals of John Mills and others for 16 estates of 2740 acres in all, Aug. 23, 1717, C.O. 152: 12, P 18.

41 *Ibid.*, P 19-33.

42 John Mills to Board of Trade, Sept. 19, 1717, C.O. 152: 12, P 40.

43 Rowland Tryon to Board of Trade, Aug. 30, 1717, in behalf of the several persons named, *ibid.*, P 35.

44 William Codrington to Board of Trade, Sept. 12, 1717, *ibid.*, P 41.

The number of provisional holders in 1718 was 97,[45] and in 1721 it was 137.[46] All these grants were annulled, in 1726, and three commissioners were sent to St. Christopher to sell the French lands in parcels of not over 200 acres to the highest bidders.[47] This arrangement, excluding, as it did, the smaller planters from the French half of the island, gave a further impetus to large scale production in St. Christopher.[48]

Some of those small planters who were forced out of the larger sugar islands moved off with their few slaves to a group of smaller and less productive settlements known as the Virgin Islands. Here they raised a little sugar, rum, cotton, and provisions. The canes were so poor that the raw juice from them was as a rule made into rum, which was consumed by the inhabitants.[49] The population of these islands, in 1717, was given as follows:[50]

45 Gov. Hamilton to Board of Trade, Sept. 10, 1718, *ibid.*, P 142.

46 Same to same, Apr. 12, 1721, C.O. 152: 13, Q 68, 69.

47 H. M. Commission to William Mathew, Gilbert Fleming, and Edward Mann, for managing the purchases, June 4, 1726, C.O. 152: 15, R 175. Official notices of sale and nullification of the provisional grants, St. Christopher, Aug. 6, 1726, *ibid.*, R 176, 177. It was stated, in 1764, that good lands in St. Christopher had often sold at £100 per acre. Anon., *Considerations which may tend to promote the Settlement of our New West-India Colonies*, London, 1764, p. 41.

48 Richard Coope, agent for St. Christopher, writing to the Board of Trade, May 22, 1734, on the exclusion of poor whites, in addition to the causes at work in the other islands, attributed it to "the late Sale of the French Lands which has put them into fewer hands." C.O. 152: 20, V 38.

49 Capt. Chandler of H. M. Ship *Winchelsea* to Sec. Burchett, May 12, 1717, C.O. 152: 12, P 12. Chandler, who had just visited these islands, said their sugar canes were too short to make sugar; "they make only a little Rum of them, which they drink themselves." He thought it not worth the government's while to settle these islands. They "will always be a nest for Pirates as are now."

50 Enclosures in letter from Gov. Walter Hamilton to Board of Trade, Mar. 7, 1717, C.O. 152: 12, P 68, 70, 71, 72, 73.

	Men (Planters)	Women	Children	Slaves
Anguilla	96	154	234	824
Spanish Town	53	60	204	308
Tortola	37	34	38	176
Crabbe Island	46	—	—	62

In Anguilla, it appears that the number of planters who had twenty or more slaves was only eight; the average number of negroes to each settler was 8.6. Spanish Town had six slaves to each planter; Tortola had five; Crabbe Island had a little over one, the highest number to any planter being seven. Some, who had exhausted the lands here, were moving to the Danish islands of St. Thomas, St. John, and Santa Cruz, and to the Dutch colony at St. Eustatius, where lands could be had at very low rates or for nothing.[51] Governor John Hart, who visited the Virgin Islands in 1724, said they were poor, barren, and incapable of improvement. The inhabitants, in their clothing and diet, showed every sign of poverty. "I found that the first Inhabitants," wrote Hart, "were such as had fled from Barbadoes, and the Greater Islands of this Government for Debt, or to avoid the punishment of their Crimes, and have since been increased by Pirates, who have come in upon Acts of Grace, and are Married and Settled there." He observed a general absence of government and legal forms and that piracy was encouraged.[52]

Governor Mathew stated, in 1734, that these islands produced no sugar for export, their chief products were

[51] Hamilton to Board of Trade, 1717? (undated), C.O. 152: 12 (unnumbered); "Philo Patria" to Board of Trade, Aug. 23, 1717, *ibid.*, P 42; minutes of council and assembly of Antigua, Mar. 31, 1718, *ibid.*, P 114; Hamilton to Board of Trade, June 14, 1720, C.O. 152: 12.

[52] Gov. John Hart to Board of Trade, St. Christopher, July 12, 1724, C.O. 152: 14, R 96.

cotton and provisions.[53] Cotton could be raised with very much fewer slaves than sugar, and the Virgin Islands found this their most profitable export. In 1743, Anguilla, Spanish Town, and Tortola, with 3000 slaves, made about 1000 hogsheads of sugar and 1,000,000 lbs. of cotton.[54] In 1751, they were making the same amount of sugar with about 8000 slaves in all. It was then thought that, with more slaves, they could have made three times the quantity of sugar.[55] The backward state of society in these islands was noticed by all who visited them. Postlethwayt wrote of Anguilla in 1774: "They have no great quantities of sugars upon the island but addict themselves rather to farming, in which they have had very good success; and this it is that enables them to live in the old patriarchal way, every man being a kind of sovereign in his own family, and no other Government there is in Anguilla.[56] Barbuda also was chiefly employed in raising cattle, corn, and other provisions, which generally came to a good market in the sugar islands.[57]

[53] Gov. William Mathew to Board of Trade, Aug. 31, 1734, C.O. 152: 20, V 46.

[54] Robert Dinwiddie to Newcastle, endorsed, Aug. 1743, C.O. 5: 5, f. 202.

[55] Lieut. Gov. James Purcell to Sec. Thomas Hill, July 11, 1751, C.O. 152: 27, Aa 39; Purcell to Boyd, a London merchant, Nov. 26, 1751, *ibid.*, Aa 41.

[56] Postlethwayt, *Dictionary of Commerce*, 1774, Art., "Sugar Colonies, British."

[57] *Ibid.*

CHAPTER V

THE EXCLUSION OF SMALL PROPRIETORS FROM JAMAICA

By far the largest of all the British sugar colonies was Jamaica. It was in this island that the great "frontier" of fresh land lay, and it was to this area that the British sugar consumers of the eighteenth century looked for an increase of production. In this hope, however, they were to be greatly disappointed, for the extension of sugar culture in Jamaica proved an exceedingly slow process. This slow development is to be attributed very largely to the agrarian system, the social ideas, and the policy of restricted production followed by its great planters. The attitude of the consuming classes in England first found expression in innocent hopes, but these gradually changed to mild criticism and, finally, bitter invective.

As soon as it was observed that sugar culture under slave labor required for its best results a system of extensive agriculture, the more wealthy planters began monopolizing the sugar lands. The small plantations and the little provision farms of the poor whites were purchased by the wealthy and consolidated into great estates.[1] These planters came from the rural aristocracy of England, and the transformation of the landscape in

[1] Richard Harris told the Board of Trade, Mar. 20, 1724/5, that "the decrease of Small Freeholds, was by reason of the greater Eating up or buying out all the lesser planters & keeping vast tracts of Land unoccupied." C.O. 137: 16, R 8.

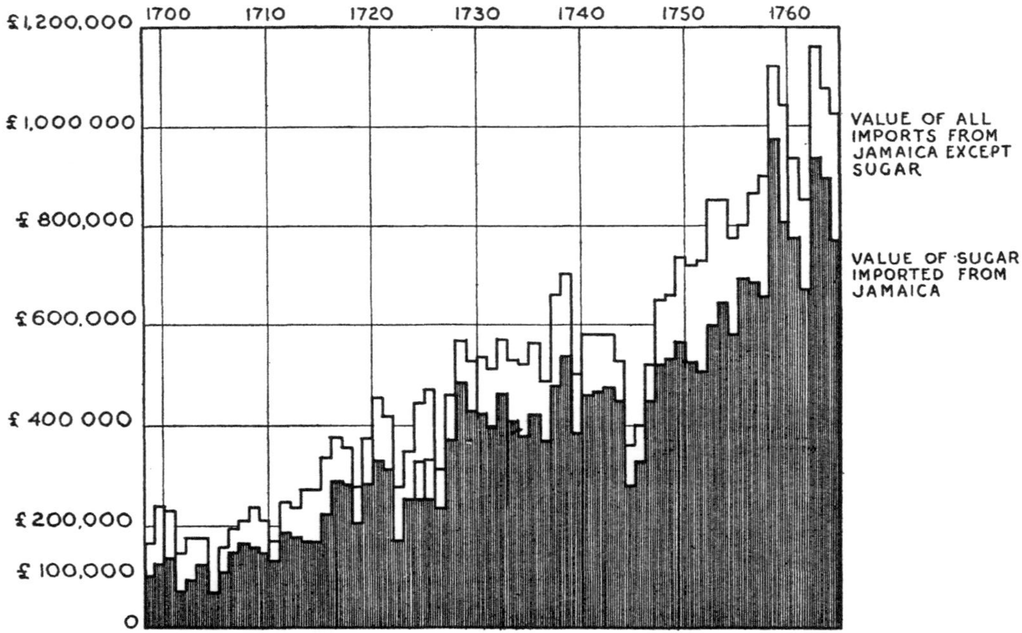

CHART II. VALUE OF THE YEARLY IMPORTS INTO ENGLAND FROM JAMAICA

The shaded portion of the above chart represents the total value of imports of sugar into England from Jamaica; the upper area which is not shaded represents the total value of all other imports into England from Jamaica. The diagram is based on tables compiled from Custom House Accounts, Ledgers of Imports and Exports.

Jamaica was largely in conformity with the social ideas of that class. Poor white districts gradually disappeared and in their places appeared cattle and sheep pastures or vacant tracts of land. But the exclusion of the whites, in the interests of large scale agriculture, was attended with disadvantages that greatly embarrassed the sugar planters. It was found unprofitable to employ slaves in the production of provisions which the whites had formerly raised. In consequence, prices of food and the general cost of living were greatly enhanced.[2] The colony became increasingly dependent on North America for its flour, grain, and lumber, but as Northerners preferred to buy their sugar, molasses, and rum from the French, who sold it cheaper, this tended to drain Jamaica of specie. Then, the absence of whites exposed the island to a chronic state of negro insurrection, which was not terminated till 1739, and which seriously retarded the development of the colony.

[2] The expense of living in Jamaica was a common complaint from travelers and prospective settlers in the colony. Gov. Lawes said in 1719: "If it were possible to regulate the Markets, so that Poor Men and their Families might be able to live and settle among Us, it would be one good Way to multiply our People, who are much too thin, and the Rich and Wanton forsake Us every Day." Speech to the council and assembly, Oct. 20, 1719, C.O. 137: 13, P 65. Again, in 1720, Lawes spoke of "The Impossibility of their [poor whites] Living here, under the Prices that govern your Markets; which, to my Knowledge, has Obliged several Families to leave Us, for the sake of living cheaper in the Northern Colonies." Speech of June 24, 1720, C.O. 137: 13, P 89. "I'll assure you half a Crown in England," wrote an army officer, "will go farther than a Pistole (about £1) here; I shall think my Self well off if this Expedition costs me only a Thousand pounds Extraordinary; I have taken a Little house here at a Place called Sigony the Pleasantest Part of the Island, but no Better than an English Barn (for) which I am obliged to pay two hundred pounds a Year; for my Cook which is a very indiffert one fifty; & every thing Dear in proportion, you may Judge what a pleasant Situation I am in." Col. Robert Hayes to Maj. Soule, Mar. 11, 1730/1, C.O. 137: 19, S 118. Taxation per capita in Jamaica, in 1754, averaged 16s. 8d., Gov. Knowles to Board of Trade, Jan. 12, 1754, C.O. 137: 27, X 200.

In its early stages this monopoly of land, in weakening the defensibility of Jamaica, led those in authority to view it as an abuse. Governor Hamilton, who still had, in 1715, a certain amount of Crown land at his disposal, thus described the demand for fresh soil and the agrarian situation in general: "Certain it is many have Considerable Tracts of which great part is uncultivated; Notwithstanding which I have dayly Applications for further Grants, which upon Surveys taken thereof not Exceeding Five Hundred Acres in One Grant is by Custom Esteemed as of right. Tho' indeed often there is little Intention of Settling the Same; but rather to form themselves a Sort of Barrier against an approaching Neighbour. This I have thought a great abuse and have upon the occasion I have mention'd, refused giving the Warrant of Survey, and I shall Still be more Cautious for the future in giving grants of Land but upon very good grounds, That there may be a Sufficient Quantity remaining ungranted to be disposed of in the best manner for the good of the publick. But the greatest part of the Valuable Lands Unsettled has been long Since patented and now in hands who neither Cultivate nor care to dispose of it, I shall endeavour all I can to procure proper remedys which I'm affraid will meet the greatest opposition."[3] At this time, many proposed that all tracts of land exceeding one hundred acres, whereof no part was planted or inhabited with whites, should be taxed or be surrendered to the Crown, and regranted only on condition of being cultivated within a limited time.[4] The Jamaica council was aware of the great need of a community of small provision farms. But the only land

3 Hamilton to Board of Trade, Aug. 30, 1715, C.O. 137: 11, O 10.

4 Memorial from Jamaica planters and others to Board of Trade, rec'd Mar. 25, 1715, C.O. 137: 10, N 127. The motive stated for better settling the colony was "thereby securing that Island against the Insurrections of Negroes and Invasion of Enemys."

it had to grant for this purpose lay "mostly in uninhabited parts of the Island very distant from any markets." Furthermore, the expense of settling had so increased that small farmers could not make the start without the loan of a slave or two, tools, and provisions. Such assistance the English government could not be prevailed upon to give, and Jamaica itself was not yet ready to make the attempt.[5]

To counteract the disadvantages arising from a system of great plantations, Jamaica began, about 1720, to experiment with a number of land schemes which sought to bring back and reëstablish the small farming class. None of these measures, which extended over a generation, met with the desired results. They call attention, however, to the peculiar problems to which the agricultural system of the island gave rise.

The poor whites of the Virgin Islands with their slaves were invited, in 1720, to come and settle in the eastern end of Jamaica.[6] Governor Lawes hoped to get about two hundred families in this way. Each family was to have ten acres of Crown land and other encouragements. The district was well timbered and it was believed the lumber, staves, and provisions from such farms would find a good market in the island. It does not appear, however, that the Virgin Islanders availed themselves of the offer. Lawes, at the same time, called attention to about 60,000 acres, owned by about 150 persons, in the northeastern part of the Island, "Patented Forty-five

[5] Memorial of Jamaica council to Board of Trade, Mar. 13, 1715/6, C.O. 137: 11, O 47. It was often suggested to the Board of Trade that England should supply to pioneers a part of their initial capital, as the French were doing in Hispaniola. Gov. Lawes to Board of Trade, Dec. 6, 1719, C.O. 137: 13, P 62; representation of London merchants to Board of Trade, 1724, C.O. 388: 24, R 155; Gov. Robert Hunter to Board of Trade, Dec. 24, 1730, C.O. 137: 19, S 124.

[6] Lawes to Board of Trade, Aug. 24, 1720, C.O. 137: 13, P 88.

Years ago, and not one Settlement or Stick fallen, and very little or any Quit-rent paid to His Majesty.'"[7] The Jamaica government, in 1722, undertook to reinvest all this land, whereon the quit rents were unpaid, in the Crown, indemnifying the holders in 1719 at the rate of five shillings per acre.[8] The succeeding governor, the Duke of Portland, sought to overcome the tendency toward monopoly by securing the execution of this scheme. "Open a free Passage," he urged, "into those vast Tracts of uncultivated Lands which are now shut up under the Notion of Property, tho' they are useless, Unprofitable, and of no Service."[9]

Governor Hunter, in 1730, reported that the island was undergoing a considerable decrease in white people. This was not owing, he said, to the want of land to grant new settlers, but to the great expense of starting a plantation. Proposals to give credit to new settlers for a number of negroes, utensils, and provisions had often been suggested, but without success.[10] Some had hopes that the culture of coffee would provide a profitable industry for the smaller planters. Coffee had been first cultivated in Jamaica by Sir Nicholas Lawes in 1728.[11]

7 Lawes' speech to council and assembly, Oct. 4, 1720, C.O. 137: 13, P 100; Lawes to Board of Trade, Apr. 20, 1721, C.O. 137: 13, P 126; Lawes' speech to council and assembly, June 24, 1721, C.O. 137: 14, Q 7.

8 Lawes to Board of Trade, Aug. 28, 1721, C.O. 137: 14, Q 16; same to same, Dec. 10, 1722, *ibid.*, Q 50; Duke of Portland, governor of Jamaica, to Board of Trade, Mar. 4, 1723/4, advocating the measure, C.O. 137: 14, Q 84; Duke of Bolton's opposition to a forced sale of his wife's claim of 1000 acres at 5s. an acre, B. T. Jour., Dec. 8, 10, 1724, C.O. 391: 33. The act of 1721 was entitled "An Act to Encourage the Settling the North East Part of this Island." It was not confirmed till Aug. 5, 1727. Order in Council, C.O. 137: 17, S 13.

9 Portland's speech to council and assembly, Oct. 1, 1723, C.O. 137: 14, Q 77.

10 Hunter to Board of Trade, Dec. 24, 1730, C.O. 137: 19, S 124; same to same, Apr. 21, 1731, *ibid.*, S 128.

11 Benjamin Moseley, *Treatise on Coffee*, 1785, p. xiii.

In 1732, several Jamaica planters and merchants petitioned parliament for an act to encourage coffee raising.[12] "The Planting of Coffee," said the petition, "will be chiefly carried on by the Midling sort of People, who are not able to bear the great Expence necessary for Erecting and Carrying on a Sugar Plantation, or for raising of Indigo, Cotton, or Ginger." It was thought that a planter with only two slaves might profitably carry on the culture of coffee. A reduction of the duty in England from 2s. to 18d. per pound was granted, but foreign coffee still competed with the English grown product.[13] It is probable that the introduction and development of this industry in Jamaica saved a good number of the small planters from being excluded altogether.

The disadvantage from the loss of the poor whites which was, perhaps, most keenly felt at the time and which materially retarded the development of Jamaica, was its effect upon slave insurrections. The wild, mountainous interior of the island offered a secure retreat for

[12] Petition to H. M. from Jamaica planters and merchants, referred to Board of Trade, Jan. 19, 1731/2, C.O. 137: 19, S 142. See also Moseley, p. xiv.

[13] Moseley, *Coffee*, p. xvii. The average annual production of coffee in Jamaica, in 1743, was estimated at 100,000 lbs. Robt. Dinwiddie to Newcastle, Aug. 1743, C.O. 5: 5, f. 202. In 1775, the export of coffee from Jamaica was 440,000 lbs. Moseley, p. xxi. The comparative values of the exports from Jamaica for 1753 are

Sugar	£1,000,000 sterling
Rum	90,000 sterling
Molasses . . .	16,239 sterling
Coffee	5,400 sterling

Gov. Knowles to Board of Trade, Jan. 12, 1754, C.O. 137: 27, X 200.

On coffee culture in Jamaica, its advantages to the poor farmer, with poor land, etc., see Journal of the Board of Trade, C.O. 391: 44, ff. 239-240; James Knight, *History of Jamaica* (ca. 1746, in MS. only in Brit. Mus. Add. MSS. 12419), II, Pt. 8, f. 146; Long, *Jamaica*, I, 513, 527; *Report of Jamaica Committee on Sugar and Slave Trades*, 1792, pp. 14-16; Brit. Mus. Add. MSS. 12432, f. 13; Daniel McKinnen, *Tour through British West Indies*, 1804, pp. 79-80, 89, 90.

runaway negroes. At the time of Cromwell's conquest of Jamaica, a number of the old Spanish slaves, commonly called the Maroons, had betaken themselves to the mountains where they formed a fortified community. Three parishes in northeastern Jamaica were infected by their presence: the parishes of St. George, St. Elizabeth, and St. James. Their principal center was Nanny Town. About the beginning of the eighteenth century, this negro settlement found a capable leader in one Cudjoe who was their headman or chief. They held lands in common, and cultivated sugar and provisions. It was generally believed they secretly traded with Jews in Kingston who supplied them with guns and powder. It was alleged, also, that the free negroes in Jamaica assisted runaway slaves to join the Maroons by supplying them with powder and arms in exchange for stolen goods.[14] Their number was a matter of conjecture. It was supposed there were, in 1733, 300 fighting men at Nanny Town, and 200 at Carrion Crow Hill, a community of probably over a thousand in all.[15] At any rate they were sufficiently powerful through their methods of guerrilla warfare to hold at bay the white militia and the royal troops.[16] The most effective method of holding the rebels in check was the employing of Mosquito Indians to fight them.[17]

14 Gov. Hunter to Board of Trade, May 10, 1730, C.O. 137: 18, S 102. *Cf.* also examination of Charles Spooner before a committee of the Privy Council, in 1788, on the demoralizing influence of the free negroes. B. T. 6 (Africa): 9, f. 205.

15 Examination of the Negro Sarra *alias* Ned by order of Gov. Hunter, Oct. 1, 1733, C.O. 137: 21, T 14. C.O. 137: 18-22 contain a good deal of material relating to the rebel slaves during the decade 1730-1740, when they were of greatest danger to the whites.

16 "A Short State of Jamaica with respect to the rebellious and runaway negroes," July 25, 1730, C.O. 137: 18, S 99.

17 Thus, in 1720, Gov. Lawes made a contract with Jeremy, king of the Mosquito Indians, the latter to furnish fifty Indians for six months to search out and destroy the rebellious negroes lurking in the mountains of Jamaica.

The desertion of slaves from the large plantations was occasioned in a great degree by the absence of the proprietors. The overseer's chief aim was to secure a maximum production of sugar, which was too often achieved at the expense of the contentment of the slaves. The governors frequently lamented the prevalence of absenteeism because it deprived the island of the best talent for public service and fostered rebellion among the slaves.[18] For these reasons the Deficiency Laws of the Island, as has been pointed out, sought to discourage absenteeism by an additional tax upon non-resident planters.[19]

The period of greatest danger from the rebel negroes was from 1730 to 1739. This was also the period, it should be noted, during which the profits from sugar greatly declined and during which the exodus of poor whites from all the English islands was most marked. It is not unreasonable to suppose that the condition of both the sugar and provision markets, at this time, inclined overseers to make greater demands upon the slaves for labor and cut down their supplies of food and clothing. The negroes deserted the plantations in great numbers in 1730.[20] The official correspondence of the succeeding eight years is filled, largely to the exclusion of other subjects, with accounts of depredations on the

Lawes agreed to pay the Indians 8 pieces of eight or 40s. per head for the negroes. Lawes to Board of Trade, Aug. 24, 1720. C.O. 137: 13, P 93.

18 Hamilton to Board of Trade, Aug. 30, 1715, C.O. 137: 11, O 10. *Cf.* also Gov. Lawes' speech to council and assembly, Oct. 20, 1719, C.O. 137: 13, P 65; also his speech of June 16, 1720, C.O. 137: 13, P 86.

"In the absence of the Masters, and sometimes from the Cruelty of Overseers they [the slaves] are driven to revolt thro' Dispair, where they have no Master Resident to Resort to." Address of the assembly of Jamaica to His Majesty, C.O. 137: 25, X 44, enclosed in a letter from the Duke of Bedford to Board of Trade, Mar. 19, 1749/50.

19 Trelawny to Board of Trade, Oct. 14, 1747, C.O. 137: 24, W 141.

20 Hunter to Board of Trade, Mar. 12, 1729/30, C.O. 137: 18, S 82, 83.

interior estates and unsuccessful expeditions against the rebels. The white militia of the island proved useless. It was composed mainly of white indentured servants on whom no great reliance could be put. Two additional regiments of royal troops were sent over in 1731. These seem to have prevented a general insurrection by overawing the slaves. In the opinion of the great planters, wrote Hunter, "The arrival of these two Regiments at that time was so seasonable, that it was look'd upon as a Special Stroke of providence in their favour, for their plantation Negroes were grown to that degree of Insolence that They durst hardly order them out to work, and have been very tractable ever since."[21] But even the regular soldiers were of little use in reducing the rebel bands in the northeast and along the central ridge. This was owing to the great mortality among the troops in a tropical climate and to the guerrilla methods of fighting. It was supposed, in 1733, that there were at least two thousand rebels throughout the island. Even "Our most Trusty Slaves," declared a Representation from the legislature, ". . . who now remain in Seeming Subjection to us wish well to their Cause, and only Wait for an Opportunity of joining them."[22] The royal troops, in

21 Hunter to Board of Trade, Nov. 13, 1731, C.O. 137: 19, S 146.

22 Representation of Council and Assembly of Jamaica to Board of Trade, 1734, C.O. 137: 21, T 17. The general disorder in Jamaica is reflected in an extract from a private letter of Richard Henings to his aunt in London, dated Jamaica July 5, 1734: "The insecurity of our Country occasioned by our Slaves in Rebellion against us, whose insolence is grown so great that we cannot say we are sure of another day and Robbings and Murder so common in our Roads, that it is with the utmost hazard we Travel them, The Methods hitherto taken to Suppress them, have been attended with unsuccess, and so vast an Expence that I can safely Say two thirds of the inhabitants are already ruined, and Tax's from this Calamity so high that it is impossible we can long stand under them." C.O. 137: 21, T 40.

1735, were reinforced by six more companies.[23] In none of the campaigns against the rebels were the results commensurate with the expense and loss of life. A formal peace was agreed to in 1739 between the English commissioners and Cudjoe, on behalf of the negroes. The rebels received their freedom and 1500 acres of land in northeastern Jamaica which they might cultivate. Cudjoe and his successors were given a limited autonomy within this territory. Runaway slaves in the future were to be returned from this settlement. The Maroons might also have licenses to trade to the white markets.[24]

This period of disorder gave a serious check to the extension of sugar culture in Jamaica. In 1739, Governor Trelawny wrote: "The chief reason of this Island's being so thinly inhabited, is because there is hardly any Good Land which has been hitherto safe from the Incursions of those Rebels unoccupied, at least unpatented; there is enough & upon all Accounts as good as that already patented which has remained desert for fear of those Incursions, and many who have begun Plantations expos'd to that danger have been forced to abandon them on that account."[25] Again, in 1741, Trelawny reported that raids of wild negroes "occasioned the throwing up of almost all the small Settlements in the inland parts of the Country, and which Settlements to this day are over ran with Trees and Bushes, & perhaps the former Proprietors are dead, or gone off the Island, or not able to recultivate them. . . . It seems to me," he added, "that the great hindrance to the settling the Island before, was

23 Ayscough, president of the council, to Board of Trade, April 16, 1735, C.O. 137: 21, T 48.

24 Peace signed Mar. 1, 1738/9, C.O. 137: 23, W 4. An order in Council of May 21, 1741, confirmed an act of Jamaica embodying this treaty, C.O. 137: 23, W 38. R. C. Dallas, *History of the Maroons*, etc., 2 vols., London, 1803, is an entertaining narrative by a novelist but inaccurate.

25 Trelawny to Board of Trade, Mar. 30, 1739, C.O. 137: 23, W 3.

the fear of the rebellious Negroes, who made frequent incursions upon the old Plantations.''[26] During the remainder of our period, though the parish of St. George remained but little cultivated owing to the proximity of the Maroon quarter,[27] negro rebellions were more quickly suppressed[28] and ceased to be an important hindrance to the industrial development of Jamaica.[29]

The need of a larger poor white population, which the slave rebellion of 1730-1739 emphasized, led Jamaica to adopt new and more radical measures to recover and settle the uncultivated lands. The situation was discussed by the home government[30] and a principle of

[26] Trelawny to Board of Trade, Nov. 21, 1741, C.O. 137: 23, W 47.

[27] Gov. Knowles to Board of Trade, Dec. 31, 1754, C.O. 137: 28, Y 43.

[28] For example, the insurrection in St. Mary's parish, April 8, 1760. Gov. Henry Moore to Board of Trade, Apr. 19, 1760, C.O. 137: 32, Bb 1; June 9, 1760, Bb 3; July 24, 1760, Bb 4.

[29] Trelawny wrote the Board of Trade, Aug. 15, 1752, that, since the peace of 1739 with the rebels, ''people may now settle with safety in any part of the Island,'' C.O. 137: 25, X 101.

[30] The Board of Trade, in a Representation to the House of Lords, Jan. 14, 1734/5, stated:

''In the Island of Jamaica there are very large Tracts of Lands proper for bearing Sugar Canes, and capable of most other American Productions, which have not yet been cultivated; And we are sorry to observe to your Lordships, that the most fertile and best situated Lands in this Colony have been formerly granted to private Persons in such Exhorbitant Quantities, that at present there remains very little or no land for the reception of new Comers, unless they purchase it at a very high Price, except in such Parts of the Island as are very much exposed and lye under such Disadvantages as may justly deter Men from settling upon them.

''We conceive the best means of promoting the Prosperity of Jamaica, and of securing the possession of it to Great Britain, would be to take all the Possible Methods to People it with White Inhabitants, and to encourage every kind of Agriculture proper for the Soil, and capable of being carried on by the People of small Substance, But tho' we have long understood this to be the principal Interest of Jamaica, we conceive it impossible to invent Methods of attracting new Inhabitants thither, whilst the Lands of the Country remain confined as they are at present, in the Possession of a few wealthy Planters.

''The People of Jamaica have appeared so sensible of these Truths, and

reform agreed upon. The Committee of the Privy Council, in 1735, ordered the Board of Trade to draw up a bill proper to be passed by the legislature of Jamaica "for the dispossessing the Proprietors of all such Extensive Tracts of Land as lye uncultivated."[31] Governor Cunningham was instructed that the Jamaica law must oblige all proprietors to plant their lands or dispose of them to those who would cultivate them, in default of which, such lands should revert to the Crown to be regranted for that purpose. Also the governor was forbidden to grant more land to any person already having 1000 acres, or

of the ill Consequences that might arise from them, that in the Year 1722, they made a Law for vesting all such Lands in the Crown for which the Proprietors had paid no Quit Rents within a certain time, upon Condition that the Lands so forfeited should be regranted in small parcells to new Inhabitants, under certain Conditions mentioned in the Act.

"Four other Acts were afterwards passed for purchasing more Land to the Publick for the Use and Encouragement of New Comers for building a Town to be called by the Name of Portland, and for forming a Settlement of Port Antonio, which is a Port of Consequence in the North East part of Jamaica.

"But either these Acts were insufficient to effectuate the Purpose for which they were designed, or the Execution of them hath been evaded, or the Intention of them weakened by subsequent Laws, because these lands have not been purchased by new Inhabitants, but for the most part lye still uncultivated, and the Island is more destitute than ever of white Inhabitants.

"We are humbly of Opinion therefore, that if the People of Jamaica will not be induced to frame an Act which may divest particular Persons of those extensive Tracts which now lye uncultivated, this may be a proper Subject for the Consideration of the British Parliament, by whose Authority an Effectual Method may be taken to reassume those Ancient Grants that have hitherto been useless even to their owners as well as the Publick or otherwise to put the Proprietors under a necessity of cultivating them." C.O. 5: 5, ff. 114-118. A memorial from 128 merchants of London, Bristol, Liverpool, and other Jamaica traders to the king pointed out the evils in the Jamaica situation and suggested similar reforms. Referred to Board of Trade, Oct. 27, 1735, 14 folios, C.O. 137: 22, V 9.

[31] Order of the Lords of the Committee of the Council for Plantation Affairs, July 2, 1735, C.O. 137: 21, T 51. The draft of a bill was prepared but does not appear to have reached parliament, B. T. Jour., July 18, 1735, C.O. 391: 44, f. 126.

hereafter to grant more than 1000 acres to any one person.[32]

An act of Jamaica, in 1736, vested in the Crown 15,000 acres of uncultivated land at Manchioneal, and 15,000 acres at Norman's Valley for the use of new settlers.[33] The land was to be allotted to a family in an amount proportioned to its numbers in the following manner, namely, to the master of a family fifty acres, to his wife fifty, to each child twenty acres, to each servant fifteen acres, and to each slave ten acres; the whole not to exceed three hundred acres to one household. The transportation of each family was to be paid, and provisions provided them for one year. Upon the encouragement of this act Trelawny reported, August 15, 1752, that ninety-seven families came from the Leeward Islands and took up these lands. Another law passed in 1743 gave the same allowance to new comers that should take up Crown land and settle it in the parish of Portland. The total amount of land here, which had been reinvested in the Crown by acts of 1722 and 1737 amounted, in 1737, to about 60,000 acres.[34] There remained, in 1752, over 5000 acres to be

32 Order in Council, July 30, 1735, approving the instruction to Gov. Cunningham, C.O. 137: 22, V 1.

33 An Act for Introducing of White People into this Island, for subsisting them for a Certain time and providing them with Land that they may become Settlers, passed May 15, 1736; confirmed by order in Council, Feb. 20, 1741/2, C.O. 137: 23, W 51. See also Trelawny to Board of Trade, Aug. 15, 1752, C.O. 137: 25, X 101; same to same, Nov. 21, 1741, C.O. 137: 23, W 47, 49. Trelawny wrote, in 1741, that these encouragements "have been Effectual with Men who had Slaves or Mony, but they will not do for very poor People." He also gave an account of the efforts the island had taken to establish barracks near the Maroon community in order to encourage the settlement of the adjacent lands.

34 Order in Council, Feb. 20, 1741/2, confirming (1) an Act of 1738 for effectually settling the Parish of Portland by vesting all the unsettled Lands in the said Parish in the Crown, and (2) an Act of 1743 "giving the same allowance to New Comers . . . within the Parish of Portland . . . as those . . . at Normans Valley . . . and Mantioneal Harbour," C.O. 137: 25, X 121; renewal by order in Council, Feb. 7, 1753.

taken up in this parish. Another act, passed in 1749, sought to induce the great planters themselves to reinforce the number of small farmers and artificers.[35] This act provided that the passage of each family should be paid, and twenty acres of land given to them, four acres of which were to be ready planted with Indian provisions; also a house was to be built worth £50, and a negro slave worth £35 and £20 in money provided. Any Jamaica planter, who would settle a family on these conditions, should be paid £145 besides allowing each person in the said family to save a deficiency to the planter for four years. For artificers the act provided that their passages should be paid, and £10 be given to each after his arrival in the island. On the effect of this act, Trelawny wrote: "Several families have come over from Great Britain, upon the encouragement of that Act, but few of 'em being husbandmen, and in course not understanding any thing of Agriculture, few Planters have car'd to settle them; so that the Country has been at a vast expence in paying their Passages and maintaining many of them a long while in the Towns, and has at last been oblig'd to buy Land near the Bath in St. Thomas in the East, and allotted it out to such as were willing to settle on it, (for some have refus'd to be settled in the Country)."[36]

In pursuance of these three acts, Jamaica spent, from 1739 to 1752, £17,300 15s. 4d. The increase in the white population since 1734 was about 1500, while the slaves had increased by 35,000.[37] The act of 1749 was renewed in 1752, and it was stated that, from December 2, 1749,

35 "An Act for the better and more effectual Encouragement of White Families to become Settlers and for giving a bounty, to certain Artificers to come over and exercise their Trades," passed 1749; repealed Nov. 18, 1758. Acts of the Assembly of Jamaica, Gov. Henry Moore to Board of Trade, Jan. 25, 1759, C.O. 137: 30, Z 39.

36 Trelawny to Board of Trade, Aug. 15, 1752, C.O. 137: 25, X 101.

37 *Ibid.* These figures, Trelawny stated, were based on lists from all but two or three parishes and upon inquiries.

to November 20, 1754, there had been brought in under it 347 men, women, and children at an expense to the island of £14,908 18s. 4d., besides £1400 remitted to Barclay and Fuller in London for transportation, amounting in all to £16,308 18s. 4d. The list of these immigrants brought at the public's expense include shoemakers, carpenters, barbers, a surgeon, tailors, bricklayers, a distiller, a sugar baker, a cooper, a harness-maker, a peruke-maker, an engraver, and a millwright.[38] The number of grants of Crown land to new settlers from 1734 to 1754 was 208.[39]

But this increase in population seems to have been made largely at the expense of the other sugar islands. "From the Temptations thrown out from Jamaica," wrote Governor George Thomas of the Leeward Islands, in 1753, "we are daily loosing Numbers of our Inhabitants." It was the military danger to Antigua, St. Christopher, and Montserrat, occasioned by the loss of their white servants, that caused complaints against Jamaica.[40]

[38] Gov. Knowles to Board of Trade, Dec. 31, 1754, C.O. 137: 28, Y 47. See also "Address and Representation of the Council and Assembly of Jamaica to His Majesty," Nov. 20-22, 1752, wherein it is stated that the Act of 1749 brought in 32 families and 17 artificers, which cost the island £7000, C.O. 137: 25, X 115. According to Gov. Knowles, £20,000 had "been thrown away" up to 1754 in trying to get settlers by the Act of 1749. Knowles to Board of Trade, Jan. 12, 1754, C.O. 137: 27, X 200.

[39] Copies of dockets of patents of Crown land for the use of new settlers made from 1734 to Oct. 9, 1754, signed Thomas Hay, sec. of Jamaica. Enclosed with letter of Knowles to Board of Trade, Dec. 31, 1754, C.O. 137: 28, Y 53. The "Address and Representation of the Council and Assembly to H. M.," Nov. 20-22, 1752, states that the number of families that were brought in from 1735 to 1752 to take up these grants was 103, at a cost to the public of £11,425 16s. 10d. "And there is still much room for settlement on these lands vested in the Crown by these Acts." It was the "Accidents and Prejudices attending the Climate," some thought, that prevented emigration from Great Britain in spite of these inducements. C.O. 137: 25, X 115.

[40] Gov. George Thomas to Board of Trade, Antigua, Oct. 8, 1753, C.O. 152: 27, Aa 84; same to same, May 22, 1754, C.O. 152: 28, Bb 16.

All these colonizing schemes of Jamaica accomplished nothing toward ameliorating the disadvantages arising from large scale production. Even the industrious small planters who did come in could not be protected against the forces which were operating to exclude the small producer. The general results were thus summarized by Governor Knowles in 1754: "The several Laws passed since the Year 1734 relating to Grants, and for enforcing the Cultivation of Lands and encouragement of new Comers, have been so far from answering the purposes intended by them, that they have done more hurt than good to the Country, by encouraging a number of idle disolute wretches to resort here, the Majority of whom by their irregular abandoned Course of Life, dye almost as soon as Landed, and the few who have shown some inclination toward Industry, have met with too little encouragement, wherefore they have been obliged to seek their Bread in another Country, after Strugling for some time, with the inclemency of this; This being the Case and the Laws proving defective, the Commissioners for executing them seem to have laid aside all further hopes of success, and therefore the sooner all these Laws are repealed, the better."[41] The act of 1749 had proved the most expensive of all. It "has constantly cost the Island £6000 per Ann:", wrote Governor Moore in 1759, "without ever answering the purposes for which it was intended." After ten years' operation, it was finally repealed.[42]

While these artificial remedies were being applied to retain a small planting class, especially on the areas exposed to rebellion and invasion, the monopolizing of sugar lands went on unchecked throughout most of

41 Knowles to Board of Trade, Dec. 31, 1754, C.O. 137: 28, Y 43.

42 Gov. Henry Moore to Board of Trade, Jan. 25, 1759, C.O. 137: 30, Z 39. The repeal of the act, Nov. 18, 1758, did not take effect till one year later.

Jamaica. Only in small districts in the northeast was this tendency restrained. Many believed that the venality of public surveyors played no small part in the creation of large holdings.[43] The total acreage patented, in the year 1754, was 1,671,569. This area was held by about 1620 planters, the average size of an estate being over 1000 acres.[44] On the troubles of a planter in obtaining an estate, at this time, Governor Knowles remarked: "It is inconceivable the difficulties that attend finding a piece of good Land to settle upon, without purchasing and at a most extravagant Rate, for after a setler has obtained a Patent for a piece of good Land, purchased Negroes, cleared and planted the ground, it often happens that the son or Grandson of these oppulent Planters with a prior Patent starts up, and lays claim to the land, brings an ejection and at once disposses the poor setler with the loss of many years labour and expences. Not only new Comers are discouraged by the overgrown Landholder in this manner, but Merchants and Gentlemen desirous of becoming Planters have been deterred from settling by the precariousness of Titles and the scarcity of Land."[45]

The amount of uncultivated sugar land in Jamaica attracted much attention in England from the consuming classes. Petitions from the refiners and grocers of London and Bristol were laid before parliament, in 1753. Their aim was either to oblige the great proprietors of land in Jamaica to produce larger quantities of sugar, or to gain leave for the petitioners to import sugar from other countries, when the price of British sugar rose above a certain price.[46] From the most reliable surveys

[43] Knowles to Board of Trade, Dec. 31, 1754, C.O. 137: 28, Y 43.

[44] "List of Jamaica Planters and their Estates," 1754, C.O. 142: 31 (one volume).

[45] Knowles to Board of Trade, Dec. 31, 1754, C.O. 137: 28, Y 43.

[46] *Parl. Hist.*, XIII, 1293; *Account of the Late Application to Parliament from the Sugar refiners, grocers, and others dealing in sugar in Lon-*

of the time, the disposition of the Jamaica lands appears to have been as follows:[47]

The total area was estimated at	3,840,000 acres
Of this two-fifths was supposed to be mountainous, inaccessible, rocky, or barren land . .	1,706,664 acres
There remained, therefore, of good, plantable land probably	2,133,336 acres
Of which there was patented	1,500,000 acres
There remained of plantable land not yet taken up	633,336 acres
There remained of the patented land still uncultivated it was estimated	1,000,000 acres
Total amount of uncultivated plantable lands, exclusive of woodlands and pastures, was probably	1,633,336 acres

Trelawny thought the greater part of this land fertile enough and well situated, the absence of good roads being, however, a drawback to its settlement.

The publication of these facts in England, at a time when the prices of sugar were nearly double what they were on the continent, stimulated a great deal of criticism of the aims and methods of Jamaica planters.[48] Parliament, during the spring of 1753, received petitions

don, Westminster, and Southwark, 1753, *passim*. *Commons Journal*, XXVI, 611, 659, 664-665, 703, 709, 775-840 *passim*. Nothing was done by parliament.

[47] Trelawny to Board of Trade, Aug. 15, 1752, C.O. 137: 25, X 101. The account sent home by Gov. Knowles, Dec. 31, 1754, is substantially the same as this account, the total amount of uncultivated plantable ground being estimated at 1,600,000 acres. Knowles to Board of Trade, Dec. 31, 1754, C.O. 137: 28, Y 43. Knowles thought that, of the ungranted lands, there were fit for canes 100,000 acres.

[48] *A Short Account of the Aims and Methods of Jamaica Planters*, 1754; J. Massie, *Summary of the Sugar Trade*, 1759, *passim*; Postlethwayt, *Great Britain's Commercial Interest Explained*, London, 1757, I, 435-437. Postlethwayt (*ibid.*, I, 494) also declared the non-cultivation of land and its monopoly to be the fundamental condition that favored the illicit trade from the Northern Colonies to the French sugar islands.

from sugar consumers complaining of the high prices of that commodity and urging the need of forcing Jamaica planters to cultivate more land. The matter was discussed at some length, and a bill introduced (May 9) for the better peopling of the island, etc. But the planting interest, by this time, was strongly intrenched in parliament, and the bill was smothered.[49] English consumers of sugar were at the mercy of the West India planters, and remained so for many years.

[49] See *Commons Journal,* XXVI, 611-840 *passim,* Feb. 23-Sept. 27, 1753.

CHAPTER VI

CAPITAL, CREDIT, AND CURRENCY

The West Indies, like all frontier communities, were in great need of capital to develop their unusually rich natural resources. But, in the eighteenth century, the loan of capital by Europeans was by no means as common a practice as it has become in more modern times. Nevertheless, the British West Indies profited by the loan of probably more European capital than did any other colonies in the new world. Comparing the English islands with the French in this respect, Adam Smith observed that the capital "which has improved the sugar colonies of France, particularly the great colony of St. Domingo has been raised almost entirely from the gradual improvement and cultivation of those colonies. It has been almost altogether the produce of the soil and of the industry of the colonists, or, what comes to the same thing, the price of that produce gradually accumulated by good management, and employed in raising a still greater produce. But the stock which has improved and cultivated the sugar colonies of England has, a great part of it, been sent out from England, and has by no means been altogether the produce of the soil and industry of the colonists. The prosperity of the English sugar colonies has been, in a great measure, owing to the great riches of England, of which a part has overflowed, if one may say so, upon those colonies. But the prosperity of the sugar colonies of France has been entirely owing to the good conduct of the colonists, who must therefore have had some superiority over that of the English; and this superiority has been marked in nothing so much as in the good management of their slaves.'"[1]

One justification offered for the mercantile system, whereby English sugar was given a preference in the British market and British manufactures a monopoly of the colonial market, was that it encouraged rich men to loan their money to the colonists. "It is this," said Long, "which has given a confidence to the Merchant in the Loan of his capital; and to the Planter in the application of that Loan, to Industry and Improvement.'"[2]

By the loan of capital is to be understood, not so much the supply of money, as the delivery to the planters of utensils, dry goods, supplies of various sorts, and slaves upon long time credits. For the supply of negro slaves, especially, the planters were always deeply in debt to the Royal African Company. The correspondence from

[1] Smith, *Wealth of Nations*, ed. *World's Classics*, II, 89. Smith cited from Anderson's *Annals of Commerce* the assertion of the West India merchants and planters, in 1775, that there was capital worth £60,000,000 in the sugar colonies, and that half of this belonged to residents in Great Britain. In the same year, it was stated that London merchants and others had no mortgages on rural property in North America except on the tobacco and rice plantations. In New England only those engaged in trade with Great Britain enjoyed the loan of British capital. *American Husbandry*, 1775, pp. 166-167. Compare with Smith's statement about the French colonies the following by the British governor of St. Vincent: "Their Credit was small, and their progress in making Settlements was much slower than the progress of the English planter on Account of this want of Credit. Hence very trifling Debts are owing by them, at the same time that the Estates of the English are heavily incumbered. From this mode of Cultivating their Lands by degrees they become habituated to a plan of Eoconomy, altogether unknown to the English Settler; and perhaps for these reasons, they can Afford to send their Sugars to Market cheaper than the English." Gov. Seton to Lord Sidney, July 1, 1788, B. T. 6: 11.

[2] Papers of Edward Long, Brit. Mus. Add. MSS. 12407, f. 25.

the agents of that company in the West Indies reveals the enormous extent of such indebtedness and the difficulties the agents often had in recovering it. In Jamaica and Barbadoes slaves were always paid for partly in silver and bills of exchange. Elsewhere payments were made almost always in sugar—usually in a mortgage upon the next ensuing crop. The terms upon which the company furnished slaves to the Leeward Islanders, together with the risks it ran, are clearly put forth in a representation to the Lords of Trade in 1687. The communication was as follows:

May it Please yor Lordships

The Royal African Company of England Having had a great Regard to the Comands that yor Lordshipps have been Pleased to lay upon them fformerly That they should supply his Maties Plantations in America wth such Quantityes of Negroe Slaves as the said Plantations could well pay for, which they have done so largely as That they have very great debts owing them there, —and Having lately made Inspection into their Trade the last Year They do find they have been very greate loosers by such Negroes as they have sent unto the Islands of Nevis Antegua, Mountserrat and St Christophers, In wch Islands their Sale hath been always for Pounds of Sugar (The accompts in those Islands being kept in Suger only, & not in mony as in the other Plantations) and Whereas Formerly the £100 [one cwt.] of Sugar in those Plantations was Estimated at 12^{s} 6^{d} Sterling mony & did for many Years together, Yield neare soe much Cleare of Charges in England, It now happens That what the Company received all last yeare comes to yield them Cleare of Charges but 6^{s} & 5^{d} the 100^{li} weight of Sugar, according to the acco^{t} they herewith lay before yor Lordshipps (The truth of wch they are Ready to prove upon Oath if Required)

And although from the other Plantations they doe also loose Considerably by the sugers they receive from them the Company Yitt in regard have a greate part of their Returnes from Barbados and Jamaica in silver and bills of Exc^{a} on wch the losse is less then upon sugers and somewhat helpes to support the Trade of those Places, But here in the aforesaid Foure Leeward Islands all the losse by the abatements of the Prizes of Sugars Falls upon the Company all their Payments being in suger yett the Prizes of Negroes are but as formerly they were sold, By wch meanes the Company doe find the Negroes that have of late Gone to those Islands have not yielded by a very Considerable Proportion, what they have realy cost the Company. Besides wch they have very bad Payments made them in those Islands when at this day are owing to them the value of 22000li sterling

The Company doe further humbly offer to yor Lordships that whereas they did in the Yeare 1672 Publish a Declaration wch was sent to the severall Plantations, in America the Prizes they would Furnish a Ships Lading of Negroes in those Plantations, Giving them Security here in England for the Payment of the mony here at Three Payments One Third at Two Months One other Third at Foure months and the Remainder at Six months after Producing Certificates of the delivery on wch tearmes the Prizes then offered for the Leeward Islands was 16li Pr head, The Company doe now (for the Encouragemt of any of the Inhabitants of those Islands) offer to Furnish them on the tearmes of their said Declaration at 15li Pr head paid in England, To any that shall soe Contract and give good Security, and they doe humbly beg of yor Lordships, that this may be Transmitted to his Maties Governors of those Islands to be made Publique there That they may be Encouraged to Contract, as those that Trade to Barbados Jamaica & Virginia Have frequently done and that they may have timely Notice not to Expect The Company can supply them soe much to their Detriment and losse as of late they have done unlesse they advance the Prizes of Negroes answerable to the fall of sugers.

African House the 19th of Aprill 1687

By order of the Royall African Compa of England
Sam Heron secr[3]

3 Papers of Royal African Co., Public Record Office, Treasury, 70: 169, ff. 84-86. In Virginia, slaves were sold for tobacco; "its impossible to sell a ship of Negroes for Good bills." Col. Garvin Corbin, agent to Com-

It appears that in the larger islands, Barbadoes and Jamaica, where trade with the Spaniards brought in specie, the African Company was paid largely in coin. Their ships returning to England carried not only sugar but large quantities of pieces of eight.[4] But in the smaller islands the company's agents seldom received anything but sugar and that more often in future deliveries. Moreover, misfortunes would frequently prevent the planter from meeting his obligations, whereupon the company would attach his land, negroes, and utensils. But the property was usually appraised by freeholders of the neighborhood who were likely to be prejudiced against the company. Such a plantation was worked under a receivership, and the annual income paid to the creditor till the debt was discharged. If the income failed to be paid, then the creditor had to institute a new suit in order to gain actual possession of enough property to meet the debt. During the last quarter of the seventeenth and first half of the eighteenth centuries a considerable amount of the company's assets were thus tied up in the estates of insolvent planters. It was often complained that the capital and talents of creditors, which were designed for trade, were occupied with the management of decrepit plantations for which they were never intended. The courts were crowded with litigation. The short-lived land bank of Barbadoes, in 1706, aggravated the situation by attempting to force creditors to accept legal tender paper money as payment. Here, however, the British government came to the rescue of creditors

pany, Virginia, June 25, 1707, Treas. 70: 8, f. 45. In the Carolinas they were sold for rice and pitch. Thomas Broughton to Company, Oct. 2, 6, and Nov. 20, 1707, Treas. 70: 8, f. 50.

[4] Abstracts of letters from agents for the years 1706-1719, Papers of Royal African Co., Treas. 70: 8.

and suppressed the scheme within a year.[5] At other times, colonial legislatures attempted to help debtors by passing acts that made sugar at a fixed price per hundredweight legal tender for all obligations. Thus, when the market price for sugar fell below the legal rate, creditors might be cheated. Other devices to embarrass creditors in foreclosing mortgages were sometimes adopted. It is surprising that English merchants should have had the courage to make loans at all under so many precarious conditions. In the first half of the eighteenth century a considerable number of West India estates were in the possession of the Royal African Company and the amount of its mortgages was extensive. What was true of this company must have been the case, in a lesser degree, with many other creditors.[6]

Following the peace of Utrecht, investments in the sugar industry rapidly increased. Many planters, who began in a small way with fifteen or twenty slaves for the

[5] *Acts of the Privy Council, Col.*, VI. §§ 152, 181; *C. S. P. Col.*, 1706-1708, §§ 542, 546, 578, pp. 348-349.

[6] An extensive investigation of the great collection of Papers of the Royal African Company would throw much light upon the economic history of all the plantation colonies. Such an investigation is now under way by Professor Zook of Pennsylvania State College. See, on debts to the African Company in 1664, *C. S. P. Col., 1661-1668*, no. 689; petition of Roy. Afr. Co. against the debtor's act of St. Christopher, 1683, Papers of Roy. Afr. Co., Treas. 70: 169, ff. 15-16; on the Antiguan debtor's act, 1685, *ibid.*, ff. 51-53; on debts owing the company, impediments to their collection, etc., Treas. 70: 8, ff. 5, 22-24, 26, 32, 36, 40, 44, 46, 48, 49, 50, 51, 56, 59, 61, 81, 82, 88, 94, 102, 107, 108, 109, 111, 113, 119, 130, 143, 159, 171, 178, 185, 187; 170, ff. 23-24, 148. The agents' letters are characterized by such statements as the following: "By delays of promised payments from Debtors in Generll," an agent in 1706 "almost dispairs of getting in any Debts to make returns to the Ballancing yor accots this year," *ibid.*, 8, f. 22; "Are pressing their Lawyers to prosecute for those Debts wch they hope are recoverable," *ibid.*, f. 36; "Few or no debts are paid without suing for" (1707), *ibid.*, f. 44; "Are fearful of having lands put upon them appraised by the Neighbours in like Circumstances at double what they can sell for," giving an instance of one Haggart's house and land which was appraised to the company at £50 per acre and would not sell for £30, where-

production of provisions, ginger, and cotton, which were inexpensive to raise, soon became ambitious to engage in sugar culture. This they attempted through credit and the assistance of friends, paying interest rates that often amounted to fifteen or twenty per cent, for bonuses were frequently included in the principal of the loan.[7] "I presume 'tis well known to every Gentleman here," said Henry Peers, speaker of the Barbadoes assembly in 1728, "that the Generality of the Inhabitants are very much indebted, & that many of them are sinking under the weight of the heavy interest they now pay."[8] Meanwhile, in the neighboring island of Santo Domingo, French activity was laying the basis for the ruin of many such English planters. In the decade 1720 to 1730, the total sugar output far surpassed that of any previous period, while its demand in Europe did not propor-

upon they petitioned for a reappraisement, *ibid.*, f. 82. The agent sends a list of all mortgages held by the company in Jamaica in 1710, *ibid.*, f. 107. "Without the Planters have Credt for slaves its impossible that ever they should pay their debts," *ibid.*, 24. The governor of Barbadoes is accused of encouraging debtors against the company (1710), *ibid.*, f. 102. The debts outstanding from one ship load of negroes was £3500 in 1711, *ibid.*, 109. Antiguan debts that were made payable in 1707 with sugar at 20s. per cwt. might have been paid with silver at one-half that price, *ibid.*, f. 46. In 1710, at Nevis and St. Christopher, slaves brought from 7000 to 8000 lbs. of sugar apiece, *ibid.*, f. 108. "Mr. Parsons has promised to make All Accompts clear the next Crop,"—a typical planter's promise, *ibid.*, f. 32. Badly cured sugar, and rum at 21s. per gallon, were received for slaves in 1713, *ibid.*, ff. 131, 132. In 1714, the agent in Jamaica "Has taken a slight view of the Mortgages in the Secrys office, and is apprehensive the Company have suffered Much from their lying so long dormant there remaining nothing of several of them but tracts of waste land," *ibid.*, f. 143. In 1715, he declares he "shall shortly make an advertisement of the compe's Estates, in order to sell them to the best Bidder," *ibid.*, f. 187. The Barbadoes agent, in 1715, writes: "Most of the Old Debts are bad," *ibid.*, f. 185.

[7] James Knight, "History of Jamaica," Brit. Mus. Add. MSS., 12419, f. 141; Long, *Jamaica*, I, 557. Knight's *History* is in manuscript only, among the "Long Papers" in the British Museum.

[8] Minutes of Barbadoes Assembly, Aug. 5, 1728, C.O. 31: 18.

tionately increase. A decided slump in prices resulted.[9] Continental markets for British sugar were seriously cut into by French competition, while the English market became overstocked. This period of depression, which began about 1728, lasted until 1739. During this time a large number of English planters, who operated in the expectation of a narrow margin of profit, were ruined. Unable to satisfy their creditors, they often made hasty departures from the islands, taking with them their slaves and movable property.[10]

9 Average prices per hundredweight for sugar sold at the London Custom House at this period are as follows:

YEAR		MUSCOVADO SUGAR		CLAYED SUGAR	
Xmas to	Xmas	Shillings	Pence	Shillings	Pence
1727	1728	24	10¼	37	2½
1728	1729	24	5½	39	3
1729	1730	21	8½	32	4½
1730	1731	19	8¼	31	11¼
1731	1732	17	10	31	0
1732	1733	16	11¼	31	1½
1733	1734	25	8½	32	9¾
1734	1735	18	9½	31	4
1735	1736	19	5½	29	9¾
1736	1737	24	9	33	¾
1737	1738	21	7¾	34	5¼
1738	1739	25	8¼	37	6
1739	1740	32	½	38	6¾
1740	1741	30	5	38	7¼
1741	1742	29	1	41	1¾
1742	1743	27	3½	35	¾
1743	1744	30	7	42	0

Treas. 64: Bundle 274. For prices in the years 1744-1758 see p. 186 *n.* 64.

10 While the running away of debtors occurred at nearly all times, its extent from 1730 to 1739, especially from Barbadoes, appears to have been greater than ever before. This drew forth a considerable amount of discussion from the assembly, governors, and such writers as Ashley and other pamphleteers. See William Gordon to Board of Trade, London, July 14, 1720, C.O. 28: 17, V 36; *Present State of the British Sugar Colonies*, 1731, pp. 10-11; Gov. Lord Howe to Board of Trade, Nov. 7, 1734, C.O. 28: 24, Aa 28. Howe stated that English, Irish, and North American merchants complained that they "have Sufferd very much of late Years in their

The legal rates of interest were lowered in most of the islands during this critical period. Barbadoes, in 1729, reduced the rate, which had been ten per cent since 1668, to eight per cent; Montserrat, in 1735, did the same; Antigua, in 1728, reduced it from ten to six per cent; and Jamaica, in 1739, reduced the rate, which since 1681 had been ten per cent, to eight per cent. Among the reasons assigned for this action by Jamaica were the low prices of the island's produce, the impoverishment of its land, and the high price of imports.[11] In 1752, the rate on Jamaican and Barbadian loans was reduced to six per cent.[12] For loans made by Englishmen resident in Eng-

Fortunes by the running away of the Inhabitants of this Island, who are indebted to them, with all their Slaves and Effects to South Carolina where they have been protected by a Law of that Province from any Arrests or Suits for Such Debts as they have here contracted.'' See also James Dottin, president of council of Barbadoes to Board of Trade, Aug. 2, 1735, C.O. 28: 24, Aa 32. ''Our produce of late Years,'' wrote Dottin, ''has scarce Sold for more than it's cost.'' John Ashley's scheme for operating estates of insolvent planters under receiverships (25 MS. pages, 1737), is given in C.O. 28: 25, Aa 60. See also Gov. Byng to Board of Trade, Feb. 21, 1739/40, C.O. 28: 25, Aa 82; Dottin to Board of Trade, Jan. 21, 1740/41, C.O. 28: 25, Aa 106; Gov. Thomas Robinson to Board of Trade, May 10, 1744, C.O. 28: 26, Bb 11. James Knight, formerly colonial agent for Jamaica, writing of speculative planters, said: ''But seeing the prosperous Condition of some of their Neighbours, they were tempted to take the same measures, without considering their Ability, or the difference of their Soil, and other Advantages; for want of which they involved themselves in debt, and paying such high Interest, have not been able to Extricate themselves: nay some have had their Estates eaten up by that means, and the lowness of their produce in Europe. This ought to be a Caution to all Young Planters how they proceed, and not to attempt things, that are not within their power and Ability.'' ''History of Jamaica,'' f. 141.

11 C.O. 139: 17, no. 1; C.O. 28: 45, f. 39: a report to Gov. Worsley on the rate of interest; C.O. 152: 25, Y 101; *ibid.,* 23, X 37; objections to the Antiguan act by merchants and planters of the island are in *ibid.,* X 61, whereupon an order in Council, Nov. 22, 1739, ordered that the act should ''lye by for the present,'' *ibid.,* 24, Y 8. Six per cent was evidently below the market rate for loans by island merchants at that time. The Jamaican act of 1739 is in C.O. 139: 15, no. 11.

12 Long, *Jamaica,* I, 555; Hall, *Acts of Barbadoes,* 1764, no. 199: C.O. 30: 1.

land, the West India colonies would not, however, legalize a rate higher than the legal rate in England. The British legal rate from the 21st year of James I until the 12th of Queen Anne was eight per cent, after which, in the eighteenth century, it was five per cent.[13] Planters of good standing who were acquainted with English factors or merchants were able, in this century, to obtain loans at five per cent. Those who could not obtain English credit were forced to pay the higher rates demanded by Jews and other merchants or factors resident in the islands. In many cases bonuses had to be given, so that, actually, rates as high as twenty per cent were commonly paid.[14] At the close of the Seven Years' War, Grenada, Tobago, St. Vincent, and Dominica were added to England's West India possessions, and a new field for speculative investment was opened. At the same time, however, the war debt and the increasing industrial development of Great Britain itself tended to diminish the amount of capital available for improving the colonies. If the islands had legalized a higher rate of interest on British loans, say six per cent, they might have attracted a larger share of capital.[15] But instead of this, parliament passed, in 1773, an act "for encouraging the subjects of foreign states to lend money upon the security of freehold and leasehold estates in any of his Majesty's Colonies in the West Indies, &c." at five per cent.[16] The purpose was to attract Dutch capital for the development especially of the islands ceded by the peace of Paris. But the outbreak of the American Revolution,

13 Smith, *Wealth of Nations*, Bk. I, Ch. ix; ed. *World's Classics* I, 99.

14 Long, *Jamaica*, I, 558. Long's residence in Jamaica from 1756 till 1769 gives his description of money lenders in the West Indies the value of a contemporary source.

15 *Ibid.*, I, 556-557.

16 13 George III, c. 14.

soon afterward, threw the islands into the turmoil of war and, temporarily at least, checked their development.[17]

While the larger part of the capital for the development of the West Indies was advanced by British merchants,[18] a considerable amount that was laid out in improvements and slaves represented the accumulated savings of the islanders themselves. Settlers might borrow money for their lands, buildings, and initial stock of slaves, but they depended upon the profits of agriculture to increase the number of negroes. It was said that "All the overseers and tradesmen, and a few of the bookkeepers, save something out of their salaries; and they have no other way of laying out their money but in the purchase of slaves; whereby the tradesmen, if they are industrious, will in time be able to set up for themselves in business, and the overseers procure settlements to retire to when old and infirm."[19] Under the system of slavery in the West Indies, as in the Southern States of America, but little capital was saved for anything but the demands of agriculture. Absentees in England, after reinvesting a part of their profits in fresh lands and slaves, spent the rest in an expensive mode of life in the mother country. As for good roads, fine buildings, schools, private libraries,—the West Indies were sadly lacking in these and many other possessions that belong to a civilized society.

As there were no gold or silver mines in the British

17 Edwards, *West Indies*, I, 386-387, 444.

18 Next to soil and climate, wrote Robert Dinwiddie to Newcastle in 1743, it was "the great Credit the Merch^{t} has Continually Supplied them with, which I really believe has been the greatest Advantage & help, to their present Flourishing Condition, which I wish may long Continue." C.O. 5: 5, f. 205.

19 *Report of Jamaica Committee on Sugar and Slave Trade*, 1792, pp. 13-14, 17.

West Indies, metallic money could be extracted only from trade. From an early period commerce with the Spanish Main, Cuba, and the Spanish settlements in Central America brought into Barbadoes and Jamaica a considerable amount of Spanish American coin. British manufactures and slaves were clandestinely sold for gold and silver. In 1700, this trade employed about 4000 tons of English shipping and the British goods sold amounted to a million and a half pounds sterling. Also the logwood traders between Jamaica and Honduras introduced a more or less constant supply of money. During the wars with Spain when trade was interrupted there were periods of great scarcity of coin in the English islands. The islands came to rely very largely upon Spanish money, in the eighteenth century, to pay for their supplies of lumber, provisions, and livestock from North America and for a certain amount of their European imports. British imports and slaves were most always paid for in sugar, though the slaves were purchased partly in coin. As the century progressed, North Americans were less and less disposed to sell their supplies for English-grown sugar, molasses, and rum. Instead, they took coin and bought their return cargoes in the French islands where prices were generally lower. After 1713, this created in the British West Indies a strong resentment against North Americans for draining the islands of their coin. The acts of parliament in 1733 and 1764, discouraging North Americans by heavy duties from importing foreign-grown sugar, molasses, and rum, were designed largely to prevent the money of the West Indies from being drained away. Nevertheless, when New Englanders disregarded the acts and insisted upon coin for their supplies, the West Indies needed the supplies so much more than the cash that they could not afford to refuse payment.

With metallic money thus chronically scarce, substitutes were constantly used. Sugar was almost always acceptable as a medium of exchange and was at times made legal tender at a fixed price. As the market price, however, often fell below the statute price at which sugar passed as money, creditors suffered loss, and made the subject one of constant complaint. Jamaica passed an act in 1683 whereby the island's produce, as appraised by churchwardens of the parish, was made legal tender for payment of all debts. But an act of 1751 admitting the uncertain value of sugar stated that the earlier act had worked an injury to creditors. Henceforth it was provided that "no other payment shall be allowed and Deemed a Good payment in Law but in Current Coin of Gold or Silver unless in such Cases where both party's agree for payment in Sugars or other Produce of this Island or in Discharge of Debts Contracted before the Passing of this Act." In the other islands so sound a financial measure was not passed during our period, and creditors had to exercise great care in forming contracts. Indeed, the colonists showed some disposition to cheat them by inflating the currency with cheap money. This is revealed in the Barbadoes Land Bank notes of 1706 and in the Jamaican act of 1758 to raise the value of coin.

Owing to the scarcity of coin in Barbadoes, sugar was often proclaimed or enacted a legal tender for all debts. The legal value of sugar in English money was repeatedly fixed by statute. Fees, taxes, fines, and salaries were payable in sugar.[20] "This noble island wants a money trade, for here all things are bought and sold in the way of truck or permutation to the much damage of the inhabitants." Thus wrote Nicholas Blake from Barba-

[20] See *Acts of Assembly of Barbadoes*, Act of 1661, no. 41, pp. 39, 40, 61-63; nos. 61, 66, 77, 102, 116; Act of 1670, no. 145, p. 77, no. 218; *C. S. P. Col., 1661-1668*, p. 111, no. 49, p. 374, no. 1576; *ibid., 1669-1674*, p. 9.

does to the king in 1669.[21] A writer in 1695 said, " 'Tis but of late years that any other way of buying and selling was us'd in those islands but for sugar, all their accounts being computed by pounds of sugar, and the Law of the Countrey, allowing all debts to be paid in that Specie.' "[22] Rev. James Cruickshank of Montserrat, in 1699, was allowed 20,000 lbs. of sugar per annum, provided he preached every Sunday.[23]

As early as 1661 the chancellor of the exchequer, Lord Ashley, recommended that a scheme for a land bank, which had been submitted to him, be tried out in Barbadoes. Accordingly, December 9, 1661, a warrant was issued to the attorney or solicitor general to prepare a bill authorizing Thomas Elliott, groom of the bedchamber, Sir John Colleton, and Francis Cradock, and others, from time to time appointed by the governor, council, and assembly of Barbadoes, to erect and manage a bank or banks in that island, "founded on the security of lands and goods, with sole power to give credit and transfer the same from one months account to another, as is done by the ownership or credit of money in foreign parts." No more than six per cent interest should be charged on loans to planters. The government of Barbadoes was to appoint a person to inspect the whole management and determine the value of all lands and the credit to be laid on them, and to appoint reasonable rates for warehouse room. Records of estates were to be kept at the bank. The loans upon land or goods were to be made in bank notes, counterfeiting of which was to be severely punished.[24] Though the notes probably were not to be declared legal tender, yet it was expected they

21 Feb. 28, 1669. *C. S. P. Col., 1699*, Addenda, no. 1113, p. 591.

22 *A Discourse of the Duties on Merchandize*, 1695, p 5.

23 Council Minutes, Montserrat, Mar. 14, 1699, *C. S. P. Col., 1699*, no. 173, p. 102.

24 *C. S. P. Col., 1661-1668*, pp. 60, 62, and no. 194.

would be generally acceptable and provide the island with a currency for internal trade.

While the scheme of 1661 for providing Barbadoes with an inland currency did not materialize at the time, it was not forgotten. As in Jamaica, a considerable amount of Spanish coin found its way into Barbadoes and was used in part payment for slaves and imported goods. Soon after the outbreak of the War of the Spanish Succession, however, returns in Spanish money became impossible. "Pieces of 8 are not to be had," wrote the African Company's agent in 1705, and added that "So much money has been carryed off some think that the trade will revert into a sugar trade."[25] At this juncture the legislature of Barbadoes passed an act (1706) establishing a land bank. The scheme provided "That every person inhabiting and having an Estate of Inheritance in Barbadoes, may have a Bill or Bills of Credit, signed and sealed by John Holder Esqr nominated in the Act for that purpose, to the value of the fourth part of the Island as ready Money, to the full value of the sum therein mentioned; and no person to refuse the same, under the Penalty of forfeiting a full moiety of the sum contained in the Bill, the Bills to pass but for one year, but renewable from year to year till the last year before the Expiration of the said Act, which is to continue five years & no longer. The end and intent of the Act is declared in the Preamble, to remedy or supply the want of cash in the Island and to help Creditors to pay their Debts." No provision was made for redemption of the bills into specie. The Board of Trade recommended the disallowance of the measure on the ground that the depreciation of the bills would

[25] James Aynsworth to Royal African Co., Nov. 15, 1705, Treas. 70: 8, f. 5; same to same, Jan. 22, 1705/6, *ibid.;* Benjamin Bullard to Royal African Co., Apr. 8, 1706: Coin is so scarce, he states, that their trade is near lost for want of cash, *ibid.*, f. 7.

work a hardship to creditors and raise prices. It would also obstruct the trade in foods from North America since the Northerners would refuse to take anything but coin.[26]

The enactors of the Bank Note Law were debtors, some of whom, it was said, paid their debts with the bills.[27] In the colonial council the supporters of the act were William Sharp, William Cox, William Cleland, James Colleton, and Alexander Walker, all of whom were much in debt.[28] In England all the leading Jamaica merchants signed a memorial against the scheme.[29] Also Mr. Bromley and other absentees in England presented a memorial against the act. "Since there is not silver sufficient," it read, "to negociate the Trade of the Island, sugar should be made use of instead thereof, as it was about thirty-

26 Representation of Board of Trade to the Queen, Oct. 17, 1706. It was signed by Robert Cecil, Ph. Meadowe, Mat. Prior, William Blathwayt, and John Pollexfen, C.O. 29: 10, ff. 144-150.

27 C.O. 28: 38, f. 167: list of the council who voted for the act. This bundle (38) contains many papers on the subject.

28 *Ibid.*, 9, N 66.

29 "Memorial against the Credit Act," *ibid.*, 29: 10, ff. 134-140. The Barbadoes merchants in England who signed this memorial were

Rd. Scott	Edw. Chilton	Edw. Lascelles
Jas. Gohier	Rich. Bate	Mathew Matson
David Miln	Guy Bale	Abraham Mendez
Edw. Alanson	John Saile	R. Hallett
Tho. Ward	Jonath. Leigh	Rd. Haynes
Geo. Lillington	John Dersley	John Donaldson

A defense of the act by one of its officers, John Holder, is entitled "The Rise, Progress & Determination of the Bank of Barbadoes," dated, March 28, 1707, 10 1/3 pp. in manuscript. *Ibid.*, 28: 10, N 63. The author confesses to a great outflow of silver from Barbadoes during the operation of the bank. See also "Observations upon a Paper Entitled The Rise, Progress, and Determination of the Bank of Barbadoes," received from Mr. Walters by the Board of Trade, Oct. 18, 1707, *ibid.* A memorial against the act from the Royal African Company was received by the Board of Trade, Oct. 10, 1706, and is in *ibid.*, 9, N 37, and in Treas. 70: 170, f. 118.

five years agoe, and is of an intrinsick universal value.'"[30] In the face of so much opposition the act was naturally disallowed by order in Council October 21, 1706.[31]

Although the Land Bank had but a few months of actual existence, it caused a serious upheaval in the economic life of Barbadoes. "Indeed the late fattall paper act," wrote Governor Crowe in 1708, "has reduced many men that were in great Possessions, to a Low Ebb, Especially Judge Burke.'"[32] It was within the power of the treasurer of the bank to issue "by a Modest Computation" over £100,000 in notes. Prices of commodities almost immediately rose "Ocationed by their cursed Bank Notes," wrote the African Company's agent, "which are esteemed 20 p Ct worse then ready Cash, if this Bank had not been Sug^{r} had not exceed[ed] 14: or 15^{s} and it was made for no other end, then to advance the growth of this Island.——What Negroes are sold is for Bank Notes and the Planters make this difference in their Commodityes Vizt Paper is at leest 20 p Cent Disadvantage to Trade.'"[33] A month later the agents reported "That the Paper has so raised the Price of Sugars that in the difference between the Sugars p the first & Second fleets the Comp^{a} have lost £500.'"[34] Not only the African Company but all creditors who were forced to take payment

30 Nov. 7, 1706, C.O. 28: 9, N 50. The only means to resupply the island with coin, said the memorial, was to renew trade with the Spanish Main.

31 C.O. 29: 10, f. 153. See also *ibid.*, 28: 9, N 39; *Acts of the Privy Council, Col.*, II, § 1007; *C. S. P. Col., 1706-1708, passim.*

32 Gov. Crowe to the Earl of Sunderland, May 18, 1708, C.O. 28: 38, f. 190; *C. S. P. Col., 1706-1708*, no. 1483.

33 Benjamin Bullard to Royal African Co., July 10 and Aug. 14, 1706, Treas. 70: 8, f. 22. He "ffears there will be a sufficient loss among the Merchants when this Act is repealed."

34 Bullard, Bate, and Stewart to Royal African Co., Sept. 19, 1706, Treas. 70: 8, f. 28.

for debts in paper suffered by its rapid depreciation.[35] Within eighteen months of the original issue the notes were circulating at a discount of forty-five per cent.[36] The royal disallowance of the scheme was received with rejoicing by the creditor class.[37]

Barbadoes was thus rescued in the early stages of a cheap money mania. But the situation that prompted it continued, and was even more distressing to the indebted planters. For practically all specie had now been driven from the island. "This Island is so drained of Cash," wrote Governor Crowe November 5, 1707, "that there is little to be procured for the minutest occasions, which is one great reason for the poorer sorts leaving us; and last week one Harrison a planter, (being much in Debt notwithstanding the strictest Ord^rs^ given to the Forts) run off in a Sloop with above Sixty Negros, leaving his land to the Creditors. I wish this may be the only instance of that kind, considering the miserable and low Condition of the Inhabitants, which will be insupportable Except her Majesty be Graciously pleased to take this place under her Consideration and Compassion.[38] Remittances from the African agents at Barbadoes and Antigua as

[35] In their memorial, Oct. 10, 1706, to the Board of Trade against the act the African Company said: "It must inevitably occasion a Discount on said paper Credit & consequently in buying & selling of Goods the Payments in paper must be considered & cause a difference in prices betwixt that payment & money whereby it will be a very great loss to us the Royal African Company & to all those who shall be forced to take their present Debts in Paper." Treas. 70: 170, f. 118.

[36] Bullard, Bate, and Stewart to Royal African Co., Dec. 7, 1707, Treas. 70: 8, f. 55.

[37] Same to same, Jan. 3, 1706/7, Treas. 70: 8, f. 36. The assembly, in 1708, was said to have been attempting a new paper credit measure which aroused fear in the agents. But nothing came of it, Raynes, Bate, and Stewart to Royal African Co., Aug. 30 and Sept. 2, 1708, *ibid.*, 8, f. 74. *Cf.* f. 78.

[38] Gov. Crowe to Board of Trade, Nov. 5, 1707, *C. S. P. Col., 1706-1708*, no. 1176.

well as Jamaica, in the absence of gold and silver, were for the most part in sugar, indigo, ginger, and pimento.[39]

The period of the War of the Spanish Succession revealed in their most aggravating forms the inconveniences that Barbadoes suffered for the want of a sound and adequate currency. In the years following the peace of Utrecht, trade with Spanish America somewhat improved the monetary situation, but scarcity of coin continued a topic of constant complaint. The growing tendency of North Americans to insist upon payments of money instead of molasses and sugar for their supplies undoubtedly drained Barbadoes and the Leeward Islands of much of their coin. The Barbadians came to believe that the Northerners were chiefly to blame for the distress and decline of the island. The movement of 1730-1733 which culminated in the Molasses Act was in part, therefore, designed to secure a metallic currency for Barbadoes and the other sugar islands. The hopes of John Ashley and the other champions of this policy were not, however, realized. Coin continued scarce and sugar at market prices was everywhere accepted as money.[40]

39 Peter Beckford and Lewis Galby, Jamaica, Dec. 31, 1707, and Jan. 16, 1707/8, Treas. 70: 8, f. 57; Bate and Stewart, Barbadoes, May 23, Sept. 23, and Oct. 12, 1708, *ibid.*, 8, ff. 66 *et seq.;* Edward Chester & Son, Antigua, Mar. 6, 1707/8, *ibid.*, f. 60.

40 See John Ashley's Memoir of 1744 in Brit. Mus. Add. MSS. 12404. Ashley on several occasions urged that England should adopt the French practice of coining small silver for the particular use of the sugar islands. It might have been put into circulation by paying it in salaries to government officers and to the troops. This would have relieved the constant distress that arose from the want of an internal currency. Long recommended in like manner the introduction of small copper coins into Jamaica. *Jamaica,* I, 571.

In Antigua, St. Christopher, Nevis, and Montserrat there was little coin in circulation. "There is another currency in all these islands viz: their own produce, sugar, rum, &c., which they compel people to take, for it is a fact that if a man demands to be paid sterling money or other gold and silver money he lent conditioned to be paid in the same species it is in the

In Jamaica the scarcity of money was a subject of perpetual comment. "So great is the scarcity of money," said a Jamaican, in 1665, "that unless there be free trade or war with the Spaniards, the colony will never flourish or hardly be kept."[41] War might, at least, lead to the capture of silver laden galleons. Barter of commodities for commodities was, according to Blome in 1672, the commonest mode of trade.[42] It was in 1683 that an act of Jamaica made sugar, ginger, anatto, indigo, cocoa, cotton, or pimento, as valued by churchwardens of the parish, legal payment for all debts. This act was not repealed until 1751.[43]

Among the coins most common in Jamaica were the Spanish doubloon (20s.), French pistole (20s.), French crowns (5s.), Mexican or Seville pieces of eight (5s.),

option of the debtor as the law now stands to pay in country produce." "General Observations upon the Raising of the Coin in the Leeward Islands," 1740, *Cal. Treas. Books and Papers, 1739-1741*, CCCII, no. 3. In an investigation of Lascelles' administration of the customs of Barbadoes before the Board of Customs, in 1743, "Mr. Lascelles produced the affidavits of Mr. George Newport & Mr. George Knight Merchants, who have both made oath that during the time they lived in Barbadoes, which was many years they never knew or heard that Mr. Lascelles at any time insisted or desired money for the dutys of sugar, and Mr. Newport says, that when he was in Barbadoes, he had often desired Mr. Lascelles to take money instead of sug[r] but he refused it, and they have both made oath that whenever the collector took money instead of sugar they always understood it to be a favour, and at the request of the merchant, and it was allowed by the complainants and all the persons that attended the examination that it had been a practice for many years before Mr. Lascelles was collector for the duty on small parcells of sugar, at the desire of the merchant, to be paid in money." The Customs Board in their report to the Lords of the Treasury stated this to be the true practice in the sugar islands. The collector had been suspected of making a personal profit arising at times from the difference between the price at which produce was credited to the government and its actual market price. Brit. Mus. Add. MSS. 33028, ff. 378-383.

41 John Style to Sec. Arlington, Jamaica, July 24, 1665, *C. S. P. Col., 1661-1668*, no. 1023.

42 Blome, *Description of Jamaica*, p. 9.

43 C.O. 139: 17, no. 1.

and Peru pieces of eight (4s.).[44] Although it was against the British Navigation Acts for Spaniards to trade in their own Ships with the English colonies, British West India governors were instructed to grant licenses for Spaniards to trade in the islands for slaves and English manufactures, for such trade brought in coin and yielded colonial customs revenues in pieces of eight.[45] An attempt of the colonial government to increase the value of pieces of eight, even light ones that were worth five shillings to six shillings, was made in an act of 1688 to run two years. This device was employed more than once in the West Indies. It tended to cheat creditors, and at this time the African Company asked for its disallowance, inasmuch as the company had over £100,000 out in debts.[46]

An example of the mode of trade in slaves with the Spanish may be illustrated from an asiento of 1689. In that year Don Santiago del Castillo, "Commissioner Generall for the Introduction of Negroes into the Spanish Indies" in London, entered into an agreement with the Royal African Company for buying slaves in Jamaica and Barbadoes from the company and individuals. Castillo agreed to import specie, bullion, jewels, and other products from Spanish America. In this case the asiento, or agreement, was sanctioned by both the Span-

44 *Laws of Jamaica,* I, no. XIX (Act of 1681).

45 *Acts of the Privy Council, Col.,* II, no. 587: An order of Mar. 11, 1662/1663; Royal Instruction, Whitehall, Feb. [?], 1663, *C. S. P. Col., 1661-1668,* no. 415. Pieces of eight were said to be worth 4 shillings. *Acts of the Privy Council, Col.,* II, no. 182: an order in Council, Apr. 29, 1685; *ibid.,* no. 183: a circular letter to all the West India governors. See also Charles II to Gov. Molesworth of Jamaica, Nov. 30, 1684, recommending encouragement to the Spanish Trade, Papers of Royal African Co., Treas. 70: 169, ff. 39-40. *Cf.* Royal African Co. to the King, July 3, 1688, in which they desire Jamaica to be restrained from hindering Spanish ships to come there for slaves. Treas. 70: 169, ff. 99-100.

46 Papers of Royal African Co., Treas. 70: 169, ff. 98, 106, July 3, 1688.

ish ambassador in England and the Privy Council. "Factories" were to be located in both Jamaica and Barbadoes. Castillo had already resided three years in Port Royal, Jamaica, and had managed the Spanish trade there.[47]

Of the extent and character of the trade between Jamaica and the Spanish in 1700, Governor Beeston wrote: "As for the money I am Well Assured from the Merch^ts that 150,000^li is the least Quantity that is Yearly sent here, most where of arises from the sale of neg^rs, flower, and other Provision, but not much now from any drapery or dry goods, that trade being lost to us by the Greate Quantitys supplyd by the Dutch and french who can under sell us 40 p Ct and therefore have Caused that trade to Ceize, and in all Appearance the flower Trade and Provisions will fayle us too, if some way be not found to hinder it, and the Profit to the Merch^ts, and Imploym^t of our sloops and vessels lost to this Island, and by Consequence much money allso."[48] Obviously the French and Dutch West Indians were quite as desirous of the precious metals from Spanish America as were the English. Prior to the outbreak of the War of the Spanish Succession, as estimated by a number of Jamaican merchants in England, at least £200,000 in Spanish gold, pieces of eight, and bullion was brought from Jamaica to Great Britain yearly, "All of which was purchased with merchandize sent from Jamaica mostly English manufacturers." This amounted to nearly a third of the total value of imports from Jamaica.[49]

So vital was the trade for bullion and coin that during the war the governors were instructed to permit the

[47] Treas. 70: 169, ff. 112-115.

[48] Gov. Beeston to Board of Trade, Mar. 14, 1700/1, C.O. 137: 1, E 2.

[49] Memorial from Jamaica merchants, rec'd Jan. 20, 1703/4, C.O. 137: 6, G 41.

colonists to trade to Spanish America.[50] In 1708, the African Company's agents in Jamaica were able to write: "Will continue to make remittances in ps ⅝ as long as they are to be had."[51]

It was during the war that French competition, already pointed out by Beeston in 1700, seriously cut off the Spanish trade to Jamaica and the inflow of coin. "The reason of it was," wrote Governor Handasyd, "by the ffrench Carrying all Comoditys to the South Sea's; and Selling them there; as Cheap as wee can sell them here; which discourages the Spaniards from buying in these parts, and so [we Jamaicans] are ready to sterve; as well as our Other Merchants here."[52] At another time he wrote: "If there can be a Method found out to prevent the ffrench Trading to Lima, and the South Seas, Trade will soon be in a flourishing Condition."[53] "As to News here," he wrote in the spring of 1711, "there is very little, Trade being very dead and like to be Worse, Except some Method coud be taken for preventing the ffrench Trading to the South Sea, and the River plato, from whence they Supply the Spaniards with Negros Slaves and other Necessarys."[54]

At the close of the Spanish War the French occupied an enviable position in the commercial centers of Spanish

[50] Circular letter from Earl of Nottingham to governors, 1705, C.O. 5: 3.

[51] Peter Beckford and Lewis Galby to Royal African Co., Mar. 26 and 31, 1708; Papers of Royal African Co., Treas. 70: 8, f. 61.

[52] Gov. Thomas Handasyd to Board of Trade, June 4, 1710, C.O. 137: 9, M 13.

[53] Same to same, July 20, 1708, C.O. 137: 8, L 31. See also his letter of Mar. 25, 1710, *ibid.*, L 99.

[54] Same to same, Mar. 20, 1710/1, C.O. 137: 9, M 48. All Handasyd's letters in this period refer to the declining trade with the Spanish. For a recent treatment of the French trade to the Plate region and the Pacific Coast of Spanish America, see E. W. Dahlgren, *Les Relations Commerciales et Maritimes entre La France et les Côtes de l'Ocean Pacifique (Commencement du XVIIIe Siècle)*. Tome I, *Le Commerce de la Mer du Sud Jusqu'a la Paix d'Utrecht*, Paris, 1909, pp. xvi, 729.

America. By special passes and permits to Frenchmen under Spanish names, they had built up a trade from Spain to the Northern colonies in South America, and through them to Chili and the other "South Sea Colonies," i.e., Ecuador, and Peru.[55] So excited did the English become that instructions were drafted, in 1715, for Methuen, ambassador to Spain, requiring him to complain to Madrid against Frenchmen obtaining passes to trade from Spain to South America, from which valuable trade Englishmen were excluded.[56] England appears to have obtained no particular satisfaction, however, from the Spanish Court. Louis XIV's Spanish policy yielded rich returns to French merchants and colonists in the western world. "In fine," wrote Governor Lowther of Barbadoes, "the French behave themselves like Lords paramount over this part of the World and treat the Spaniards just as they think fit. They have diverted the old channel of trade and have carry'd it from the North to the South Sea: heretofore Cartagena, Portobel, and Sta. Fe used to be the Chief Marts, where most of the European Commodities were vended, and the Spaniards on the southern part of the Continent resorted with their silver to those places to buy what they wanted, but the French now supply them by Panama and the other ports in the South Sea."[57]

The scarcity of money which resulted from the decline of the Spanish trade was greatly relieved after the restoration of peace.[58] The asiento which lasted from

[55] See paper on the subject sent to the Board of Trade by Richard Harris, an important English merchant, Jan. 17, 1714/5, C.O. 389: 21, ff. 502-514. Harris thought the English also ought to gain a right to trade to Spanish America for logwood, indigo, cochineal, and spices.

[56] Draft of instruction for Methuen, Jan. 17, 1714/5, C.O. 389: 22, ff. 71-72.

[57] Gov. Lowther to Board of Trade, Jan. 25, 1716/7, C.O. 29: 13, f. 358.

[58] In 1715, an act was framed in Jamaica to prevent the export of coin

1713 to 1739, whereby England alone was to supply slaves to Spanish America and gained permission to import one British ship load of English goods a year into their ports, opened Spanish America to an untold amount of illicit trade. It was not till 1728 that Jamaica was again especially distressed for want of currency. The assembly then passed an act to make sugar and other produce of the island at fixed prices legal tender for payment of debts. But Governor Hunter refused to sign the bill as it would work an injury to creditors. Sugar, considered as money, was at this period entering upon a decade of great depreciation. The assembly tried to fix its purchasing power at a rate higher than the market value.[59] Creditors were willing to take sugar for old debts but only as appraised at the time of payment.

During the period 1730-1739 most of the pamphleteers refer to the scarcity of coin. "If any Body has any," wrote a Jamaican in 1733, "they keep it by them for Fear They should never see it again."[60] The smallest coin in circulation was the bit, worth seven and a half pence. Indeed, the absence of small change was a great grievance to the poorer class of people throughout the century.[61]

The activity of Spanish coast guards in suppressing illicit trade with the British, from 1737 on, again interrupted the supply of coin to the English and brought on the war with Spain. The Spanish War of 1739 and

to any place but Great Britain or the Northern Colonies. But its passage was blocked, each house of the legislature refusing the amendments of the other. Council Minutes, Nov. 2-29, 1715, C.O. 140: 13, ff. 112, 120, 122, 127, 129, 140, 142.

59 Gov. Hunter to Board of Trade, Aug. 3 and 5, 1728, C.O. 137: 17, S 35, 36, 37.

60 A letter from a Gentleman of Jamaica to his Friend in London, Sept. 16, 1733, Brit. Mus. Add. MSS. 19049, f. 7.

61 Statement of Sir Chaloner Ogle to Board of Trade, Oct. 29, 1735, C.O. 391: 44, f. 240.

the capture of Porto Bello increased the circulation of Spanish money in Jamaica and enriched the island.[62] Although North Americans in these years, it was said, would receive nothing but money in payment for their supplies,[63] the West Indies had sufficient coin with which to pay. In 1750, however, a vigorous protest against the New Englanders for draining the island of coin was registered with the Board of Trade.[64]

Between the peace of Aix la Chapelle and the outbreak of the Seven Years' War the Spanish trade was renewed. But it appears to have been somewhat reduced in volume. Prior to 1756 "there were," according to Governor Lyttelton of Jamaica, "between fifteen and twenty vessels employed in the Spanish Coast Trade to Aruba, and Rio de la Hache who carried off some Rum and Dry Goods and in return imported about two thousand Mules and Horned Cattle annually which the War in a great measure has put a stop to. From various pretences the Spaniards used to come to this Island and bring large Sums of Money in which they never were interrupted also they used to Run over from Cuba to the North side of this Island in small Boats with Mules and Cattle for which they carried off Dry Goods but the War put a stop to it."[65] In trade with the Spanish the balance was always in favor of the Jamaicans, and was paid in money and bullion. The colony "had by that means a sufficient quantity of Specie for its internal Commerce, and also was enabled to send large Sums annually to England in

[62] Dr. J. Houstoun, *Memoirs*, 1747, p. 266.

[63] Knight, *History of Jamaica*, ff. 204-205.

[64] "Memorial of the Sugar Planters Merchants and others . . . relating to the illicit Trade carryed on from the Northern Colonys to the ffrench etc. Received from John Sharpe by the Board of Trade, Oct. 18, 1750." C.O. 323: 12, O 59. Another copy is in C.O. 5: 38. See Appendix XI. p. 414.

[65] Gov. W. H. Lyttelton to Board of Trade, July 9, 1763, C.O. 137: 33, Cc 19.

Silver.''[66] Long estimated that about the year 1750 the value of the slaves sold to the Spanish was £112,500, and the English merchandise amounted to as much, most of which was paid for in coin. The larger part of this money was spent in Jamaica by the Guinea ships for sugar. Thus the £150,000 specie, which was needed for the trade of the island, was secured.[67] So adequately supplied with coin was Jamaica that the act of 1751 which repealed the law of 1683, legalizing the payment of debts in sugar, and which made gold and silver the only legal tender aroused no opposition among the planters.[68]

In 1760, there was a very noticeable decrease in the volume of metallic money in Jamaica. This was attributed, partly to the interruption of Spanish trade, but mainly to an unprecedented drain upon the island's specie by the North Americans whose illicit trade to Monte Cristi (Santo Domingo) was then at its height. Speaking for his native island, Alderman Beckford told the Board of Trade ''That the Nature of the Trade with the Northern Colonies exhausted their Cash, the Traders from which, in return for their Lumber, take only a small part in Produce, but chiefly Cash, with which they go and

66 ''The Traffick which used to be carried on with the Spaniards, who came in small Vessels upon the Coasts of the Island, and purchased large quantities of Good's the Manufactures of Great Britain, in return for which they disposed of some Mules and Cattle, for the use of the Planters, and paid a large Balance in Money and Bullion, was deemed very beneficial both to the Mother Country, and this Colony, which had by that means a sufficient quantity of specie for its internal Commerce, and also was enabled to send large sums annually to England in Silver;'' but his orders now treat this as illegal, which he intimates is mistaken policy. Lyttelton to Board of Trade, Apr. 2, 1764, C.O. 137: 33, Cc 39.

67 Long, *Jamaica*, I, 530.

68 C.O. 139: 17, no. 1. The draining of the island's coin by North American traders continued, however, a sore point with Jamaicans. See ''An Inquiry into the Causes of the Present Scarcity of Money,'' by a Jamaican writing about 1750 (1757 is the probable date), Brit. Mus. Add. MSS. 30163.

purchase a Cargo at the French and Dutch Islands at 50 p Cent Advantage.''[69] As if to increase its cash by fiat Jamaica had, in 1758, passed an act to raise the nominal value of every Spanish dollar by two pence.[70] The abler leaders recognized, however, that such a measure ''never can answer to keep Money in the Island; That the Balance of Trade being against the Island, Money will and must go out; That it will not answer the purpose intended, and will only hurt Creditors, and alter the Rate of Exchange.''[71] The measure excited much opposition from the creditor class in England, and must have strengthened the movement for a regulation and control of North American trade such as would effectively prevent it from draining the coin of the sugar islands.

The situation in Jamaica, arising from the scarcity of coin, was very embarrassing. Even a wealthy planter might not be able to procure money enough to pay a sudden demand for £500 to a Jamaican creditor. An ill-natured creditor had it in his power to distress and damage the planter's estate to five times the value of the debt, by levying on his negroes. Furthermore, neither sugar nor rum, after 1751, would be accepted as money, unless the creditor was disposed to take them. ''He may insist on payment in coin, although there is no mint in

[69] B. T. Jour., Feb. 29, 1760, C.O. 391: 67, f. 60.

[70] An Act for Ascertaining the Value of Spanish Milled Money, C.O. 139: 19, no. 15.

[71] Stephen Fuller, agent of Barbadoes, Feb. 29, 1760, C.O. 391: 67, f. 62. William Vassal also opposed the act. Alexander Grant told the Board that ''the only way to lessen the Balance of Trade against the Island, and increase the Cash would be to lower the Price of their Commodities.'' *Ibid.*, f. 63. Mr. Morant who had just arrived from Jamaica said ''That the scarcity was so great, that not half the Taxes could be paid, nor the Troops subsisted.'' He thought some drastic measure had to be taken. *Ibid.*, f. 64. Mr. Elletstone, also lately arrived from Jamaica, declared the bill was of absolute necessity. He said its effect had been to raise the value of dollars from 6 shillings to 6 shillings and 6 pence. *Ibid.*, ff. 64-66. The Board of Trade recommended the disallowance of the act, March 7, 1760. *Ibid.*, f. 72.

the island, no coin peculiar to it, the channel by which it used to be supplied stopped up, a perpetual efflux by an ill-managed trade with North America, and so little remaining for circulation, that the inhabitants can with difficulty provide enough to pay their taxes, or purchase their daily subsistence.''[72] The inconveniences here revealed were by no means peculiar to Jamaica, but existed in practically all the British American colonies and in the United States until near the outbreak of the Napoleonic wars. To remedy the situation by discouraging North Americans from trading to the French and practically compelling them to take sugar, molasses, and rum, instead of money, from the English sugar islands was the purpose of the Sugar Act of 1764 and the administrative reforms adopted to enforce it.

[72] Long, *Jamaica,* I, 560-561. Scarcity of money was a topic of permanent interest to Jamaicans. In 1787, the same year that Shays' followers were fighting for paper money in Massachusetts, it was proposed for the first time in Jamaica to establish a bank to issue paper money in loans secured by land. In this way, it was urged, the island would obtain a currency for internal trade. Article by Senex in *Kingston Morning Post and General Advertiser,* January 5, 1787 (no. 539), Brit. Mus. Add. MSS. 12431, f. 183.

CHAPTER VII

FOREIGN MARKETS FOR BRITISH SUGAR

The demand for sugar on the continent of Europe was largely satisfied from England until near the end of the seventeenth century. Writing in 1670, an Englishman records that "By reason of this Industry of our Plantations wee have already beate out the Portugeize at Brazile from sending home any Muscavadoes, And have reduced their fine Sugers from 7 & 8 pounds the hundred, unto 50 and odd shillings, and are by this means the Sole Merchants, almost of all that Suger that is Manufactured into loafe or hard Suger either in Holland, France, or Hambrough as being all made out of our Muscovadoes."[1] The enumeration of sugar, the writer added, was likely to lose for England much of her foreign market for that commodity.[2] Of all the sugar imported into England the planters estimated in 1675 that only half was consumed in England.[3] Refined sugar as well as raw was exported, particularly to the East Country, that is the Baltic ports.[4] A decline in the export of refined sugar, however, set in after the increase of import duties on sugar from

1 "The true state of the Manufacture of Sugers within our Plantations," etc., British Museum, Egerton MSS. 2395, f. 636. For earlier shipments to the continent, *cf. C. S. P. Dom., 1668*, pp. 332, 337, for ships with sugar from Bristol, Hull, Lyme, and Pendennis bound for Rotterdam, Amsterdam, and Rouen.

2 Brit. Mus. Egerton MSS. 2395, f. 632.

3 C.O. 31: 2, pp. 179, 180, Petition to the king from Barbadoes, Apr. 16, 1675.

4 Report of the Commissioners of Customs on the Case of the Sugar Refiners, Jan. 13, 1691/2, *Cal. Treas. Papers, 1557-1696*, XVII, no. 14.

the colonies in 1685 under James II.[5] As no drawback of duty was allowed on refined sugar exported, its price abroad rose until competition with Dutch and French refineries was impossible.[6] On raw sugar the whole additional duty could be drawn back with the result that exports of raw sugar continued undisturbed. In fact, the relief from English competition in refined sugar stimulated the Continental refineries to make larger demands for the British raw product. Prices for raw sugar rose and benefited the planter. The injury fell upon the English refiner who eventually lost his export trade.[7] He was in part compensated, however, by an additional duty of three farthings per pound on refined sugar from the colonies and five farthings on foreign refined. In 1693, these additional duties on sugar were removed,[8] raw sugar paying 18d. per cwt. with a drawback of 9d. on reëxport. But during the eight years, the Europeans perfected their art and were fortified for the future.[9]

To Italy, Cary noted that England exported "Sugars both White and Brown, the last principally to Venice, but more thereof in times of Peace than we do in this time of War, Freights being high, and the Commoditie

[5] I Jac. II, c. 4; *Commons Journal*, IX, 724, 733, 737, 738; *Hist. MSS. Com. Rep.*, 1909, II, 155.

[6] *Discourse of the Duties on Merchandize*, 1695, p. 9. The Old Subsidy of 1660 was 18d. per cwt. on raw sugar, half of which was drawn back on reëxport.

[7] *Ibid.*, 15.

[8] 2 William & Mary, sess. 2, c. 5; 4 and 5 William & Mary, c. 15.

[9] Cary thus remarked on the situation in 1695: We sent to the continent "formerly great Quantities of Refined Sugars, till we gave the French and Dutch leave to undermine us, partly by the Advantages they had by the late Imposition on Muscovadoes, and partly by the Imprudence and ill Management of our Sugar-Bakers, who would not take Pains to comply with the humours of that People as the others did; but I hope if due care be taken, that profitable Trade may be recovered again." *Essay on Trade*, 1695, p. 116.

dear at home.'"[10] In return were brought raw and thrown silk, red wool, oil, a certain soap, some paper, and currants. To Holland went sugar, tobacco, indigo, logwood, fustic, ginger, and cotton. What the Dutch did not consume of these was dispersed throughout Germany.[11] Even the poorer grades of sugar were acceptable in the Dutch markets.[12] At Hamburg, English sugar, tobacco, and cloth were exchanged for German linen yarns, linens, and other commodities.[13] Sweden, besides other manufactures, took considerable English sugar and tobacco.[14] Some tobacco and sugar were shipped to Denmark. From these northern kingdoms England was supplied with pitch, tar, hemp, masts, timber, and iron.[15]

Exports of British manufacture to Portugal were exchanged for Portuguese sugar which the English transported to Italy.[16] Sicily, in 1713, was reported as usually receiving sugars at that time from England.[17]

Considerable information on the foreign trade of the English is obtainable from British consular reports preserved in the Commercial Series of the Board of Trade. In 1711, the consuls of England were requested to assemble the merchants and send to the Board of Trade a report every six months on the trade of their localities.[18] An examination of these reports down to 1765 reveals that English sugar and other colonial products were widely sold in Europe but in amounts that, for sugar at least, steadily declined. This was usually claimed to be

10 Cary, *Essay on Trade,* 1695, 122.

11 *Ibid.,* 123.

12 Oldmixon, *British Empire in America,* London, 1708, II, 152.

13 Cary, *Essay on Trade,* 1695, p. 125.

14 *Ibid.,* 126.

15 *Ibid.,* 128.

16 *Ibid.,* 138.

17 Macpherson, *Annals of Commerce,* III, 37 n.

18 B. T. Jour., Mar. 1, 1710/1, C.O. 39: 21, p. 243. *Cf.* also Board of Trade to Sec. of State Lord Townshend, Apr. 1, 1715, C.O. 388: 95, I (23).

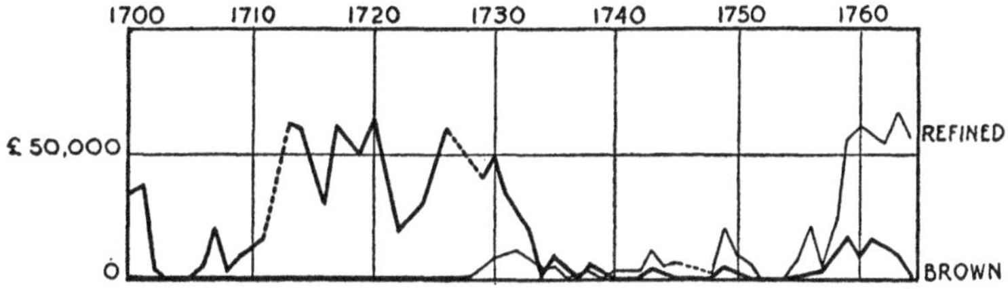

CHART III. VALUE OF SUGAR EXPORTED FROM ENGLAND TO FLANDERS

From the Custom House Accounts, Ledgers of Imports and Exports, it has been possible to compile tables of the values of brown and refined sugars exported each year from England to various parts of Europe. Charts III-VII represent the values of such exports to the principal foreign markets, the heavy line indicating raw or brown sugar and the thin line refined sugar. The lines are dotted where statistics are wanting. Similar tables compiled for Sweden, Denmark and Norway, the "Streights," Italy, Venice, and Turkey reveal so little trade with those localities that it is unnecessary to construct charts of it here. The diagrams show clearly what a decided falling off occurred in the sugar trade to Holland after 1720 and to Flanders and Germany after 1730. The effects of the Austrian Succession War and the Seven Years War in stimulating the foreign sugar trade is also well marked. Exports of refined sugar became of considerable importance only after the middle of the eighteenth century.

due to competition from the French and Dutch. The sugar distributed by the Dutch was often of French origin.

The Spanish Netherlands, according to the consul, Burchett, took from England from Christmas 1712 to Christmas 1713 of brown sugar 24,855 cwt., of white 70 cwt., and of refined 102 cwt.[19] The raw product was refined by the Flemish themselves. During the eighteenth century the sugar trade to Belgium is given but slight notice. Formerly loaf sugar from England was imported, but having established refineries of their own, that trade came to an end. Raw sugar, usually about thirty hogsheads per ship, remained with tobacco and rice a feature of each English ship's cargo to Ostend as late as 1765.[20] Their duty on brown sugar was very low compared with that upon other grades.[21] At Brussels, in 1765, many sugar-houses had recently been erected and though English sugar was imported from time to time, they preferred French raw sugar which they found cheaper than English.[22]

From Elsinore, Denmark, Robert Tigh reported that Danish refiners were protected by high duties, and that the amount of English sugar imported had declined owing to high duties in retaliation for British duties against Norwegian deals.[23] For many years, Tigh kept lists of all ships passing through the Sound bound for

[19] C.O. 388: 17, N 204, Burchett to Board of Trade, Oct. 27, 1714. *Cf.* also C.O. 388: 18, O 60, Mar. 21, 1715. See also Chart III facing p. 158.

[20] Thomas Mortimer, vice-consul at Ostend, to Board of Trade, Feb. 28, 1765, C.O. 388: 95, I (14). He noted a general improvement in British trade to Ostend since 1740 in tobacco, rice, and raw sugar.

[21] Mortimer to Board of Trade, Ostend, Mar. 13, 1765, C.O. 388: 95, I (15).

[22] James Porter, consul at Brussels, to Lord Sandwich, Feb. 10, 1765, C.O. 388: 95, I (16).

[23] Robert Tigh, consul at Elsinore, to Board of Trade, Dec. 7, 1715, C.O. 388: 20, P 66. Also Apr. 25, 1719, C.O. 388: 21, P 185.

the Baltic.[24] Many cargoes of sugar were observed going east to Stettin, Danzig, Königsberg, Eastsea, and Stockholm, but practically all were carried in Dutch ships from Amsterdam, and it is probable that a great deal of the sugar was of French origin. In 1765, the minister to Copenhagen reported that, with the exception of tobacco, rice, leather, coal, lead, and tin, the products of Great Britain were excluded from Denmark and Norway (after 1760), and that trade with England was declining.[25] In North German ports excessive tolls were encountered. Largely because of this hindrance Holland was able by land routes to control the markets of Westphalia, Cassel, Brunswick, Hanover, and Bremen,—the territory, that is, between the Rhine and the Weser.[26] Sweden discriminated in favor of sugar imported in Swedish bottoms.[27]

The consumption of sugar in Russia appears to have been slight. From the report of the British consul at St. Petersburg, it appears that only small amounts of English raw and refined sugars, coffee, indigo, logwood, and tobacco were imported. In 1764, England supplied St. Petersburg about one-seventh of its sugar.[28]

Austria, Bohemia, and Hungary, prior to 1743, were supplied with sugar refined at Hamburg from French

[24] Tigh's lists, forwarded to the Board of Trade yearly between about 1714 and 1740, give the names of all ships passing through the Danish Sound, their home ports, destination, and character of their cargoes. These lists constitute little books, and are in C.O. 388: 25-38.

[25] Dudley A. A. Cosby to Earl of Sandwich, Copenhagen, Mar. 16, 1765, C.O. 388: 95, I (20).

[26] Elkin to Board of Trade, received Dec. 6, 1726, C.O. 388: 25, S 91.

[27] R. Jackson to Board of Trade, 1728, C.O. 388: 28, T 86; same to same, Oct. 29, 1715, C.O. 388: 18, O 93. For a Swedish design on Tobago, see Finch to Harrington, Stockholm, May 5, and Aug. 11, 1731, C.O. 388: 30, V 82, 83.

[28] Samuel Swallow to Board of Trade, St. Petersburg, Aug. 8/19, 1765, C.O. 388: 95, I (22).

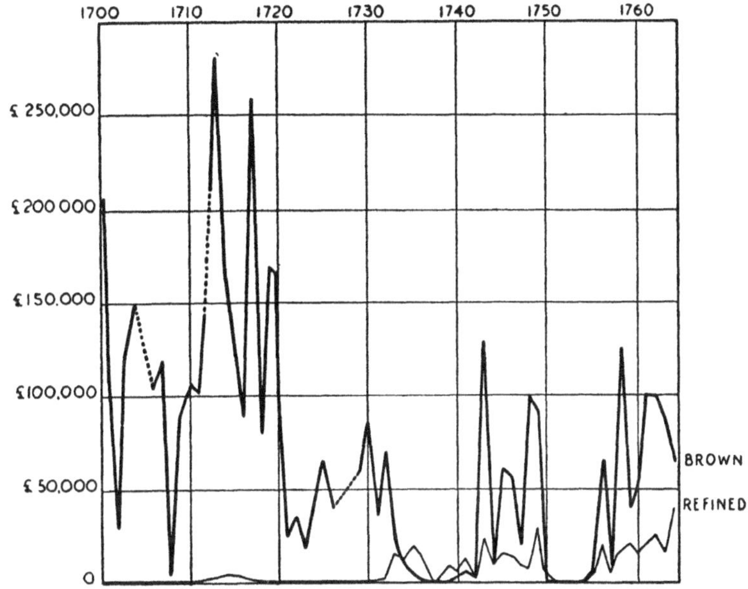

CHART IV. VALUE OF SUGAR EXPORTED FROM ENGLAND TO HOLLAND

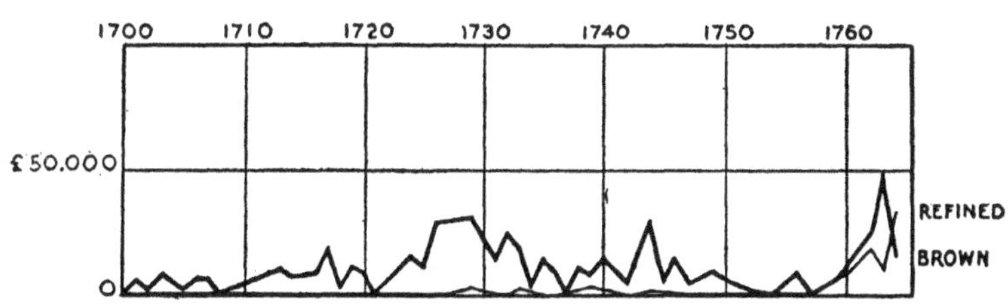

CHART V. VALUE OF SUGAR EXPORTED FROM ENGLAND TO THE "EAST COUNTRY" (BALTIC PORTS)

raw sugar. But, in 1743, Austria and Hungary gave English sugar an eight per cent advantage. The English agents in Vienna expected that this would open a great market to England. "The article of Sugar is of vast importance, the consumption very great."[29] "Sugar and pepper are two immense articles, on which certainly we shall secure the greater part of the consumption."[30] New tariffs giving proportionate advantages to England were arranged with Styria, Carinthia, Carniola, and Tyrol. The port of entry for all this region was Trieste. Freight to Austria via Trieste was estimated much lower than via Hamburg.[31]

Hamburg, in the early part of the eighteenth century, was supplied with raw sugar from England alone. In 1764, however, the British merchants wrote that "Latterly the French have imported here from Fifty to Sixty Thousand Hogsheads per Annum; . . . the English for many years past next to none at all. . . . From hence it appears but too plain that the French from their Industry and Increase of their Plantations have entirely cutt us out of that Trade by importing better and cheaper Sugars, thereby putting it out of the power of the English, to act against Them, without Loss."[32] The Seven Years' War marked a brief revival of the English sugar trade to Hamburg. But the French soon after the war recovered their control of that market.[33] Whereupon the

29 Thomas Robinson and J. Porter to Lord Carteret, Vienna, Feb. 2, 1743, C.O. 388: 42, Bb 4.

30 *Ibid.*, Dec. 27, 1743, C.O. 388: 42, Bb 24.

31 *Ibid.*, Bb 31.

32 Report (Sept. 12, 1764) of the Company of Merchant Adventurers upon the State of their trade to Hamburgh. Annexed to the consular report of R. Woodford to Board of Trade, Hamburg, Mar. 12, 1765, C.O. 388: 95, I (17).

33 R. Woodford to Board of Trade, Hamburg, Feb. 14, 1766, C.O. 388: 95, I (19). Woodford gives figures to show that since the close of the war

English merchant adventurers there offered to the Board of Trade their explanation of why England was unable to maintain its hold on the Continental sugar market. This document really gives the clue to the whole situation in the sugar trade at the close of our period. "The acquisitions, We have made," so reads the report, "and an Attention to the Cultivation, not only of our New, but old settlements, may answer good purposes. . . . It is presumed in some of our islands there is much monopolized, uncultivated land; . . . the chief Proprietors and Planters seeking nothing further, than to supply our Consumption at home, when, if by keeping the market bare in England, They can sell ten thousand hogsheads dear, they do not incline to grow twenty thousand. . . . This, however, seems rather calculated to answer private Ends, and a disadvantage to the Publick. —The increase of the use of Sugar in England, even of late Years, is undoubtedly considerable, and one excuse alleged for not sending abroad is, that our plantations do not furnish more, than a sufficiency for home consumption. . . . If a considerable tax was laid on all uncultivated land in any of our Sugar Islands, Time would shew, if we could not grow more, than are wanted at home. . . . If it had no other effect, the consumer would be supplied on easier terms. . . . In the year 1743 or 1744 an Act of Parliament[34] passed, permitting the exportation of sugars, coffee and ginger &c direct to Hamburgh and other foreign markets; and a Cargoe or two came here about that time, since discontinued for reasons already mentioned. The legislature no doubt had then a

English shipping to Hamburg had fallen off and the French increased. However, "the Broker of the English Company, a very intelligent Person in Trade, assures me, that We maintain still the superiority over the French in the Sugar Trade." *Cf.* also Macpherson, *Annals*, III, 431.

[34] Act of 1739, 12 Geo. II, c. 30.

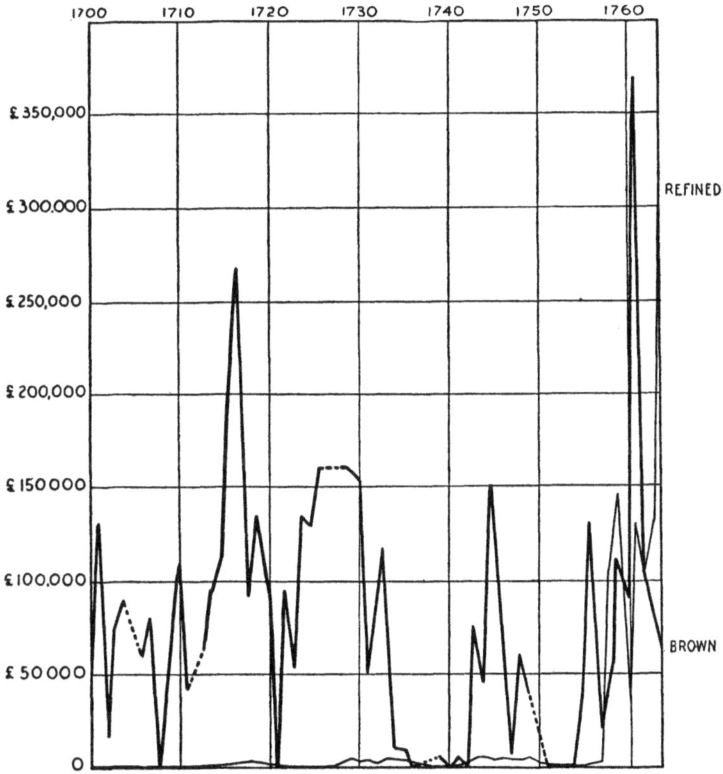

Chart VI. Value of Sugar Exported from England to Germany

View of increasing the quantity and of interfering with the French, at this and other Markets. . . . The Means however being found insufficient, it may be necessary to give further Encouragement.'' In conclusion, the report recommended that to revive and maintain the trade to Hamburg it might be beneficial if bounties were granted on all sugar, coffee, etc., exported from the British West Indies directly to Hamburg.[35]

Occasional mention is made in the consular reports of the English sugar trade to Italy. It was of considerable importance. The fish trade from the Northern Colonies attracted more attention, but sugar, raw and refined, pepper, ginger, and indigo were shipped from the plantations to Leghorn, Naples, Sicily, and Venice. Portuguese and French sugars, however, competed with the English and were in more general use.[36] Sicily had discriminating duties against English and in favor of Turkish sugar.[37] The same was true of Naples.[38] Ancona, however, was a somewhat more hospitable port for British sugar.[39] Italy, who once directed the commerce of Europe in spice and sugar, was herself now dependent for these very commodities upon the Western nations. Their struggle for commercial supremacy here was intense and, as elsewhere, England seemed to be losing. Spain produced, in the main, her own supply of sugar,

[35] Report of Merchant Adventurers at Hamburg, Sept. 12, 1764, C.O. 388: 95, I (17).

[36] References to the trade to Italy are in C.O. 388: vol. 19, P 9; vol. 20, P 74, P 94; vol. 21, Q 23; vol. 23, R 44; vol. 95, I (4); and in C.O. 5: 1227, shipping returns for New York, June 15, 1748.

[37] William Chamberlaine to Board of Trade, June 28, 1728, C.O. 388: 28, T 98; same to same, Oct. 31, 1729, C.O. 388: 29, U 6.

[38] Edward Allen to Sec. Popple, Nov. 1, 1719, C.O. 388: 29, V 41. See also Robinson to Harrington, Vienna, Dec. 10, 1732, C.O. 388: 32, W 104. See also W 137.

[39] C.O. 388: 33, X 35. On the prospects for English trade at Ancona see Addison's *Travels in Italy*, 3d ed., printed for Tonson, London, 1726.

though some came from France. No English sugar appears to have been used in that country.[40] Turkey and the Levant were chiefly supplied with sugar by the Portuguese from Brazil, and from France via Marseilles. Scarcely any British sugar found its way so far east.[41]

Of all the markets to which sugar was reëxported Ireland was the most important. The consumption of foreign sugar, however, was very considerable in Ireland during the eighteenth century. This was especially indicated by a decline in the export of English sugar to Ireland during periods of war with France.

From the ledgers of imports and exports it is possible to construct tables giving the valuation of refined and brown sugar exported each year from England to the markets on the continent of Europe and to Ireland,[42] and a table, printed by Bryan Edwards, gives the total amount of sugar, refined and brown, exported from England to all points abroad for the period 1699-1772.[43] It may be observed from these tables that when prices were high in England the amount of sugar exported tended to decrease, and when prices were low in England the amount exported increased. Also during the wars, when large amounts of prize sugar came to England, shipments to the Continent were heavier than in time of peace.

The decline of the foreign sugar trade undoubtedly brought losses to many British merchants. The period of depression, however, was only temporary, lasting from about 1720 until 1739. The growing demand for sugar in England ultimately more than compensated planters and

40 C.O. 388: 22, Q 38, Q 41; C.O. 388: 23, R 51.

41 Gee, *The Trade and Navigation of Great Britain Considered*, 1729, ed. 1738, p. 64; Anthony Haynes, consul at Smyrna, to Rt. Hon. Gen. Seymour Conway, Jan. 2, 1766, C.O. 388: 95, I (11).

42 See opposite pp. 158, 160, 162, 164 for charts of the trades to Flanders, Holland, the Baltic ports, Germany, and Ireland.

43 See Chart VIII, facing p. 168 for total exports of raw sugar.

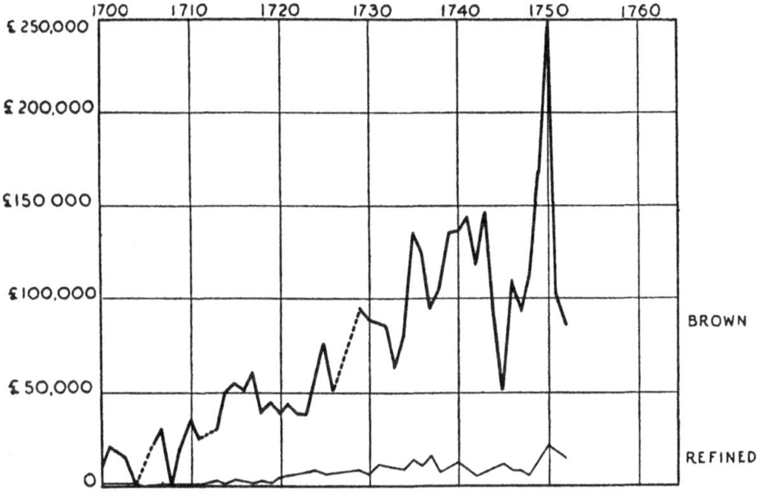

CHART VII. VALUE OF SUGAR EXPORTED FROM ENGLAND TO IRELAND

merchants for the loss of foreign commerce. Moreover, the measure which was finally adopted in 1739, ostensibly to revive foreign trade, proved in reality a scheme of the planting interest to exploit British consumers to an extent unparalleled, perhaps, in previous history. The effect on English opinion and commercial policy of the decline of foreign commerce will be considered in the succeeding chapter.

CHAPTER VIII

THE AGITATION FOR DIRECT TRADE BETWEEN THE COLONIES AND EUROPE

In the Navigation Act of 1660, the colonial commodities "enumerated" for shipment to English dominions alone were, with the exception of tobacco, all British West India products. They were sugar, tobacco, cotton, indigo, fustic or other dyeing woods.[1] In 1673, intercolonial trade in these products was discouraged by subjecting them to customs duties. Thus the duty upon sugar exported from a British West India colony to another British colony was five shillings per hundredweight on white sugar and one shilling six pence per hundredweight upon brown or muscovado sugar.[2] A further limitation upon the marketing of sugar was laid in 1663, when a duty payable to the crown of four and a half per cent was imposed upon all dead commodities exported from Barbadoes and the Leeward Islands.[3] All shipping to and from the colonies was, after 1660, confined to British or British colonial vessels. Such were the fundamental restrictions upon the sugar trade.

In the latter third of the seventeenth century, we

[1] 12 Car. II, c. 18. sec. xviii.

[2] 25 Car. II, c. 7. sec. v.

[3] Sept. 12, 1663, Hall, *Acts of Barbados*, London, 1764, no. 36; C.O. 5: 4, f. 253.

observed frequent protests against these requirements. It was alleged that "enumeration" increased the cost of freight and handicapped the English sugar merchant in competing at Continental markets. The four and a half per cent duty still further enhanced the price of Barbadian and Leeward Island sugar. The "enumeration dues" of 1673, in the same way, put a brake upon the sugar trade to the British Northern Colonies and tended to encourage them to purchase their tropical supplies from the foreign West Indies.[4]

Barbadians and others were often reported for selling sugar directly to the Dutch who took it to Europe as the product of St. Eustatius. In return the Hollanders sold directly to the British West Indians, or at St. Eustatius, European manufactures.[5] This was in violation of the Navigation Act of 1663 requiring the colonists to obtain all their manufactures through England.[6]

In 1689, the author of the *Groans of the Plantations* regarded the "English ships" rule as no longer a grievance, but was sure that requiring the passage of sugar through England had gone a great way in destroying the British sugar trade to Europe.[7] It is a fact that the export of sugar from the islands to England had fallen

[4] For criticisms of the Navigation Laws by West Indians in the latter part of the seventeenth century, see Cunningham, *Growth of English Industry and Commerce, Modern Times*, Pt. I, p. 360; *C. S. P. Col., 1663*, nos. 578, 792; *1661-8*, nos. 85, 127, 129, 156, 149, 989, 1023, 1895, 1726; *1669-1674*, no. 48.

[5] Dalby Thomas (1690) in *Harl. Misc.*, II, 382. Thomas estimated that about 1500 hogsheads of sugar per year were sold to Dutchmen by Leeward Islanders. At the time of the American Revolution (1778), St. Eustatius produced not more than 600 barrels of sugar a year; its agricultural population was then 120 whites and 1200 blacks. J. F. Jameson, *American Historical Review*, VIII, 683-684.

[6] 15 Car. II, c. 6.

[7] Littleton, *Groans of the Plantations*, 1689, p. 5.

off in the last decade of the seventeenth century.[8] In 1699, the merchants of Barbadoes and the Leeward Islands petitioned the Board of Trade for a reduction of duties on sugar and an opening of European markets to direct trade from the islands so that English sugar might compete with French.[9]

The decline of the sugar trade from England to the Continent was most marked in the decade 1713-1723.[10] The situation attracted much attention and led the Board of Trade to make an extended investigation of the condition of the sugar and tobacco trades. Hearings and reports on the subject occupied the board during most of the year 1724.[11] A large number of merchants conversant with trade to Barbadoes, the Leeward Islands, Jamaica, and the European Continent were summoned before the board and questioned. Upon request, a few submitted written reports on the condition of trade.

[8] Brit. Mus. Sloane MSS. 2902, p. 115: "General Estimate of the Present State of the Trade of England, January the 14th, 1697."

[9] C.O. 28: 4, no. 93; *C. S. P. Col., 1699*, p. 585.

[10] Raw sugar exported from England:

1713	184,609 cwt.
1714	158,996 cwt.
1715	143,337 cwt.
1716	161,941 cwt.
1717	290,171 cwt.
1718	124,375 cwt.
1719	167,622 cwt.
1720	121,778 cwt.
1721	66,743 cwt.
1722	83,609 cwt.
1723	63,479 cwt.

See also Chart VIII, opposite.

[11] B. T. Jour., Apr. 30, 1724 *et seq.*, C.O. 391: 33. An order in Council of Apr. 30, 1724, required the Board of Trade to prepare and lay before his Majesty a "State" of the Sugar and Tobacco Trade, C.O. 388: 24, R 135.

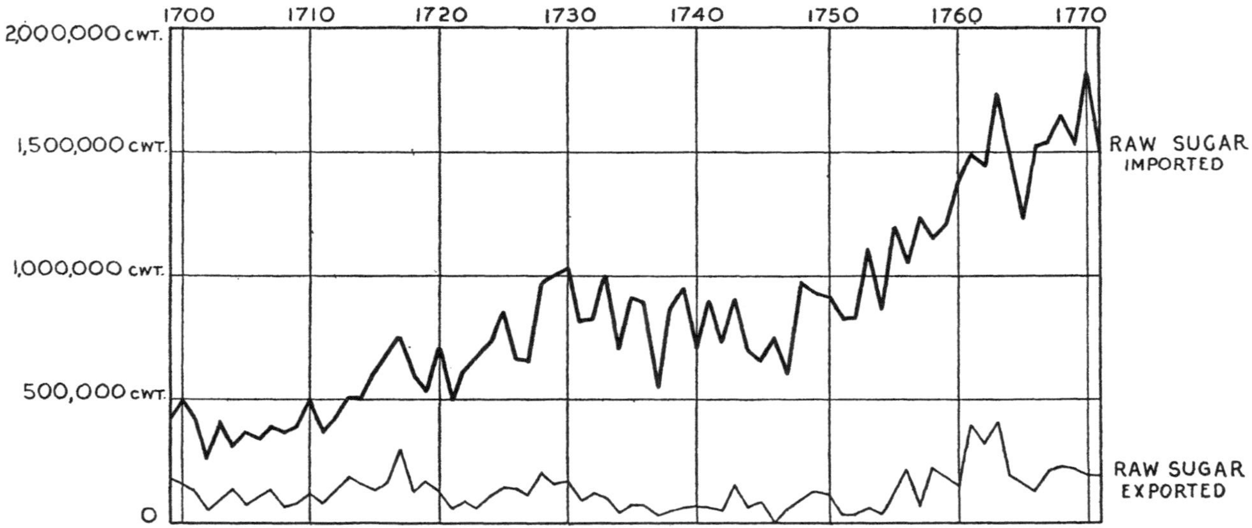

CHART VIII. TOTAL QUANTITY OF SUGAR FROM BRITISH PLANTATIONS IMPORTED INTO ENGLAND AND EXPORTED FROM ENGLAND, 1699-1772

The heavy line in the above chart represents the total quantity of sugar from British plantations imported into England between January 5, 1699, and January 5, 1755, and thereafter into Great Britain to January 5, 1772. The thin line represents the total quantity of British raw sugar exported from England or Great Britain during the same periods. The area enclosed between the heavy and thin lines represents the quantity of British sugar consumed in England or Great Britain from 1699 to 1772. Based on accounts printed in Edwards, *West Indies*, II, 498.

These testimonies and reports may be summarized as containing first, reasons for the loss of the European market, and second, recommendations for its recovery.

It was generally admitted by all that England had lost her standing in the foreign sugar markets. Of first importance among the causes for this was the enormous increase in home consumption. "The consumption of Sugar in England," said a London merchant, Drummond, "by the great use of Tea and Coffy is very much encreased, of late, especially by the cheapness of Tea which will alwise enlarge the Consumption."[12] It is, indeed, a question whether the British West Indies were ever, after the second decade of the eighteenth century, prepared adequately to supply even the English market. The magnitude of Dutch, French, and Portuguese competition on the Continent was dwelt upon at length. Whereas in 1700 the import of Dutch sugar from Surinam into Holland was from 4000 to 6000 hogsheads a year, in 1724 it amounted to between 36,000 and 50,000 hogsheads. Its quality had also much improved. "In 1701," said Drummond, "I found it generally worse than our leward Island Sugars and its now equal to our Barbados Muscovadoes & a fine round grain." Furthermore, the Dutch East India Company had recently improved its sugar plantations in Batavia, whence they exported great quantities of sugar to Mokha, Surat, the coast of India, and Persia, besides about 7000 tons yearly to Europe. "Of a quality almost equall with the Basill or Lisbon blanco Sugar, this sorte of sugar has much lessened our consumption both for Amsterdam and Hamburg it being of a rich nature fit for the Scale without any refining." Being transported from the East Indies in great Dutch fly-boats of 700 to 800 tons the cost for freight was

[12] Memorial to the Board of Trade, received June 16 and read June 24, 1724, C.O. 388: 24, R 142; B. T. Jour., June 24, 1724, C.O. 391: 33.

small.[13] But "No Nation," said Drummond, "runs down a European Market so much as the French, for they are so impatient for quick return, that at Amsterdam I observed for many years, that when either Suriname Sugars arrived in the winter, or that English plantation Sugars were consigned to us in that season, we laid them up in warehouses to wait for an advantagious market & the Spring Consumption, whereas the French never housed their Sugars but sold them from the lighters in lotts by auction at any time." Lisbon sugars, the produce of Brazil, were in great use through all Italy and the Levant. "They are a rich Sugar," continued Drummond, "come home clean in chists, and the best sorte called blancos are used as they come home without refining, the coarser sorte called Muscovados come in large quantity to holland cheefly for refining."

Thus England's competitors furnished sugar both in sufficient abundance to control the European market and of a quality that won the respect even of British merchants.

Minor causes alleged to have discouraged the British sugar trade were the burdensome custom duties, such as

[13] C.O. 388: 24, R 142. The following account of the amount of sugar imported by the Dutch East India Company into Holland is among the Papers of James Knight, Brit. Mus. Add. MSS. 12431, f. 142.

1700......	102,502 lbs.	(Amsterdam)	1713......	1,871,737 lbs.
1701......	1,104,771 lbs.		1714......	2,638,545 lbs.
1702......	778,243 lbs.		1715......	1,621,845 lbs.
1703......	728,582 lbs.		1716......	1,842,394 lbs.
1704......	2,704,771 lbs.		1717......	3,617,800 lbs.
1705......	1,988,352 lbs.		1718......	2,477,124 lbs.
1706......	1,750,997 lbs.		1719......	4,770,937 lbs.
1707......	1,422,580 lbs.		1720......	3,927,425 lbs.
1708......	997,492 lbs.		1721......	3,011,292 lbs.
1709......	1,326,295 lbs.		1722......	2,905,641 lbs.
1710......	1,830,437 lbs.		1723......	3,320,393 lbs.
1711......	2,334,701 lbs.		1724......	4,778,896 lbs.
1712......	2,001,957 lbs.		1725......	5,912,964 lbs.

the four and a half per cent export duty in Barbadoes and the Leeward Islands, the "enumeration dues," excessive customs in England, and the lack of complete drawbacks on reëxported sugar. It was urged that the four and a half per cent duty be abolished, or else, that its proceeds be allowed to remain in the islands, that "enumeration dues" be repealed, English customs reduced, and complete drawbacks paid on reëxported sugar. Aside from the effect such reforms might have on the revival of foreign trade, these proposals were undoubtedly reasonable in the interests of planters and North American and British consumers. Other reforms in the customs administration itself were justly advocated. Too many holidays, exorbitant fees, favoritism, venality, and a mass of red tape in the service were great hindrances to efficiency in modern commerce.[14] Also better naval protection against pirates and meddlesome Spanish patrols, said the merchants, would prevent the loss of time occasioned by the necessity of sailing in fleets and reduce insurance and other charges.[15]

[14] Stephen Godin, a London merchant, to Board of Trade, May 30, 1724, C.O. 388: 24, R 140; B. T. Jour., June 9, 1724, C.O. 391: 33; Worsam, a merchant, to Board of Trade, June 24, 1724, C.O. 388: 24, R 143; B. T. Jour., June 24, 1724, C.O. 391: 33; *ibid.*, Chertsey, July 2, 1724, C.O. 388: 24, R 153; memorial of merchants trading to Jamaica to Board of Trade, May 31, 1724, C.O. 388: 24, R 145; John and David Lequesne to Board of Trade, July 21, 1724, C.O. 388: 24, R 150; anonymous report to Board of Trade, received and read July 22, 1724, C.O. 388: 24, R 155.

[15] Richard Harris and other Jamaica merchants stated that in the English custom-houses there were forty-seven holidays besides Sundays. Thus business stagnated during nearly one-sixth of the working days in the year. This was a handicap to which no other Protestant merchant in Europe was subject. The practical difficulty was thus stated: "For Instance, when 30 or 40 Sugar Ships are unloading at the same Time as is often the case Half that number of Foreign ships lye waiting to Load of the said Sugars Tobacco &c, let the consequence be what it will they must all stand still till the Holydays are past, in the meantime come French or Portugeze Sugrs to Hamburgh Holland &c, and then truly there be Orders by next Post to forbid buying any Sugars here so we loose the Market

Of all the memorials received by the Board of Trade on the decline of foreign commerce, that of Nathaniel Torriano,[16] a merchant of London, appears to have made the deepest impression, and was referred to on several occasions during the succeeding twenty years. Torriano's chief interest lay in opposing the suggestion, proposed by many at the time, that sugar, rice, and tobacco be removed from the "enumerated list" for European ports south of Cape Finisterre. His objections were: (1) Sailors would demand their wages at these ports in southern Europe and expend money there to the detriment of England and the hurt of their families. (2) After so long a voyage ships would have to refit in southern Europe instead of England—to the detriment of England. (3) Spain and Portugal would offer no market for sugar since they have enough of their own. Mediterranean ports of Spain would prefer Spanish or Portuguese sugar. Southern France would consume only French sugar. Italians were now supplied with Portuguese sugar and would probably continue to buy that. (4) Such a trade would be prejudicial to English navigation for at present sugar ships are small pinks and hackboats lightly manned. Whereas a direct trade to the Mediterranean would require them to be "Act

abroad & other Nations gain it from us this being one instance of many of the like kind which often happen by Reason of these Holydays." Debenture fees, though not large, caused frequent delays and were a great nuisance. Favoritism was shown in the dispatch of this business and tips to clerks were an abuse. Fees of various kinds, twenty-five in number, were enumerated whereby it was shown that the public charges on a hundred ton ship for a round trip between London and Jamaica amounted to £44 11s. 2d. All this involved too much red tape and delay. Harris and others to Popple, July 6, 1724, C.O. 388: 24, R 148. The letter is signed by R. Harris, Wm. Parrott, Tho. Sandford, Anth. Chamberlain, James Knight, John Gibbon, Edw. Watkinson, Edw. Lloyd, and Joseph Long.

16 Nathaniel Torriano to Board of Trade, July 20, 1724, C.O. 388: 24, R 152. See also Torriano's letter to Board of Trade, Aug. 8, 1724, C.O. 388: 24, R 162.

Ships'' manned with a third to a half more hands which would make freight dearer. Besides, there would be no return cargoes, which would make it impracticable. Warehouse rents which might be paid in Italy would detract profit that England gains by this at present. (5) Planters now sell their sugar in England for manufactures and supplies which are at once returned to them. A Mediterranean trade would make their returns uncertain and long delayed. Further, a cargo is now usually shipped partly on the English merchant's account and partly on the planter's account, and they could never agree to send to the same place. (6) Finally, to permit colonists to ship off clayed or refined sugars would be detrimental to English refiners and also to English navigation since fewer ships are needed to transport refined than raw sugar.

These arguments were read to the Board of Trade on August 11.[17] On September 15 they were considered again and a representation to his Majesty on the subject of a direct trade was agreed to and signed. The board opposed the idea of removing sugar, etc., from the ''enumerated list,'' and consequently the policy of enumeration for sugar was maintained for another fifteen years.

But the disadvantageous position of England in the Continental markets continued to attract attention. In the year 1730, rice was removed from the ''enumerated list'' for ports south of Cape Finisterre.[18] This fact

[17] B. T. Jour., Aug. 11, and Sept. 15, 1724, C.O. 391: 33. Oct. 28, an order in Council referred the Board of Trade's representation with a copy of Torriano's letter of Aug. 8 to the Committee of the Council. C.O. 388: 24, R 167. Torriano's letter, said John Ashley in 1735, ''must no doubt have influenced several Gentlemen of the British Legislature'' as well as merchants and factors. Ashley, *Observations on a Direct Trade,* 1735, pp. 3, 5.

[18] Between Sept. 29, 1730, and Feb. 29, 1731, 35 vessels took licenses to carry rice direct from the Carolinas to southern Europe. Treas. 38: 364.

encouraged the West India planters, who in the same year undertook to secure a similar advantage for the produce of their islands. The Jamaica council and assembly resolved[19] in October that a representation to the king be prepared asking liberty to send their produce to foreign ports, and promising to pay to a Crown officer on exportation the same sum as would accrue to England should the sugar pass through her ports. East India goods were to be excluded, and the wearing of British manufactures encouraged. Charles Dunbar and Governor Mathew made the same request and promise for the Leeward Islands.[20] Some thought that the "enumeration" of sugar was the only reason for England's decreased sale of sugar in Europe, for British planters produced as cheaply as the French.[21] A number of the colonial pamphlets of 1731 recommended direct trade to Europe as vitally important and their authors printed elaborate tables to show how largely transshipment through England added to the cost of sugar in foreign markets.[22]

The agitation for direct trade to Europe also involved the desire that direct trade in sugar between the West Indies and Ireland might be restored.[23] This trade had been prohibited in 1670.[24] For a time the prohibition benefited Bristol, Chester, Liverpool, Whitehaven, Lancaster, and other western ports from which Ireland was supplied. But recently the French had discovered methods of supplying Ireland with sugar thirty per cent

19 *Journal of Jamaica Assembly*, II, 742, Oct. 24, 1730.

20 Gov. Mathew to Board of Trade, Oct. 3, 1730, C.O. 152: 19, T 142.

21 *The Importance of the Sugar Plantations*, 1731, pp. 38-39; *Present state of British Sugar Colonies*, 1731, p. 8.

22 *Present state of British Sugar Colonies*, pp. 11-12. It was estimated that "enumeration" added 30 per cent to the price of sugar. See also *Remarks upon a book*, &c., 1731, pp. 8-9.

23 *Present state of British Sugar Colonies*, pp. 16-18.

24 22 and 23 Car. II, c. 26, sec. 11.

cheaper than the English could do it. Since 1696, the trade in molasses and rum to Ireland had been under the same restriction,[25] and both of these commodities were imported from the French islands. In return for all of which, Ireland supplied the French with great quantities of beef, provisions, and horses.

Practically all observers of economic conditions agreed that the sugar trade was burdened with too many duties.[26] Whether or not direct trade to Europe were granted, a reduction of duties would bring great relief to a shackled commerce. The duties which came in for the most persistent censure were the four and a half per cent export duty in Barbadoes and the Leeward Islands, the duties on the export of enumerated commodities from one colony to another, and the inadequate drawbacks on sugar reëxported from England. Upon entering England, brown sugar paid a duty of 3s. 6d. per hundredweight; if reëxported all but about 8d. per hundredweight of this was drawn back. But sugar refiners in England successfully resisted the plan to draw back the whole import duty; they knew it would lessen the supply of their raw product and raise prices. The "enumeration dues," collectable in the sugar islands upon sugar shipped to the Northern Colonies, were less defensible. The export duty of 1s. 6d. per hundredweight on brown sugar passing from one colony to another equaled a specific tax of seven and a half per cent. This, in addition to the four and a half per cent in all the sugar colonies except Jamaica, was certainly an excessive burden. It helps to explain the extensive commerce which developed between New England and the foreign West Indies. British West India rum, besides the four

[25] 7 and 8 William III, c. 22, sec. 14.

[26] The duties were discussed by many writers. See for example *The Importance of the Sugar Plantations*, pp. 38-39.

and a half per cent in Barbadoes and the Leeward Islands, paid in England a custom and excise amounting to something more than 4s. per gallon. Rum paid more than brandy imported from Dunkirk. And, what seemed a great hardship, the excise of 3s. 8d. per gallon was never drawn back, thus making it practically impossible to reship rum to Ireland. However, the French, in New England ships and often via New England, supplied Ireland with rum as well as sugar.[27] The object of the high duties on rum was to protect malt liquors. In the face of so many objections it may seem a little strange that the sugar trade was not relieved of some of its burdens. Such relief, however, would have benefited no one but Continental, Irish, and North American consumers, and would have impaired a lucrative source of revenue. It might have stimulated enterprise in the sugar islands, but, it was argued, the planters were sufficiently encouraged by possessing a monopoly of the English market which after all was the best in Europe and offered the highest prices. And so the fiscal arrangement was left practically undisturbed during our period.

Representations and memorials, in behalf of direct trade to Europe and lower duties, arrived frequently during the years 1731-1739 from colonial legislatures and their London agents.[28] The planters of St. Christopher said that "enumeration" of sugar enabled the

[27] *The Importance of the Sugar Plantations*, pp. 39-40.

[28] Address and Representation of Council of Jamaica, Nov. 27, 1731, C.O. 137: 19, S 152; Representation of the President, Council, and Assembly of St. Christopher to the House of Lords, Sept. 24, 1731, C.O. 5: 4, f. 220; Memorial from John Yeamans, agent for Antigua, to Board of Trade, rec'd Jan. 21, 1734/5, C.O. 152: 21, V 52; Petition and Representation to the King from the Council and Assembly of Barbadoes, Nov. 13, 1733, C.O. 28: 40, f. 43; a petition from the same to the House of Commons, Nov. 13, 1733, C.O. 28: 40, f. 44; Representation to Board of Trade from Yeamans, agent for Leeward Islands, Dunbar, surveyor gen'l of Leeward Islands, and Mr. Souleger, rec'd Apr. 2, 1734, C.O. 152: 20, V 8, 9.

French to undersell them in Europe and Ireland by at least twenty-five per cent. In 1733, Governor Lord Howe of Barbadoes forwarded to Newcastle petitions signed by a great number of planters, merchants, and others praying for direct trade to Europe. He indorsed their programme, remarking that "The people here are generally in a most miserable Condition and if not soon relieved it will prove too late."[29] Governor Mathew of the Leeward Islands also favored the movement and thought it would promote the claying of sugar by English planters.[30] John Ashley of Barbadoes wrote two pamphlets in behalf of the cause, one in 1735 and another in 1737.[31]

In April, 1734, the House of Lords sent an address to his Majesty praying for further information upon the subject of direct trade in sugar to Europe and a reduction of certain duties. In answer to which, the Board of Trade, January 14, 1735, made a representation to the House of Lords.[32] They observed that during the past thirty years the use of sugar in England had greatly increased, owing to the increased consumption of tea and coffee. On the other hand exportations of sugar from England had of late years diminished very considerably. This was due to French, Dutch and Portuguese sugars

[29] Lord Howe to Newcastle, Dec. 3, 1733, C.O. 28: 45, f. 157. He enclosed the popular petition and petitions from both houses of the legislature and Ashley's printed *State of the Sugar Trade*. See also Howe to Board of Trade, Nov. 7, 1734, C.O. 28: 24, Aa 28.

[30] Gov. Mathew to Board of Trade, Aug. 31, 1734, C.O. 152: 20, V 46.

[31] John Ashley, *Some Observations on a direct Exportation of Sugar, from the British Islands, with Answers to Mr. Torriano's Objections to it. In a Letter from a Gentleman in Barbadoes, to his Friend in London.* London: Printed in the Year MDCCXXXV. 23 pages, 4to. "*The Miserable Case of the British Sugar Planters Considered In a Letter from a Wellwisher to the Trade of Great Britain,*" indorsed: "Mr. Ashley's Second Memorial upon the State of Barbadoes and the other Sugar Islands." In manuscript, 27 pages. C.O. 28: 25, Aa 62, rec'd and read July 20, 1737.

[32] C.O. 5: 5, ff. 114 *et seq.*

being sold in Europe much cheaper than sugar from the English colonies. As to a reduction of duties "It would be impossible to give our Traders any Relief in this particular, without breaking thro some established Customs, and making great Alterations in several Laws, by which many general Charges have been imposed upon Shipping for the Repair of Peers and Light-Houses. But they have long been desirous of the Liberty of carrying their Sugars directly to all the European Markets to the Southward of Cape Finisterrē, And we would humbly submit it to your Lordships, whether such a Liberty might not be granted under proper Restrictions." A reduction of import duties upon rum should be made. This would not interfere with British liquors but rather with French brandy and foreign spirits "which carry great Sums of Money out of the Kingdom, whereas Rum is paid for in our own Manufactures." Finally, as to the advantage which the French derive from the "freshness" of their sugar plantations, they considered it impossible to prescribe any remedy which would put English planters on an equal footing with the French.

Many attempts were made to determine why French sugar sold so much cheaper in Europe than English. Two considerations were generally admitted: (1) that the soil of the French plantations was fresher and more productive; and (2) that the use of sugar in England was more universal and the demand greater than on the Continent. But besides these natural factors there were fiscal arrangements which elevated the price of English sugar to an artificial level. A statement of 1734 among the Treasury accounts[33] thus summarizes the comparative charges upon French and English sugar: (1) One hundred hogsheads of brown sugar might be imported into France cheaper than it could be into England by

[33] Treas. 64: 273.

£134 : 13 : 0½ or fifteen and a quarter per cent. (2) The whole duty, measured specifically, on sugar imported into France was three and a half per cent plus a charge for the "King's Weights" of 6 sous 3 deniers per hundredweight. Whereas the duties on British sugar imported into England were four and a half per cent in the plantations and fifteen per cent in England—in all nineteen and a half per cent. Furthermore, French duties were collected on an abnormally low valuation, while English duties were collected on a valuation of 30s. per hundredweight though sugar might sell for only 15 to 20s. per hundredweight. (3) All other charges on sugar, except freight, were nearly three-fifths dearer in England than in France.

The sugar trade received some attention in parliament in the spring of 1735. On April 3, the House of Commons asked for a copy of Nathaniel Torriano's objections, made in 1724, to a direct trade to Europe. The Duke of Newcastle requested copies of the Board of Trade's recommendations on the question in that same year.[34] But, a reconsideration of the subject from the conservative standpoint appears to have given a setback to the cause of direct trade, for parliament took no action on the matter till four years later.

The special interests of the planters were sustained during the years 1735-1739 by further petitions[35] from

[34] B. T. Jour., C.O. 391: 44, ff. 51-52; C.O. 388: 34, X 44; Newcastle Papers, Brit. Mus. Add. MSS. 33028, p. 341: the Board of Trade's representation of July 24, 1724, was laid before the House of Commons Feb. 25, 1735/6.

[35] Memorial to Board of Trade, May 31, 1734, from J. Sharpe, Peter Le Heup, and George Lewis Teissier, agents for Barbadoes, C.O. 28: 24, Aa 10. Representation of the President, Council, and Assembly of Jamaica to the Board of Trade, May 2, 1735, C.O. 137: 21, T 75; *Journal of the Assembly of Jamaica* (May 2, 1735), III, 302. Yeamans, agent for Antigua, visited the Board of Trade Sept. 16, 1735, and presented a letter from the speaker of the assembly of Antigua, B. T. Jour., C.O. 391: 44, f.

the islands, by pamphleteering, and by continuous lobbying on the part of colonial agents. Englishmen were dissatisfied because so much land in Jamaica remained uncultivated. The only cure for this evil, said Ashley, lay in the stimulus to industry that would come from direct access to Continental markets.[36] Ashley revealed one of the real motives that underlay the propaganda when he stated that under the existing régime planters were "compelled to send their Sugars to a glutted and restrained Market, now, in a manner, under the sole Government of the united Companies of Grocers and Sugar-Bakers, who must buy all our Sugars while we remain under our present Circumstances."[37] In other words the planters were looking for a weapon with which to fight the British sugar trust of the time. November 23, 1736, an address to the king from the assembly of Jamaica bewailed the low estate to which the planters had been brought by confining them to a single market.[38] The removal of rice from the "enumerated list" had been followed by a revival of Carolina trade with southern Europe. This was cited on several occasions as a precedent for removing sugar from the same list. But that the same happy issue would follow in the case of sugar was far from certain.[39]

219. James Dottin, president of council of Barbadoes, to Board of Trade, Oct. 21, 1735, C.O. 28: 24, Aa 34. Torriano's arguments for the old régime were all refuted by Ashley in statements that amount to mere guesses about what might happen if a great trade to southern Europe were created. Ashley's optimism was at times ill-founded and absurd. *Observations on a Direct Export*, etc., 1735, pp. 11 *et seq.*

36 *Ibid.*, 9.

37 *Ibid.*, 10. "When a large crop comes crowding home to a glutted and restrained Market, the present Buyers will give just what they please, which is generally a losing price. Since they are of a Society or Community and well known We have now Scarcely any Re-exportation but to Ireland and therefore must Sell to them." *Ibid.*, p. 7.

38 *Journal of Assembly of Jamaica*, III, 395.

39 *Cf.* Ashley, *Observations on a Direct Trade*, p. 22.

Direct trade to northern Europe received less attention than that to the Mediterranean. It was generally conceded that to open the manufacturing countries of northern Europe to direct colonial shipping would harm seriously the sale of English manufactures in America and the West Indies. Nevertheless, it was estimated by James Knight, who was deeply interested in the West Indies, that carrying sugar directly from the English plantations to Holland in the same ship without landing in Great Britain would mean a gain to the owner of from fifteen and four-fifths to fifty per cent.[40] On March 14, 1739, John Sharpe, agent for Barbadoes, attended by several planters, called at the Plantation Office, and presented a petition for direct trade in sugar not only to the Mediterranean but to Flanders, Holland, Hamburg, and the Baltic.[41] The same day, John Yeamans, agent for the Leeward Islands, with several merchants and planters also paid a visit to the same office to enlist the assistance of the board in a petition for direct trade that was about to be laid before the House of Commons.[42]

Parliament, during the spring of 1739, gave its attention once more to the sugar trade. The Commons and Lords reviewed the matter from all angles, paying especial attention to Torriano's letter of 1724.[43] April 4, a bill was introduced and read the first time to permit the direct shipment of sugar from the plantations to Europe.[44] Petitions against the bill were presented from

40 Papers of James Knight, date about 1738, Brit. Mus. Add. MSS. 12431, f. 164. This is a single folio sheet, printed.

41 C.O. 152: 23, X 52.

42 B. T. Jour., C.O. 391: 47, f. 23.

43 *Ibid.*, C.O. 391: 47, pp. 27, 47, Mar. 23 and May 29. The board supplied the Commons and Lords respectively with information on the subject. The *Parliamentary History* contains no reference to the matter. The planters' petition was presented and read in the House of Commons, Mar. 14, 1739, *Commons Journal*, XXIII, 284. See *ibid.*, 298.

44 *Ibid.*, 316-317.

merchants of Bristol,[45] Liverpool,[46] London,[47] Chester,[48] Lancaster,[49] and Whitehaven.[50] All contended that it would hurt British navigation, as well as the middlemen's trade and manufactures. A petition from the sugar refiners of London stated that many refiners were then idle for want of sufficient sugar, and that the passage of this bill would lead to an increase in the price of sugar and would hurt the export of the refined product.[51] But the planting interest denied that the passage of the bill would raise prices, and asserted that they could produce three times as much sugar as they did.[52] Hearings were given for and against the measure April 30, May 4, and May 7. The vote for committing the bill was 76 yeas and 19 nays. May 23, the bill was passed and carried to the Lords.[53]

In the House of Lords, the fight against the sugar bill was renewed. Torriano's objections of 1724 were reviewed, petitions for and against were received, and a number of witnesses on either side examined. Among those against the bill, who were heard, were Niel Buchanan, and a Mr. Lloyd; and of those who supported it Thomas Hyam, Solomon Merritt, John Ashley, William Beckford, and James Knight presented their case. The measure was before the Lords from May 24 till June 12 when it was passed.[54]

The passage of the Sugar Act of 1739[55] marks an im-

[45] *Commons Journal*, XXIII, 318, Apr. 5, 1739.

[46] *Ibid.*, 329, Apr. 12, 1739.

[47] *Ibid.*, 338-339, Apr. 19, 1739.

[48] *Ibid.*, 340, Apr. 26, 1739.

[49] *Ibid.*, 343, Apr. 30, 1739.

[50] *Ibid.*, 345, May 1, 1739.

[51] *Ibid.*, 340, Apr. 26, 1739.

[52] *Short Account of the Interest and Conduct of Jamaica Planters* (1754), pp. 11-12. See *Hist. MSS. Com. Report*, XIV, Pt. 9, p. 247: the Bishop of Chichester's comment on the opposition to the bill, in a letter to his son, May 31 and June 14, 1739.

[53] See *Commons Journal*, XXIII, 344, 349, 351, 365, 372, Apr. 30-May 23, 1739.

[54] See *Lords Journal*, XXV, 395, 399, 401, 402, 405, 411, 414, 416.

[55] 12 Geo. II, c. 30.

portant breaking down of British colonial policy in the interest of West India planters. The act granted permission to ship sugar directly from the plantations to any foreign ports in Europe. Ships bound for northern Europe must, however, touch at some British port; those destined to points south of Cape Finisterre might sail directly without touching in England. In both cases sugar bound for Europe was relieved of all duties formerly collectable in England. The privilege was accorded only to ships built and owned in Great Britain, though in 1742 the act was extended[56] to English colonial ships. Licenses must be purchased for the privilege in England at the beginning of the outward voyage and bonds furnished to secure enforcement of the law. In all cases ships must return within eight months to England first of all before sailing again to the colonies. This would secure to England the advantages of the "Staple Act" of 1663. The act did not open direct trade from the plantations to Ireland.

After so long an agitation for the removal of sugar from the "enumerated list," it is, at first, astonishing to discover how little actual trade to the Continent resulted from the act of 1739. The first shipment under the act was when the ship *Walpole*, John Ephinstone, master, cleared from Barbadoes, December 1, 1740, with 253 casks of sugar, bound for Gibraltar and Leghorn.[57] "It is with great pleasure," wrote William Beckford from Spanish

[56] 15 Geo. II, c. 33, sec. 5. Many objected to thus encouraging New England shipping, but colonial ships operated 15 per cent to 20 per cent cheaper, and planters insisted that their use was necessary to regain foreign trade. William Perrin, *Present State of the British and French Sugar Colonies*, 1740, pp. 9 *et seq.*

[57] "An Account of the Number of British Ships laden with Sugars and the Quantity of the Sugars exported from the British Sugar Colonies to any Port in Europe south of Cape Finisterre since the year 1739 distinguishing the time and place of Exportation, Dated Custom House, London, March 24, 1753. Wm. Wood, secretary." Treas. 64: 274.

Town, Jamaica, to James Knight in London, "that I have heard of the success of the Leghorn Ship & I am determined to push the Direct Exportation to the Utmost of my power and it is the utmost satisfaction to me to find all the Gentlemen here so well Inclined to forward any Undertaking of that kind, so that I make no Doubt (If Ships offer) to be Able to Send Two or three either to Hamborough or up the Streights, we shall rely greatly on your own Endeavour[58] to get an Amendm^{t} or Clause in the Export Bill to suffer Licences to be taken out in the Colonys & that the Owners may be resident out of Great Britain the want of such Liberty has been the Only reason why I have not Exported any Sugars."[59]

The exclusion of colonial shipping from participation in this trade was regarded by the planting interest as a grievance. It was declared that New England ships could carry freight from the West Indies to Europe fifteen to twenty per cent cheaper than the French, and thirty per cent cheaper than English ships.[60] It is a tribute to the increasing strength of the planters that, in the face of opposition from British maritime interests, this disability of colonial shipping, in 1742, was removed by parliament.[61] But the restrictions under which licenses, etc., were to be obtained were such that only forty-eight licenses altogether were granted to ship sugar from the colonies directly to the south of Cape Finisterre between 1739 and March 24, 1753. Of these only five were used.

58 James Knight and William Beckford had been associated in testifying in behalf of the Act of 1739 before the House of Lords. *Lords Journal*, XXV, 405 (June 1, 1739). Knight was for a time colonial agent for Jamaica.

59 William Beckford to James Knight (London), Spanish Town, Jamaica, Nov. 21, 1741, Brit. Mus. Add. MSS. 12431, f. 122.

60 William Perrin, *Present State of the British and French Sugar Colonies*, 1740, pp. 4-8.

61 15 Geo. II, c. 33.

One was in the case of the Leghorn ship of 1740 cited above. Two other vessels sailed from Barbadoes; the *Robinhood,* David Littlejohn master, cleared May 2, 1746, with 300 hogsheads, 90 tierces, and 125 barrels of sugar, for Leghorn; and the *Upton,* Thomas Rice master, cleared also for Leghorn, July 12, 1749, with 10 hogsheads, 61 tierces, and 1 barrel of sugar. The other two ships were from Jamaica; the *Robinhood,* with the same master as above, cleared for Leghorn June 16, 1744, with 167 hogsheads, and 33 tierces of sugar; and the *Italian Galley,* William Axford master, cleared for Leghorn, September 7, 1747, with 300 hogsheads of sugar. The total direct shipment of sugar from the West Indies to southern Europe up to 1753, under the act of 1739, amounted to only 777 hogsheads, 184 tierces, 253 casks, and 126 barrels. No shipments were made from Antigua, St. Christopher, Nevis, or Montserrat.[62]

The effect and significance of the act of 1739 lay in its power to raise the price of sugar in the British market and to circumvent the refiners' association, if there was one. Said William Beckford, in 1743, "I know of but one Remedy that Can in any wise keep up the Price of Sugars against the Combinations of the sugar Bakers &c^{a} and that is Giving a Bounty or Praemium of 1^{s} p Ct. on all Sugar Exported from this Island [Jamaica] to foreign Markets."[63] The mere possibility of sending

[62] William Wood's Account, Mar. 24, 1753, Treas. 64: 274. See also for statements of no trade under the Act of 1739, Gov. Mathew of Leeward Islands to Board of Trade, July 6, 1745, C.O. 152: 25, Y 139; Gov. Henry Grenville of Barbadoes to Board of Trade, Feb. 8, 1748/9, C.O. 28: 29, Cc 28. Grenville added that no sugar had been sent to any port of Europe north of Cape Finisterre. Also Gov. Thomas Robinson of Barbadoes to Board of Trade, Feb. 20, 1746/7, C.O. 28: 27, Bb 57.

[63] William Beckford to James Knight, Spanish Town, Jamaica, June 18, 1743, Brit. Mus. Add. MSS. 12431, f. 125. The utility of the act of 1739 as a means of raising prices in England was candidly declared by the Jamaican

sugar abroad had its effect upon prices as may be seen from the quotations on sugar in London from 1727 to 1758.[64] Muscovado or raw sugar which, in 1733, sold at

planter, Rose Fuller, in the following passage from one of his letters dated Spanish Town, June 15, 1743. The cause of low prices, said Fuller, lay in the fact that English sugar bakers "have it in their power to meet and agree what they shall give for our commodities, but all these combinations can never prevail if we can bring it about that sugars may not be quite so plenty in England as they have been of late years. . . . In order to bring this scarcity about or at least to prevent a too great abundance we propose that every ship that comes from London to load here shall bring a licence to carry sugars directly for foreign markets either those to the southward or to the northward and as this licence can be had for fifteen shillings if they make no use of it it will be but a small cost to the ship and if it has no other effect it will at least alarm the bakers and keep them in a state of suspence what sugars will be sent to foreign markets; several of the most considerable shippers are resolved if things continue so bad as they now are to make very large consignments either to Hamburg or other ports and will use their utmost endeavours to persuade others to do the like, and in order to incourage the same it seems resolved by the members of the legislature tho not now sitting, to procure a meeting about Christmas and give a bounty of one shilling per hundred[weight] att least for all sugars that shall be sent to such foreign markets." Long Papers, Brit. Mus. Add. MSS. 18960, p. 13.

[64] Average prices per hundredweight for which sugar sold at the London Custom House from 1745 to 1758 are as follows:

YEAR		MUSCOVADO SUGAR		CLAYED SUGAR	
Xmas to Xmas		Shillings	Pence	Shillings	Pence
1744	1745	39	11	50	¾
1745	1746	39	5	51	1
1746	1747	42	9½	55	7
1747	1748	31	7¼	50	1¼
1748	1749	28	11¾	42	8
1749	1750	27	9½	39	6
1750	1751	30	6	40	5¾
1751	1752	38	7¾	48	4¾
1752	1753	33	..	48	5¼
1753	1754	35	8¾	50	3½
1754	1755	35	8¾	49	8¼
1755	1756	34	3¼	48	10½
1756	1757	37	1	48	6
1757	1758	42	5¾	51	2¾

Treas. 64: Bundle 274.

an average of 16s. 11¼d. per hundredweight rose, by 1747, to 42s. 9½d.; clayed sugar, or sugar partly whitened before export from the West Indies, increased from 31s. per hundredweight in 1732 to 55s. 7d. in 1747. Well might Beckford of Jamaica congratulate the Jamaica Coffee House Club in Cornhill upon its usefulness in West India politics.[65] In 1763, Governor William Henry Lyttelton of Jamaica stated: "When the Act of the 12th of His late Majesty to permit sugar to be carried directly to a foreign Market was made, that Commodity was at a very low Price scarce worth the Planters making it but immediately upon the War breaking out with Spain in the year One thousand Seven Hundred and thirty nine it rose in Value which has Continued ever since, and Great Britain having been found the best Market for our own Sugars the Merchants have not been induced to try foreign ones."[66]

The act of 1739 marks the elevation of the organized West India planting interest into a position of great power in British politics.[67] It did nothing in the way of

65 William Beckford to James Knight, Spanish Town, Aug. 19, 1741. Beckford added: "I am of the Clubb's opinion that we have been Deficient in promoting a Direct Exportation." Brit. Mus. Add. MSS. 12431, f. 120.

66 William Henry Lyttelton to Board of Trade, July 9, 1763, C.O. 137: 33, Cc 19.

67 The following episode is illustrative of the power of West India planters in parliament at this period. In 1744, Henry Pelham, chancellor of the exchequer, proposed an additional duty on sugar amounting to a farthing in the pound to the consumer. Pelham's motion was at first carried by a small majority. "Nor was the smallness of it matter of surprize to those who considered how many were either by themselves or their friends, deeply concerned in one [branch] or other of the Sugar trade, and that the cause itself was always popular in the House of Commons." Pelham's friends "did not express great forwardness to support Mr. Pelham on this occasion. . . . Col. Bladen, whose wife has a great estate in the Sugar Islands, made an elaborate speech against the duty. On the planters' side were Sir H. Lyddel, Jennings of Cambridgeshire, Pitt, Alderman Heathcote, Sir J. Cotton, Alderman Calvert, and Sir John Barnard. Those against the

recovering the Continental markets, but in raising the price of British sugar to the exorbitant level that subsequently prevailed it proved a most useful instrument in the hands of West India planters.

planters on this occasion were Winnington, Fox, and Sir W. Yonge. The sugar duty was ultimately defeated, with the assistance of Lord Carteret. *Parl. Hist.*, XIII, 640-641, 652-655.

CHAPTER IX

INTERNATIONAL TRADE BETWEEN BRITISH NORTH AMERICA AND THE FOREIGN WEST INDIES, 1686-1730

The trade restrictions imposed on British colonial commerce by the Navigation Acts of 1660, 1663, 1673, and 1696, affected the export trade of the Northern Colonies and of the English sugar islands in opposite ways. These laws directed that the chief products of the West Indies should be carried in fixed channels whereby British subjects could share in the monopoly of their production. Accordingly, all sugar exported from the West Indies had to be shipped either to England or to British colonies on the American continent; at no time prior to 1739 was it legal to ship sugar to any foreign port outside of Great Britain or her colonies. On the other hand, the chief articles of export from the New England and Middle Colonies, consisting of building materials, provisions, and live stock, were in times of peace under no such restriction. Their import into England, however, was discouraged in the interest of British agriculture, but in the West Indies and southern Europe, as far as English law was concerned, these commodities could be sold wherever markets existed.

Some reflection on the comparative extent, both in area and population, of the British colonies in the temperate and tropical zones impresses one with the inequality of these regions as complementary trade areas. The purchasing power of the Southern and West India colonists was always limited in comparison with the enormous supply of Northern products. The empire was overbalanced on its temperate zone side. On the contrary, the French empire in the same period was overweighted on its tropical side. New France was sufficiently extensive in territory but sadly deficient in population, productive capacity, and situation to answer the needs of the French West Indies. It is very natural, therefore, that we should find from the beginning a tendency toward economic equilibrium which paid no heed to political boundaries. Such were the conditions which supplied the basis for that ever increasing trade between the Northern British Colonies and the French planters in the tropics.

Illegal trade was a permanent feature of colonial commerce during the whole period of restricted trade down to the American Revolution. Were it possible to plot its curve, it is probable that we should observe marked ascents, first during the war of the Spanish Succession, then again just after the peace of Utrecht with a steady rise to 1731, then a slight descent, followed from about 1738 to 1761 by a steady rise of the line. In 1686, England and France agreed upon a treaty of neutrality[1] promising to abstain from commerce with each other's possessions in America. It is interesting that there appears to have been no objection to this arrangement from the North Americans. But in the years immediately after the peace of Utrecht, the northerners had come to regard French markets in the south no longer as mere fields for speculative adventure, but as natural outlets for Northern produce. Certain it is that, by 1730,

[1] Jean Dumont, *Corps Universal Diplomatique*, etc., Amsterdam, 1726-31, vol. VII, Pt. ii, pp. 141 *et seq.* Naval office lists of entries and clearances at Boston, 1686-1688, show practically no trade from Boston to the foreign West Indies. C.O. 5: 848.

New England, New York, and Pennsylvania felt that their economic life was vitally connected with the slave communities of Santo Domingo, Guadeloupe, and Martinique. Attempts from this time to check a free intercourse between the Northern British colonies and the foreign West Indies always provoked declarations that the population and standard of life in North America were absolutely dependent on free trade with the French West Indies.

In the French islands, also, a corresponding body of opinion grew up in consequence of the trade with Englishmen.[2] The importation of rum and of molasses, from which it was distilled, was discouraged in France in the interest of French wines and brandy. In the French West Indies, therefore, rum and molasses were abundant and cheap. Santo Domingo was in a position to purchase English supplies on terms far more attractive than Barbadoes or Jamaica could offer. It is also quite probable that the plantation duty,[3] laid in 1673, of one shilling and six pence per hundredweight on English brown sugar, and five shillings on white, shipped to the Northern Colonies, acted as a brake on British intercolonial trade. French law, however, during the whole period was against any trade whatever with foreigners, either export or import.

The first third of the eighteenth century witnessed the rise to great magnitude of a commerce between British North America and the foreign West Indies. The period is also characterized by much uncertainty about the legality or illegality of this trade. Those to whom such intercourse was a detriment pointed to the treaty of 1686

2 See Pierre de Vaissière, *Saint Domingue,* Paris, 1909, pp. 32-35.

3 25 Chas. II, c. 7. The planters' opinion of this duty was expressed by Littleton of Barbadoes as "a grievous clog to our commerce with those plantations." *Groans of the Plantations,* p. 4.

as containing its absolute prohibition. But the English government, on the other hand, while admitting its power under the treaty to restrain foreigners from trading to the British colonies, finally refused to acknowledge authority under the treaty to prohibit its own colonists from engaging in such trade provided the navigation laws were not violated. Consequently, in 1730, British opponents of free trade in the West Indies organized the movement which secured from parliament a law of no uncertain meaning designed to destroy all colonial commerce between British and foreign subjects. The remainder of this chapter is an attempt to trace down to 1730 the growth and character of trade between British North America and the foreign sugar settlements and the development of opinion concerning it.

Toward the end of the seventeenth century this trade was becoming quite common. In time of peace it does not seem to have occasioned serious alarm. But in war-time its existence was brought to the notice of the government, and measures taken to suppress it. Thus at the beginning of the War of the Spanish Succession, a circular letter to the governors instructed them on the subject as follows:

"The Queen having been informed that great Quantitys of Provisions have been laden on some Vessels in the Ports of the Continent of America, on pretence of carrying the same to Jamaica and other her Ma^{tys} Islands in those parts but that in truth her Ma^{tys} Enemys have been supplied wth them; to the great Damage and Prejudice of her Ma^{tys} Plantations, besides the advantage her Ma^{tys} Enemys receive thereby, being furnished with such provisions by this means as they would otherwise want,"[4] it was ordered, therefore, that no vessel should

[4] Earl of Nottingham to Lord Cornbury, governor of New York, Oct. 26, 1703, C.O. 323: 5, F 20.

be cleared with more provisions than enough to feed its passengers, unless security were given that none would be sold at St. Thomas or to the enemy, and that certificates issued at shipment were to be presented on delivery in the British West Indies. Robert Quary, the surveyor general of the customs, in his travels through New England, the Middle Colonies, Maryland, and Virginia, was on the lookout for unlawful practices, and has left us his observations concerning them. From returns made in Virginia by ships from the West Indies, he inferred that there was wilful laxity on the part of customs officers in the English islands. Papers issued by officers at Barbadoes and the Leeward Islands credited vessels with taking from those colonies more produce than was actually loaded upon the ships. Also English casks were put upon ships there but, as no weight of their contents was indicated, he leaves us to infer that such were empty casks to be filled with sugar, rum, and molasses in the French islands, and to be entered in North America or England as British products.[5]

As to the prevalence of illegal trade at New York, Quary reported to the governor, Lord Cornbury, that for many years all manner of illegal and prohibited trade had been "carry'd on, Encouraged, and Sheltered" at Sandy Hook. Ships from Madagascar, Curaçoa, Surinam, St. Thomas, and other foreign plantations stopped there, and transferred to smaller boats, "there ready for the Purpose," manufactures of Europe and other prohibited goods. The ships themselves often appear to have entered port in ballast, the cargoes being landed secretly from the little boats.[6]

Four years later, in 1707, Quary visited the principal

[5] Robert Quary to Commissioners of Customs, Virginia, Oct. 15, 1703, C.O. 323: 5, F 38.

[6] Quary to Lord Cornbury, Dec. 15, 1703, *ibid.*, F 37.

ports of New England and, in an interesting letter to the Board of Trade, described his impressions of the customs officers there. Of the situation he wrote:[7]

> The next place I came to was the province of Connecticut the seate of Government is called New London, tho' not much like Old London; attended the Governor Col. Winthrope who received me very kindly and desired me not to look to narrowly into the mistakes of that Government. I quickly found that there was good reason for this caution for when I went to examine the Customs houses, I found nothing but confusion and Roguery. I was appraised of many dishonest practices acted in that place before I went, but did not expect to have found matters so very bad, the p'son that acts as Collector was one Mr. Withred a Pillar of their Church but a great Rogue wch I am sure yor Lordships will believe when I tell you that there is no villany that a man in his post could doe but was constantly practiced by him, several vessels that made a trade of running tobacco from the out pts of Virginia wth out Entry or Clearing came directly to this Government and landed their tobacco, but what was farr worse he gave false Certificates, for the Shiping off this Tobacco, to other of the Plantations in which he certifyed that the Tobacco illegally imported was legally imported and that the Queen's duty was paid. I have found several of these false Certificates filed in the Custom House of Boston where considerable quantities of this Tobacco have been sent as well as to the other Plantations. it would tire yr Lordships should I give you the history of the Illegal trade carried on and encouraged in the Government, from Curacoa, Surinam and other places.

Of Rhode Island he observed:

> Their chiefe trade is to the West Indies, but more especially they have a great trade to Curacoa and Surinam the Chief Town of Trade is New Port wch is grown in few years to be a great town mainly by Illegal trade to those places nor is it possible to prevent it while the Government is in Proprs hands.

[7] Quary to Board of Trade, Jan. 10, 1707, C.O. 323: 6, I 60; *C. S. P. Col.*, 1706-1708, § 1273.

The Dutch and Danish islands were always a source of danger to the English during all the wars of the eighteenth century. Curaçoa, St. Thomas, and, later, St. Eustatius, as neutral ports, were used constantly by North Americans and French West Indians as entrepôts for the exchange of goods. Peter Beckford, writing from Jamaica in 1702, feared that New England would send most of its provisions to Curaçoa, leaving Jamaica "disfurnished." He even hoped Jamaica might be permitted to continue its trade with Spanish America by way of Curaçoa for mules and coin.[8] There is some evidence that English traders sold slaves to the French at Santo Domingo from St. Thomas during the war.[9] In 1703, Governor Granville of Barbadoes believed that flags of truce gave an opportunity for illegal trade with the enemy and reported that the Danish Island of St. Thomas "in time of war ever has been and is the staple for all sort of indirect and illegal trade and commerce."[10] In the same year, Manasses Gilligan, an English subject, was tried at Barbadoes for trading with the enemy at St. Thomas. Gilligan pretended to have been naturalized a Dane, and during the trial remarked: "You will have many Merchants in these parts retire to St. Thomas during the war that they may do the same."[11] Attorney General Northey held that English subjects residing at St. Thomas, or naturalized there without the queen's license, might be recalled, and if they remained and traded with the enemy might be proceeded against as

[8] Peter Beckford, lieut.-gov. of Jamaica, to Board of Trade, July 20, 1702, C.O. 137: 5, F 3.

[9] Peter Smith (a trader) to Monsieur —— at Santo Domingo, dated St. Thomas, June 14, 1703, C.O. 137: 6, G 25.

[10] Gov. Sir Bevill Granville to Board of Trade, Sept. 3, 1703, C.O. 28: 6, H 50.

[11] *Ibid.*, Oct. 31, 1703, C.O. 28: 7, I 2. For a full account of the Gilligan case by the judge of vice-admiralty and the attorney and solicitor general of Barbadoes, see *ibid.*, I 4.

criminals.[12] Jamaicans were annoyed that English manufactures should be sold to Spanish America through the Dutch instead of through Jamaica, as was very largely the case in time of peace. It deprived Jamaica of that influx of Spanish specie which was always of great benefit to the island as its main source of currency.[13]

Flags of truce, or ships commissioned for the exchange of prisoners, during all the wars of the eighteenth century were very frequently used for carrying on trade between the English and French colonies. For example, in 1707, a cartel was agreed to between the governors of Barbadoes and Martinique.[14] Eight months later the governor of Barbadoes remarked that this cartel "only served to promote the ffrench Trade and Intelligence."[15] In 1709, Peter Holt, a West India merchantman, told the Board of Trade that St. Thomas and Curaçoa were supplied with provisions chiefly from Carolina, Pennsylvania, New York, and Rhode Island. It was this, he said, which enabled the French to fit out privateers from these islands.[16] Further information on this trade was supplied by Samuel Brise who resided at Curaçoa from 1705 to 1709. Brise said he had seen several ships come in there with tobacco from Virginia, and at St. Thomas he had noticed several vessels from Carolina with pro-

[12] Report of Edward Northey, attorney general, Mar. 22, 1703/4, on the Gilligan case, *ibid.*, I 23.

[13] Lieut. Gov. Thomas Handasyd of Jamaica to Board of Trade, Nov. 27, 1703, and Dec. 17 and 31, 1704, C.O. 137: 6, G 44; C.O. 137: 7, I 2, 14.

[14] William Sharpe, president of Barbadoes council, to Board of Trade, Mar. 24, 1706/7, C.O. 28: 10, O 15. Sharpe significantly stated as a reason for it that few sailors from the Northern Colonies could be prevailed on to venture as far as Barbadoes since, if captured, they would be sent to France unless there was a cartel.

[15] Gov. Crowe to Board of Trade, Nov. 5, 1707, C.O. 28: 11, O 87.

[16] Peter Holt's testimony before Board of Trade, Nov. 15, 1709, B. T. Jour., C.O. 391: 21, f. 270; f. 352. See also Holt to William Billton at Amsterdam, dated London, Oct. 26, 1709. Holt mentioned being at Curaçoa in 1706, C.O. 323: 6, I 85, 93.

visions. Also, while a prisoner at Martinique for eighteen months, he had noticed flags of truce arrive from Antigua with beef, pork, and flour for sale.[17]

The economic effect of this trade with the Dutch and Danes on the British sugar planters was to increase for them the cost of food and plantation supplies and, in Jamaica, to contract the currency, which was always an embarrassment to indebted proprietors.

Another trade which assumed large propositions during the war and was even more directly detrimental to the British sugar islands was that of the Northern Colonies with the Dutch sugar colony of Surinam, the modern Dutch Guiana. In 1707, the governor of Barbadoes informed the Board of Trade that "it would be some help to this Island if the Trade between New England and Surinam were obstructed for if I bee rightly informed, great quantitys of Rum Sugar and molasses go in returns for their horses flower and other Provisions.'"[18] In 1710, it was stated in a memorial to the House of Commons on behalf of Jamaica that the Dutch facilitated the trade all they could, and that there were then some English merchants settled there to carry on the trade, and that seldom were there fewer than four hundred English seamen at that place.[19] A pamphlet of the day voiced the awakening fear of the planters that some one had begun to undermine their monopoly of the sugar trade.

> Its indispensably necessary for the government [said the writer] to cherish and support this valuable trade, and to discourage all attempts to wrest it out of our hands; in order to do this, some inspections may be necessary into the trade from New

[17] Samuel Brise's testimony before Board of Trade, Jan. 16, 1709/10, B. T. Jour., C.O. 391: 21, f. 336. Brise was a mariner.

[18] Gov. Crowe to Board of Trade, Nov. 5, 1707, C.O. 28: 11, O 87.

[19] *Some Considerations humbly offer'd on behalf of Jamaica, to the Hon. House of Commons*, 1710, Brit. Mus. Tract. 816, m. 18 (37).

England, and the northern colonies, to S^{t}. Thomas's Curasoa and Surinam; to the last they send horses, by which they carry on their sugar making, which promotes that Dutch Colony in that manufacture; there is a law or order in Surinam, that these northern vessels shall not be admitted to trade with them, unless they bring such a number of horses; and besides they import from these colonies dry goods; by which means the consumption from Britain, and the southern plantations is much lessen'd, upon an exact inquiry some prejudice to our Trade may be found, that ought to be remedied and prevented for the future.[20]

The first earnest complaint from British planters against the Northern Colonies' trade with the Dutch came in 1713, and was forwarded to the Board of Trade by their agent, William Heysham. It was no longer upon military grounds that they objected to the trade, but because a commercial rival had encroached upon a British industry. "I am directed from Barbadoes," wrote Heysham, "to lay before Your Lordships, That of Late, Rum, Sugar, and Molasses hath been Imported to New England and other Northern Collonies from Surranam, which if not timely prevented may prove fatall to her Ma^{sts} Sugar Plantations in America and greatly discourage the Trade and Navigation of Great Brittain—Suranam being a large Collony and the Land New and fertile they can make and Vend Sugers much cheaper then any of her Maj^{sts} Plantations—This humbly Craves Y^{r} Lors^{ps} consideration. And your best and speediest assistance to prevent the Growing Evill."[21]

The Board of Trade summoned Jeremiah Dummer,

20 *Present State of the Sugar Plantations*, 1713, p. 26. The statement occurs frequently that the Dutch required the English who traded to Surinam to bring a number of horses in proportion to their ship's tonnage, or the tails, in case the horses died. *Cf.*, *e.g.*, Charles Dunbar, surveyor of customs in Barbadoes and Leeward Islands, "State of the British Sugar Colonies," enclosed in letter from Gov. Mathew of the Leeward Islands to Board of Trade, Oct. 3, 1730, C.O. 152: 19, T 142.

21 William Heysham to Board of Trade, Oct. 6, 1713, C.O. 28: 14, T 32.

the agent for Massachusetts and Connecticut, and Heysham, and talked over the matter for three days in January and February, 1714 (January 28, 29, and February 2). Dummer desired no further action to be taken till Governor Nicholson of Nova Scotia had been informed and his report heard. A representation to the queen on the trade was drawn up and signed February 9.[22] It set forth the damage to the sugar plantations from this trade. Whereupon, an order in Council was issued, February 20, 1713, directing a bill to be prepared for the relief of the sugar islands.[23] Unluckily perhaps for the planters, this order, though received March 15, 1713, was not read by the board until July 28, 1715. But it is interesting to know that a relief act of parliament for the sugar planters was suggested by the West Indians and recommended by the Privy Council at least twenty years before its final passage in the form of the Molasses Act of 1733. The projected legislation did not become law at this time, but circular letters to all the governors were sent out, under date of August 19, 1714, directing them to take care that no illegal trade be carried on between their respective governments and the French settlements in America.[24] The government stated in this letter that the trade in question was a violation of the fifth and sixth articles of the Treaty of Neutrality of

Cf. P. Harley to Board of Trade, Nov. 4, 1713, C.O. 5: 866, V 24; Lowndes to Board of Trade, Nov. 4, 1713, C.O. 137: 10, N 86.

22 B. T. Jour., C.O. 391: 24, pp. 21, 104, 108, 110, 114, 117, 121.

23 C.O. 152: 10, O 94; C.O. 137: 10, N 142. The Board of Trade, in their representation to the Privy Council, prayed that the New England trade to Surinam might be prohibited by law. The order in Council of Feb. 20, 1713/4, therefore, ordered that the attorney general and the solicitor general should prepare a bill to be passed into law by parliament "this session" for prohibiting this trade. C.O. 28: 14, T 82.

24 C.O. 324: 10, pp. 58, 293; B.T. Jour., C.O. 391: 24, p. 297. *Cf.* also Gov. Hamilton of Jamaica to Board of Trade, April 26, 1715, C.O. 137: 10, N 139.

1686 between England and France. Governor Dudley of Massachusetts had copies of these articles printed and sent to the captains of the frigates attending the government of that province with instructions that they be put in execution.[25]

Trade between New England and Surinam was, however, left untouched by these trade regulations. The lowlands along the coast south of the Orinoco were, at this time, being brought under sugar cultivation by a comparatively few Dutch planters under the direction of the Dutch West India Company. Holland, however, since the loss of New Amsterdam, had no temperate zone colony to balance or complement the industrial life of its tropical settlements, and was dependent for its supplies and live stock on trade with foreigners. Its colonial policy was in general that of the open door to all foreign merchants. These conditions obviously suited the Northern Colonies, and from this time to the Revolution their trade to Surinam was a vital and important activity. A clear description of this commerce and an estimate of its importance, at the time when it first provoked jealousy and criticism among the British sugar planters, is that given by Thomas Banister in 1715. It is included in "An Essay on the Trade of New England" which was submitted to the Board of Trade in that year. Banister's statement of the case for New England is as follows:

> I shall mention to your Lordships our Trade to Surinam by way of Prevention, because the Gentlemen of Barbadoes have openly attackt it, representing it as prejudicial to the English Islands, therefore desired an Act of Parliament to prohibit it. This Trade takes off a great number of small Horses of no Use to us and fit for no other Market: It imploys a great number of

[25] Joseph Dudley to Board of Trade, Boston, May 2, 1715, enclosing printed copy of the articles, C.O. 5: 866, V 207.

Ships and sailors. The Tradesmen feel the Benefit by the Merchandise of Soap Candles Beer, building of Ships and the great Number of Cask this Trade imploys. The landed Interest Shares with them in the Export of very much Hay, Oates, Onjons, Apples, Pork, Beef, Staves, board, butter and Flower. The Fishery by a great Export of Macheril and refuse cod. Other Parts of Merchandize by shipping thither much Wine and some Salt. And the Customs house by the Wine aforesd exported without a drawback. The return for these is Molassus, which we brew and distil, and thereby raise many good Liveings; And the Merchant finds it one of the most profitable Trades he drives. Upon the whole therefore it is no wonder we are so desireous of keeping it. However if the Islands Suffer either in equality or Superiority to our Benefits, we will own it a too great Regard for our own Interest that inclines us to preserve and pursue it. If they suffer it must be (1) either because we fetch enough from Surinam to supply all our Wants, so their Goods lie upon their Hands (2) at least enough very much to lower the Price of theirs (3) or by turning so much of our Trade to Surinam we lessen it to Barbadoes (4) Or we carry the Dutch those Commodities which Barbadoes wants and suffers by that Want. In answer to these I assure the honourable Commissioners (1) that we expend all the Barbadoes Molassus we can purchase or procure, And don't use the other when this is to be had (2) We pay 3d or 4d per Gallon more for this, than we do for the Surinam Molassus, and it is almost a[s] dear again as it Useth before this Trade was opened (3) We send them more Vessels now than formerly, and such quantities of every thing that their Markets will not take them off: But our Commodities lie there and perrish (4) the Grand articles of this Trade are unfit for the Islands or are in such quantities that we can overstock both Markets. This is really the State of the Case between us, and their Endeavours to cut us off from this Trade is rather the Effect of Picque than any Public Interest.[26]

[26] Thomas Banister, An Essay on the Trade of New England submitted to the Board of Trade, July 7, 1715, C.O. 5: 866, V 91. Banister's essay was subsequently printed in expanded form as *A Short Essay on the Principal Branches of the Trade of New England, with the Difficulties they labour under and some Methods of Improvement*, by T. B., London, 1715.

British planters, on the other hand, were alarmed and believed themselves entitled to protection against impending damage to their interests. As yet, they do not appear to have really suffered by the existence of the New England-Surinam trade, but were highly sensitive to the possibility of harm from foreign competitors. Northerners were assisting Dutch and French rivals and, therefore, the trade ought to be suppressed by act of parliament. This view was stated by Governor Robert Lowther of Barbadoes, in 1715, and found frequent repetition from then to the time of the Revolution. Lowther wrote:

> I humbly conceive it would be of great Advantage to this Place, and to all His Majesties Suger Colonies, if there was a Law made in England to Restrain His Subjects in North America from exporting Horses into any Country not under His Majesty's dominion, for the French at Martinique and Guadelupe, and the Dutch at Soronam begin to Rival us in the Sugar Trade, and this is owing in Some Measure to the great Supplies of Horses They frequently receive from New England, and other Parts of that Continent, for as We grind the Sugar-Canes with Wind-Mills, so they are necessitated to do it by an Engine that's drawn by Horses and Cattle.[27]

The year of the peace of Utrecht marks an important turning point in the trade from the Northern Colonies to the French islands. From that time it assumed large proportions and attracted constant attention. People who gave evidence on the subject in later years thought of the trade as beginning at the close of the war. Thus the English merchant, Worsam, writing in 1720, referred to a "Calculate I made in the Year 1715 When that Trade was in its Infancy and when it was believed Illegall but Since they had Sr Edward Northey's pinion they have

[27] Gov. Robert Lowther to Board of Trade, Oct. 25, 1715, C.O. 28: 15, T 101.

I believe especially in Sug. Rum and Molasses Traded for above Three times as much every Year.''[28] An experienced West India merchant, who gave evidence on the trade in 1731 before the House of Commons' committee, stated much the same view as to its beginnings:

''Q. Do you believe the Northern Colonies, before the Year 1714, carried on any Trade for Rum or Molasses to the *French* Islands?

A. I never heard of it.

Q. By what Means were they supplied before that Time?

A. They were supplied from *Surinam,* our own Islands, and with *French* Brandy.''[29]

Other writers were of the same opinion and, as contemporary evidence proves, it was substantially a true one.[30]

At the close of the war and for some years after, the French were hard pressed for all sorts of provisions, live stock, and supplies. Prices were exceptionally high and attracted Northern traders in great numbers. A sea-captain who engaged in the trade at that time but wrote seventeen years later, may nevertheless be trusted for his description in general of the French markets, though

[28] Worsam to Gordon, London, Aug. 23, 1720, C.O. 323: 8, L 14.

[29] Capt. Fayrer Hall's *Evidence before a Committee of the House of Commons, April 1731.* Printed. C.O. 5: 4, ff. 274-275, or pp. 6-7 of the printed pamphlet.

[30] A writer in 1731 states that it was ''the Year 1715, when this Trade with the French began.'' *Answers to all Objections to the Bill,* etc., p. 2. Printed (1731). C.O. 5: 4, f. 270. Oldmixon's *America* makes no mention of the trade in the first edition of 1708, but in the second edition of 1741, attention is drawn to it. Postlethwayt wrote, in 1757: ''But, soon after the Peace of Utrecht, a pernicious commerce began to show itself, between the British Northern Colonies and the French sugar Colonies, which began with bartering the lumber of the former for French sugar and melasses.'' *Great Britain's Commercial Interest Explained,* 1757, 2d ed., London, 1759, I, 485.

his figures may not be strictly accurate. Fayrer Hall described the French West India markets as follows: "As soon as the Peace of *Utrecht* was concluded, the English carried to the French, Mules, Asses, Horses, and Provisions of all Sorts, which the *French* wanted exceedingly. I knew many Hundred Mules, even whole cargoes, sold at a Time for 500 Livres a Head all round. Provisions were generally from 50 to 100 *per cent.* dearer among the *French* than in our own Islands, at least for two years after the Peace."[31] Hall declared that the French, seeing the advantage of the trade, encouraged it. French molasses, he stated, was at first bought by Bostonians at 3d. and 4d. a gallon. Formerly, the French had "never esteemed [it] more than Dung; for they used to throw it all away." Rum was selling in Barbadoes, in 1712, at 11d. to 12d. and upwards per gallon, according to Hall.[32] The French governors, he said, connived at the trade and were bribed to grant licenses to Englishmen. Santo Domingo, he added, was well wooded and, in 1714, the French were largely manufacturing their own lumber for casks. Captain Hall gave the Commons' committee in 1731 the following interesting account of some of his experiences in the French trade at this period:

I have lived in and traded for twenty Years past to the *West Indies,* and the northern Colonies. I was first there, at our Islands, in 1709, afterwards at Jamaica in 1712, and in 1714 I was Master of a Sloop, and carried a Load of Provisions and Lumber from Philadelphia to Barbadoes. There were more Vessels with Provisions and Lumber, which made the Prices of those Commodities very low at Barbadoes, so I went from thence

[31] Fayrer Hall's "Letter to a Noble Lord" in the *Post Boy* of Mar. 6, 1731, C.O. 5: 4, f. 276.

[32] Fayrer Hall's *Evidence before the Commons' Committee,* C.O. 5: 4, f. 272. *Cf.* also *Remarks upon a Book,* etc., 1731, pp. 15-16, for another retrospective view of trade in 1713-16.

down to *Martinico,* where I sold some Flour, and understanding there was a good Market, or great Demand, for Mules there, I went to *Curasso,* and took in forty-eight Mules at thirty Pieces of Eight a-piece, I bought them of the Governor; these were as many as I could carry, and I carried them to Martinico, I was nineteen Days in my Passage, and lost but two, When I came the *Martinico,* I was forced to get the People to petition for Liberty to sell the Mules, and I gave the Governor a hundred Livres for every Mule; I sold the whole Cargo at six hundred Livres a Head, and sold the Mules in Health for seven hundred, any that could stand on their Legs would sell for four hundred; I got near four times as much as my first Expence. I went from *Martinico* to *Barbadoes,* and took in Flour, and went down to *Curasso* again; and in that time there were other Vessels arrived there, two from *Nevis,* and two *French* Sloops, but I could not make anything like the same Profit; I went a third Time, but could make but little of it then by my Mules.——Great Quantities of Rum were made near Fort-Royal and *Fort St. Pierre* [in Martinique], at that Time, I bought several Hogsheads of Rum and Sugar in Martinico the last Voyage, and carried to *Curacoa.*[33]

[33]Hall's *Evidence before the Commons' Committee,* C.O. 5: 4, ff. 272-273. In a "State of New England's Trade as it stands in 1715" the imports and exports are tabulated as follows:

Rum Sugar Molasses Bills of exchange and silver	from	British West Indies Surinam Isaac Cape Cape François St. Thomas Curaçoa

To Great Britain Newfoundland Annapolis, N.B.	Some Sugar Molasses Rum	To Pennsylvania Virginia Maryland N. Carolina S. Carolina Campeachy & Honduras	Rum Molasses

To British W. Indies Cape François Isaac Cape Surinam St. Thomas Curaçoa	Candles Soap Horses Onions Mackeral Cod Beef	Pork Tobacco Staves Butter Flour Boards Bear	Salt Apples "Neats" Tongues Hams &c.

C.O. 5: 866, V 67.

As to the extent of New England's trade with the foreign sugar settlements, we are mainly dependent upon opinions expressed in letters of the governors and surveyors of the customs. To be sure, we have the naval office lists which purport to give all the entries and clearances of merchant vessels and their cargoes at each port. And these lists do record the entries and clearances of a considerable amount of shipping with all the foreign West India Islands. But it is practically certain that a great deal of the sugar that purported to come from British settlements in reality came from the French and Dutch plantations. It was believed by many persons, who were likely to know, that the practice was prevalent, both in the Northern Colonies and in the British West Indies, for customs officers to issue fictitious clearance papers to sea-captains. Consequently a very large but uncertain share of the exports of the North, that are listed in the customs returns as cleared for Jamaica, Antigua, or Barbadoes, were actually delivered in Santo Domingo or Surinam. The returns from these foreign settlements were, in the great majority of cases, recorded as entered from Barbadoes or some other British island. In this way, sugar from the West Indies escaped not only the ad valorem export duty of four and a half per cent levied in Barbadoes and the Leeward Islands, and the "Enumeration Duty" of 1s. 6d. per hundredweight on exports of brown sugar from all the British West Indies, but also the additional English import duty levied on foreign sugars in case of their reshipment into England. The shipping lists indicate that considerable amounts of sugar, rum, etc., were reshipped from the Northern Colonies to Great Britain and entered there as English produce. The belief was strongly entertained that this sugar, which escaped "foreign" duty, was in the main of French or Dutch

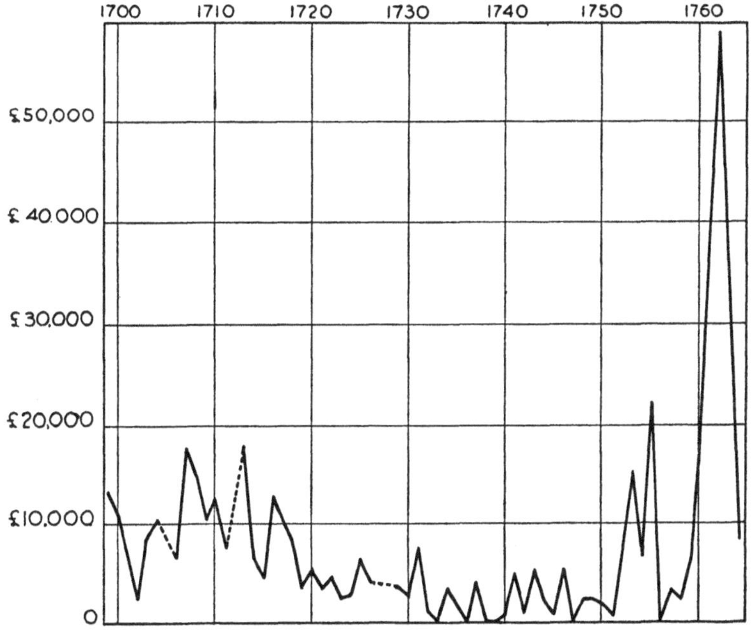

CHART IX. VALUE OF THE YEARLY IMPORTATION OF BROWN SUGAR INTO ENGLAND FROM NEW ENGLAND, NEW YORK, PENNSYLVANIA, VIRGINIA, AND MARYLAND, AND RECORDED AS ENGLISH PRODUCE, 1699-1764

This chart is based on tables compiled for each of the above districts from Custom House Accounts, Ledgers of Imports and Exports. The line is dotted where statistics are wanting.

growth. The total amount, however, of such sugar reshipped to England was small, compared with the total import of British sugar,[34] and could have had but very slight effect in depressing the prices of sugar in England.

The extent of the trade in foreign sugar at Boston, in 1716, was described by Archibald Cumings, surveyor of the customs and searcher at the port of Boston, as follows:

> Since I was appointed Surveyor and Searcher thereof . . . there is annually imported here 2000 hhds of Molasses and 2 or 300 hhds of Suger and rume from Suriname Cayan St. Domingo & St Thomas being dutch french danish ports and there being no duty more paid here then if our product: and our product in all the Islands paying 4½ p Ct. to his Majesty on the said Commodity it would seem highly reasonable these forreign Commodities should pay the 4½ p Ct. to his Majesty upon the Importation here or more to putt the Importer upon Levell in trade with those trading to our own Islands and make a revenue of 800 or 1000 pounds p annum this money which would ease the Crown of Charges here or the defraying of other Expences.[35]

The following year, Cumings made a further report on this trade and supplied figures for comparing the amount of sugar, molasses, and rum from foreign settlements with that from the British plantations. The largest item from both places was molasses to supply the distilleries, of which molasses about three quarters was of foreign production. Very little rum came from the foreign islands, but a large quantity from the British plantations. This rum was largely consumed in the fisheries, Indian barter, and the Guinea trade. The sugar came mainly

34 Tables giving the values of sugar, rum, and molasses, entering England each year, from 1698 to 1765, have been compiled from the Customs Accounts, Ledgers of Imports and Exports; a chart of the values of sugar entered from the Northern Colonies is given opposite p. 206.

35 Archibald Cumings to Board of Trade, Boston, Aug. 2, 1716, C.O. 5: 866, V 116.

from the British West Indies. Cumings' report of March 2, 1717, gives the following summary of Boston's trade with the West Indies with proposals for such a regulation of it as would discourage foreign imports and yield the Crown a revenue:

I doe observe that last year there was about 2300 hh^ds^ of Molosses imported here from foreign plantations as Cyan, Surriname, Cape Francois, & St Thomas with about 15 hh^ds^ Rume with 15 hh^ds^ 208 tierces & 595 Casks of Sugar, ten Casks of Indico, & 83 bags Cotton Wool all foreign growth which pays no more duties here then if of our own growth & often shiped for Brittain as such which is a prejudice to the revenue from our own plantations has been imported 149 hh^ds^ 208 tierces & 1107 bar^ls^ of Suger 502 bags of Cotton Wool Six Casks Indico 19 bar^ls^ 1 hh^d^ & 14 bags Cocoa 681 hh^ds^ 855 tierces & 116 bar^ls^ of Molosses 1429 hh^ds^ & 338 tierces & 180 bar^ls^ Rume most of which Rume and molosses is used here all the Cotton Wool Cocoa & most of the Casks of Suger used here Wee distil annually about 2000 hh^ds^ of Rume here of Molosses and about 1800 pipes wine imported from fyall Maderas and Canaries all which might bear a duty as a revenue for the Crown to defray the Expences that the plantations are annually to Great Brittain for Governours & officers Sallaries Support of Garrisons the Expence of the Station men of war & by Selling a stamp office in all the Islands and on the Continent for this Service for as the plantations can bear this Charge being chargeable with little or no duties so it is not reasonable they should be a burden to Great Brittain if the Government thought fitt of this I could send a perfect scheme.[36]

[36] Cumings to Cockburn, a member of the Board of Trade, Mar. 2, 1716/7, C.O. 5: 866, V 134. The total amount of foreign sugar, molasses, and rum imported into Boston during the years 1714-17 was recorded as follows:

	Molasses			Sugar			Rum		
Year	hhds.	Barrels	Tierces	hhds.	Barrels	Tierces	hhds.	Barrels	Tierces
1714	1074	55	937	53	35	59	27	20	44
1715	900	119	745	3	186	48	19	5	26
1716	800	137	778	4	521	116	23	25	8
1717	704	328	645		215 / 28 Barrels and Tierces	12	12	8	6

C.O. 5: 866, V 153, 156.

He suggested that the following duties be collected in Massachusetts and, presumably, in all the Northern Colonies:

On English produced rum and molasses . . .	20s. per hhd.
On foreign produced rum and molasses . . .	30s. per hhd.
On rum distilled in New England, an excise of .	40s. per hhd.
On English sugar	5s. per cwt.
On foreign sugar	8s. per cwt.

By such a preferential tariff Cumings believed England would obtain sufficient revenue to pay the governor's and other salaries. The governor would then be free to uphold the vice-admiralty judges and customs officers in their efforts to suppress illegal trade. In the naval office lists of ships cleared from Boston at this time, those carrying sugar are numerous and in most cases are entered as bound for London or Virginia. Vessels carrying rum were mostly cleared for Newfoundland.[37] The amount of sugar, however, whether of British or foreign growth, that reached England by way of New England was not very great; from July 15, 1717, to July 15, 1718, it amounted to one hundred hogsheads.[38]

The West Indians had no occasion, therefore, to fear that the amount of foreign sugar in the English market that came from New England could have much effect on prices there. What did alarm them was the more reasonable apprehension that, if Holland could produce its own sugar in South America, the Dutch would purchase less brown sugar in the London market. Thus, indirectly, the Northern Colonies were helping to diminish the sale of British sugar in Europe and depress its price in England. In 1718, Governor Lowther of Barbadoes again recommended that parliament should pass a law to restrain the trade from North America to the foreign

37 Naval Office Lists for New England, C.O. 5: 848, for years 1713-19.

38 Cumings to Board of Trade, July 15, 1718, C.O. 5: 867, W 12.

settlements. Lowther said that never "could they [the French] produce any Considerable Quantity of Suger if the King's Subjects did not supply 'em with Corn, Floor, Fish, Beefe, in returne of which they receive Wine, Brandy, Suger & Molosses."[39] He added that the French already were underselling the English in these products.

In 1719, Governor Shute of Massachusetts reported that the trade from that colony to any foreign plantation was inconsiderable, "except that to . . . Suriname which is carried on mostly by exporting small wild Horses not fit for Service here, nor salable in our English Plantations, for which is brought back in return chiefly Molasses, In which Trade may be employed twelve or fifteen Small Vessels."[40] He suggested that two revenue cutters and two waiters should be added to the customs service at Boston; and that another custom house be provided for New Hampshire.[41] Jeremiah Dummer, in 1720, said that the horses sent from Massachusetts to Surinam are refused by the British Islands," and that this trade did not employ more than one or two ships a year.[42] William Gordon, who had traded in the West Indies, informed the Board of Trade in 1720, that Surinam received all its mill horses from Massachusetts and Rhode Island, as well as fish and staves.[43]

39 Gov. Robert Lowther to Board of Trade, 1718, C.O. 28: 15, T 181.

40 Gov. Shute to Board of Trade, Feb. 17, 1719, C.O. 5: 867, W 99.

41 *Ibid.*, and Shute to Board of Trade, June 1, 1720, C.O. 5: 867, W 109.

42 Jeremiah Dummer to Board of Trade, Oct. 11, 1720, C.O. 5: 867, W 114.

43 William Gordon to Board of Trade, London, Aug. 17, 1720, C.O. 5: 867, W 110. He stated that Massachusetts and Rhode Island were about half supplied with their molasses by the French and Dutch.

In 1722, it was reported that New Englanders had brought coopers to set up their casks in the French Islands and that they emptied the sugar, molasses, and rum from French casks into their own. J. Burchett, sec. to the Admiralty, to Popple, sec. of Board of Trade, Aug. 11, 1722, enclosing an extract of Capt. Brand of the *Hector* to Burchett, Antigua, July 5, 1722, C.O. 323: 8, L 35.

New York also sent horses to Surinam, St. Thomas, and Curaçoa, but had very little, if any, direct trade with the French. From these places, wrote Governor Hunter in 1720, there were "No returns but gold or silver at least that are avow'd or discovered."[44] At the same time, it was alleged by William Gordon that the French colonies were chiefly supplied with flour, bacon, and some horses from New York, in exchange for sugar and cocoa, which were shipped direct to Holland. Gordon added that the quantity of French and Dutch sugar which came to the Northern Colonies was so large, that they sent a great deal of it to England as of British growth. A moderate duty was advocated on such foreign produce imported into North America.[45] In 1723, New York was reported to have sixteen distilleries "which are wholly supplied with Mollasses from Martinico."[46]

All these reports concerning the Northern trade to the foreign sugar colonies did not go unobserved by the government of the day. The Board of Trade, in 1721, in their representation to the Crown on the state of the plantations, reviewed at some length the state of this trade.[47] While it was "a very great discouragement to the Sugar Planters in the British Islands," the Northerners were not without a certain justification in undertaking it. For the exports from England to the continent of America exceeded the imports from thence about £200,000 per annum, "which Debt falls upon the Prov-

[44] Gov. Robert Hunter to Board of Trade, Aug. 11, 1720, C.O. 5: 1052, Cc 38.

[45] William Gordon, to Board of Trade, Aug. 17, 1720, C.O. 5: 867, W 110.

[46] Gov. Henry [Worsley] of Barbadoes to Secretary Carteret, Mar. 26, 1723, C.O. 28: 44, f. 17; same to Board of Trade, C.O. 28: 17, V 109.

[47] Representation to H. M. on the State of the Plantations in America, Sept. 8, 1721, C.O. 324: 10, ff. 296 *et seq.;* there is also a copy in Brit. Mus. Add. MSS. 35907 (Hardwicke Papers). It is printed in *N. Y. Col. Docts.* V, 591-630.

inces to the Northward of Maryland who probably are inabled to discharge the same by the trade they are permitted to carry on in America & to Europe in Commodities not Enumerated in the Acts of Trade." However, if naval stores were encouraged "the more Northern Colonies would be thereby enabled to pay their Balance to England without lying under the necessity of carrying on a Trade to Foreign parts, in some respects detrimental to their Mother Country."[48] The Board of Trade understood the situation in all its complexity. The interests of West India sugar producers were balanced against those of English manufacturers. The board could not see its way to recommend any restriction of the trade of the Northern colonists which would impair their ability to purchase English manufactures. A similar defense of the trade was made, in 1723, by Cadwallader Colden, the surveyor-general at New York. He pointed out that in return for the provisions sent from New York to Curaçoa, St. Thomas, Surinam, the French islands, Barbadoes, and Jamaica for transfer to Spanish America, "the Ballance being everywhere in our favour, . . . we have money remitted from every place we trade with, but chiefly from Curaçoa and Jamaica. . . . But whatever advantages we have by the West India Trade, we are so hard put to it to make even with England, that the Money imported from the West Indies seldom continues six months in the Province before it is remitted for England, The Current Cash being wholly in the Paper Bills of the Province and a few Lyon Dollars."[49]

[48] C.O. 324: 10, ff. 376, 382.

[49] An Account of the Trade of New York by Cadwallader Colden, enclosed in letter from Gov. Burnet to Board of Trade, June 25, 1723, C.O. 5: 1053, Cc 117. Colden alluded to the French trade as "some little Private Trade with the French Islands." Colden's argument was repeated by Rip Van Dam, president of the council of New York, in a communication to Board of Trade, Oct. 29, 1731, during the sugar bill controversy, Newcastle Papers, Brit. Mus. Add. MSS. 33028, f. 263.

During the decade from 1720 to 1730, it was generally held that the trade from the American continent to the foreign sugar colonies was on the increase. Prices in the French islands appear to have been always lower than in the British colonies. The English merchant Worsam, in 1724, stated that sugar could be purchased among the French four or five shillings per hundredweight cheaper than in Barbadoes. Besides, there was added to the price of sugar bought in Barbadoes or the Leeward Islands the export duty of four and a half per cent and, in all the British colonies, the "enumeration duty" of one shilling six pence per hundredweight on brown sugar and five shillings on white. This made the price of French sugar delivered in the Northern Colonies six or seven shillings per hundredweight cheaper than English sugar.[50] This view was commonly admitted at the time, and was confirmed by Governor Worsley of Barbadoes. In 1730, Worsley wrote: "Sugar at Martinico, tho' not so fine, as ours here, yet perhaps as good for the Sugar Bakers, is much Cheaper, than the Sugar of Barbados, and Melasses is there worth about 4d. per Gallon, and here tis worth 9d. and 10d., if they could not export their Mellasses, they must fling it away as they formerly did when they had no vent for it."[51] Richard Harris, a West India merchant of long experience, told the Board of Trade that the French could sell their sugars seven and a half per cent cheaper than the English could in Barbadoes or the Leeward Islands.[52] Northerners did not purchase much rum in the West Indies; they preferred to take molasses and distill it themselves.

[50] Worsam to Board of Trade, July 2, 1724, dated at "Chortsea" (Chertsey), C.O. 388: 24, R 153.

[51] Worsley to Board of Trade, July 7, 1730, C.O. 28: 21, Y 41. In Jamaica, in 1722, sugar sold from 26s. to 45s. per cwt., and rum at 1s. 10½d. per gal. Long Papers, Brit. Mus. Add. MSS. 12404, f. 308.

[52] B. T. Jour., Dec. 7, 1731, C.O. 391: 40, ff. 307-312.

For the price of rum among the English was considered very high, being from fifteen pence to two shillings per gallon for some years prior to 1728.[53] British rum, that made at Jamaica especially, was of a superior quality and the Northern people could seldom afford to purchase it.[54] Rum made in New England was much cheaper. Worsley claimed that the distilling of rum in New England had depressed the price of British West India rum by fifty per cent.[55] No such effect, however, was produced on Jamaica rum, which increased gradually from 1s. 10½d. per gallon in 1722, to 2s. 6d. in 1774.[56] In the year 1725, three thousand hogsheads of foreign molasses were imported into Boston, and distilled into rum. Cumings, the surveyor, repeated his criticism of ten years previous on the trade, again proposing a duty of 3d. per gallon on foreign molasses, and also duties on foreign rum, sugar, cocoa, and indigo.[57] The other customs officers in Massachusetts also condemned the trade and suggested that duties be imposed on it.[58]

A new aspect of New England's trade with the foreigners was revealed in 1730. Colonel David Dunbar, surveyor of woods in America, reported that many large vessels built in New England were sold to the French and

[53] Timothy Donovon, agent to the contractors for victualling H. M. ships at Jamaica, to Newcastle, April 24, 1728, C.O. 137: 17, S 23 (a copy). Large quantities were purchased for the fleet at what he considered exorbitant prices. See Chart X for quantity of rum imported into England.

[54] *Importance of the Sugar Plantations*, p. 95.

[55] Worsley to Board of Trade, Oct. 18, 1724, C.O. 28: 18, W 51.

[56] Long Papers, Brit. Mus. Add. MSS. 12404, f. 308; Long, *Jamaica*, I, 382.

[57] Archibald Cumings to Board of Trade, Jan. 28, 1725/6, C.O. 5: 869, Y 65. See also his description of the trade and similar recommendation in a letter to Board of Trade, Oct. 10, 1724, C.O. 5: 829, Y 62.

[58] "Memorial of Thomas Lechmere, Surveyor General and the several Collectors and other Officers of H. M. Customs in New England." Dated Boston, April 30, 1725, C.O. 5: 869, Y 64. Signed, Thos. Lechmere, Wm. Lumber, John Jekyll, Hibbert Newton, Nath^{n} Gay, Archibald Cumings.

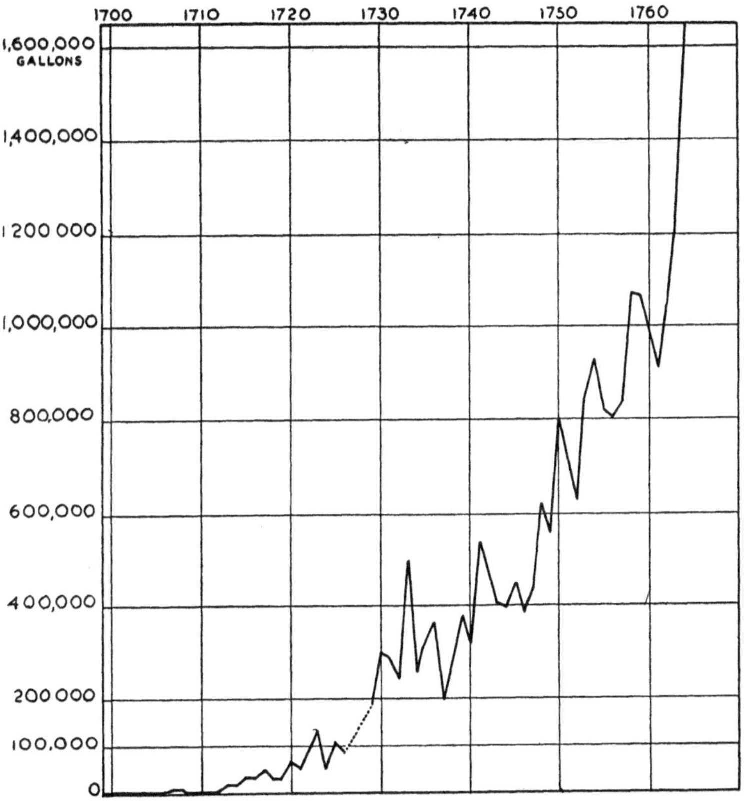

CHART X. TOTAL QUANTITY OF RUM IMPORTED YEARLY INTO ENGLAND FROM THE BRITISH SUGAR COLONIES, 1699-1764

The above chart is based on tables compiled for Antigua, Barbadoes, Jamaica, Montserrat, Nevis, and St. Christopher from Custom House Accounts, Ledgers of Imports and Exports. The line is dotted where statistics are wanting.

Spanish. They were designed for speed, rather than burden, and well protected, many of them having from twenty to forty guns, with open ports. All were "purchased with French rum & Molasses, of wch there is as general a consumption here all thro' the Continent as there was of porter and ginn in London a year agoe." These ships were registered in New England and sailed from there with an English or Irish captain and a British crew. The latter sometimes carried the ships to Cape Breton, but more often to the French and Spanish West Indies with lumber and fish. There the vessels were delivered to the foreigners, and the sailors were turned adrift or entered the foreign service.[59] Later, Dunbar gave similar information at the bar of the House of Commons, for which the Massachusetts Assembly summoned him before them and severely censured him.[60] In another letter Dunbar wrote: "This town [Boston] is full of frenchmen come to carry away large ships to their plantations, purchased here with rum & Molasses, the practice here is for the Merchant to go with the Ship beyond the Fort, & there to give a bill of Sale & the register to the french Master, & thus evade the Acts of Navigation, besides rum & Molasses the french send and bring wines and great quantities of their silks & all this connived at."[61] Dunbar recommended that Parliament

59 Col. David Dunbar to Board of Trade, Boston, Aug. 19, 1730, C.O. 5: 871, Z 130. A copy is also in Newcastle Papers, Brit. Mus. Add. MSS. 33028, f. 262.

60 Newcastle Papers, Brit. Mus. Add. MSS. 33028, f. 262. The Boston *Weekly News Letter* for July 22, 1732 (no. 1483), noted that the Board of Trade, in a representation to the House of Commons, had called attention to Dunbar's letter of 1730.

61 Dunbar to Board of Trade, Oct. 21, 1730, C.O. 5: 871, Z 141. This information was confirmed by Jeremiah Dunbar and Thomas Coram, who resided many years in New England. Representation of Board of Trade, Jan. 23, 1733, Hardwicke Papers, Brit. Mus. Add. MSS. 35907, ff. 55-56. Jeremiah Dunbar, brother of David Dunbar, also testified before the Board

lay a duty on all rum, molasses, and brown sugar imported "into the Massachusetts Government onely" and ten per cent ad valorem on all other goods.[62]

It is evident that the French were aiming to equip themselves with a merchant marine suitable for managing their own trade with New England. And though Massachusetts had been a partner in this project, she now sought to protect herself from the effect of it. In the autumn of 1730, an act passed the assembly laying discriminating duties on foreign molasses and rum, if imported by a foreigner, that is, the duty on molasses was one shilling per hogshead if imported by an English subject solely on his own account, but if imported by a foreigner it was five shillings. The duty on rum was twenty shillings per hogshead of one hundred gallons when imported by an English subject, and sixty shillings if brought in by a foreigner. No preference was shown to the British shipper in the duty on sugar, which was two shillings per hogshead. Probably but little foreign sugar came into Massachusetts at the time.[63]

The majority of West India critics of the trade of the Northern Colonies with the foreign sugar plantations agreed that this traffic damaged the British sugar islands in two ways. It depressed the price of West India produce, especially rum, and it enhanced prices of their

of Trade on this subject. When he left Boston, there were one hundred Frenchmen there importing French goods. B. T. Jour., Jan. 5, 1730/1, C.O. 391: 40, f. 2.

62 David Dunbar to Board of Trade, Sept. 15, 1730, C.O. 5: 870, Z 45.

63 Printed act of Massachusetts Assembly, C.O. 5: 872, Z 163. The preamble recited that "Whereas many Strangers and Foreigners have of late Years reaped great Gain and Profit by bringing into this Province Considerable quantities of Foreign Molasses and Rhum on their own Accompts, whereby much of the Trade that was formerly carried on with considerable Profit by the Inhabitants of this Province, altho' with the paying very high Charges for Permission, &c. in their Islands, is in a great measure, if not wholly prevented": therefore the above duties were laid.

provisions and timber. An estimate of the annual damage thus inflicted upon Barbadoes and the Leeward Islands was made, in 1730, by the surveyor of the customs in those islands.[64] This estimate by Charles Dunbar may be summarized as follows:

Antigua produced 10,000 hogsheads of Rum which, sold at 16d. per gallon, a "moderate price," would yield	£66,666 : 13 : 0
But illicit trade lowered this price to 12d. or 13d. per gallon, so that it yields only . . .	50,000.
Annual loss to Antigua from illicit trade . .	16,666 : 13 : 0
Annual loss to St. Christopher, Nevis, and Montserrat	16,666 : 13 : 0
Annual loss to Barbadoes	20,000.
Annual loss to Barbadoes and Leeward Islands on rum	53,333.
And, by the enhanced prices of provisions and lumber, the annual loss to Leeward Islands is	30,000.
A similar loss to Barbadoes	20,000.
Making the annual loss to Barbadoes and Leeward Islands on provisions and lumber . .	50,000.
Total Annual Damage to Barbadoes and Leeward Islands by the Northern trade, and that from Ireland to the foreign sugar colonies .	£103,333.

Dunbar recommended a prohibition of the export of provisions from North America and Ireland to the French or Dutch, as well as the import of foreign molasses and rum into North America or Ireland. If the

[64] Charles Dunbar, "State of the British Sugar Colonies," enclosed in letter from Gov. Mathew of Leeward Islands to Board of Trade, Oct. 3, 1730, C.O. 152: 19, T 142.

total prohibition could not be obtained, then he proposed the following import duties:

	Duty in British W. I.	In Northern Colonies & Ireland
On white or clayed sugar, per cwt.	3s.	9s.
Muscovado sugar, per cwt.	9d.	4s. 6d.
Rum, per gal.	3d.	12d.
Molasses, per gal.	1d.	6d.

The higher schedule for North America would secure compensation for revenue lost in Barbadoes and the Leeward Islands by merchants not paying the four and a half per cent export duty and the plantation duties on intercolonial trade in enumerated products.

Thus, by 1730, the British Continental colonies and the French West Indies were bound by strong commercial ties. That these regions profited immensely in consequence and that the continuance of their welfare was conditioned upon free trade between them can hardly be doubted. Public opinion both in the Northern Colonies and the French islands tended to sanction such trade as indispensable. British official sentiment, on the other hand, in accord with a long established philosophy of commerce, opposed an international trade so apparently detrimental to the British West Indies and contributing so materially to the growth of their rivals. West Indians and absentee planters naturally held similar views and associated as cause and effect the increasing intercourse between British and French colonies and the declining prosperity of their older settlements. Only rare observers such as Banister, Colden, and a few commissioners of trade appreciated the complexity of the situation and apprehended injury both to North America and England from an interference with existing conditions.

CHAPTER X

INTERNATIONAL TRADE BETWEEN BRITISH AND FOREIGN COLONIES IN THE WEST INDIES, 1699-1730

The idea of protection to the sugar industry from all foreign competition occupied the minds of West India planters in an increasing degree from the later years of the War of the Spanish Succession throughout the eighteenth century. The idea, to be sure, had been present in the seventeenth century and had found expression in the preferential duties imposed by England in 1657 on British grown sugar.[1] The desire for more protection was the result of the extensive trade between British subjects and foreign sugar settlements. This traffic was carried on by two groups of traders. There were first the Northern colonists who sold provisions, supplies, and live stock to the French and Dutch and took chiefly molasses in return, and second, the English merchants and shippers who operated among the islands themselves. The latter supplied the French and Dutch with English manufactures, provisions, and slaves, and in return took French and Dutch sugar. This foreign sugar, packed in English casks, was cleared from British islands and entered England as British grown sugar. Large quantities of French sugar were sold in England in this way. No very considerable amount of foreign sugar entered England by way of North America.

[1] By Act of June 26, 1657, see below, p. 236, note 50. The policy was sustained in 1660 by 12 Car. II, c. 18.

The alarm among the British planters aroused by this trade with the foreigners was occasioned as much by the sale of English manufactures for French sugar as by the vending of Northern produce for French molasses. It was against the former phase of the trade that the planters first attempted to protect themselves. In Barbadoes and Antigua, laws were passed to break up, if possible, the exchange of British manufactures for French sugar. The conflict was between the landed proprietors on the one hand, and merchants and shippers on the other. In the protective acts passed in Barbadoes and Antigua, in 1715-1716, are to be found the antecedents of the Molasses Act of 1733. The latter act of parliament was the application to the colonial world as a whole of a protective policy already adopted in a portion of it. The controversy of 1731-1733, which preceded the adoption by parliament of the act of 1733 and excited all parts of the American empire, was a rehearsal on an imperial scale of the controversy waged seventeen years earlier in Barbadoes and Antigua.

We have already considered the subject of trade with the foreign sugar colonies from the standpoint of North America, and have examined its character, extent, and the repeated proposals that parliament suppress it. In this chapter, we shall in like manner survey the trade between British West India merchants and the foreign settlements. With this form of the problem the planters were able to act through their own colonial legislatures. Later, they succeeded in inducing parliament to deal in the same way on their behalf with the larger phase of the problem.

There is ample evidence to show that, in the last half of the seventeenth century, the Navigation Acts were generally disregarded by the people of Barbadoes. The boldness of traders engaged in illegal traffic is comparable only with the laxity of administration. The collector of customs there, in 1698, wrote that the merchants "do not scruple the publick importing of foreign European goods & Manufacture: & also the Enumerated Plantation Commodities, without any manner of Cocketts, Certificates, or Custom House Clearings for the same . . . And have publickly Registered a French Man to be part Owner of a Vessel here; And severall other Most Notorious breaches of the said Acts they make. . . . They have had the confidence to say They will take care to make it not worth any man's while to serve the King here and that if any of the Custom house Officers shall for the future presume to put in Execution (the unreasonableness, as they call it, of) the Acts of Trade & Navigation; It shall cost them a thousand pounds sterling, but they will make an interest & get them either turned out at Home, or will ruin them here:"[2] Occasionally, some naval commander would seize a vessel importing French goods, only to find his vigilance thwarted by the collusive intervention of a customs officer. Commander Bertie, for instance, on March 31, 1702, seized the sloop *Friendship,* Thomas Balgay, master, loaded with French sugar, put aboard at Dominica by Frenchmen in sloops from Martinique. After Bertie had brought the sloop into Carlisle Bay, Barbadoes, some persons interested in the sugar showed him a permit to unload the cargo, signed by Samuel Cox, a commissioner of the customs. A few hours later, Cox and Jacob Stephens, a customs waiter, came on board, and Cox ordered Stephens to seize the sloop and affix the broad arrow to it. Then the sugar was unloaded and the ship set at liberty without any prosecution being ordered by

[2] William Sharpe to Board of Trade, Oct. 26, 1698, C.O. 28: 4, no. 6; *C. S. P. Col., 1699,* no. 476, II. On the difficulty that officers had in acting at all to enforce law, *cf.* C.O. 28: 4, nos. 6, 7, and 9.

Cox.[3] Again, on April 6, Bertie ordered the sloop *Margaret* from New York to be pursued, but before the sloop could be reached Cox came alongside in a small boat and "bad the Master make the best of his way and not take notice of the man of Warr, That the Capt. would ask him to shew his papers but to shew him none." All the naval officer could do was to protest to the council of the island against such acts of lawlessness.[4]

During the War of the Spanish Succession, flags of truce were believed by Admiral Walker and "most people in Barbadoes" to be only pretexts for collusive trade between Barbadoes and Martinique.[5] Governor Lowther, the attorney general, and the magistrates at Barbadoes all countenanced the trade with Martinique, and hindered the customs collector from obtaining writs to take up offenders. French liquors and prohibited goods were, in 1713, being daily imported from Martinique.[6] It was reported afterward that, in this year, Robert Lowther, governor of the island, together with two merchants, William Gordon and George Newport, engaged in several trading voyages between Barbadoes and Martinique, shipping horses and slaves in exchange

[3] Memorial presented to Hon. John Farmer, Esq., president of the council of Barbadoes and members thereof by Hon. Peregrine Bertie, commander of H. M. S. *Betty* now in Carlisle Bay, April 13, 1702, C.O. 28: 6, G 25.

[4] *Ibid.*, and Jno. Ransom to Popple, July 27, 1702, C.O. 28: 6, G 27. Cox was dismissed from being a commissioner of the four and a half per cent in Barbadoes by order of the Lord Treasurer. William Lowndes to Popple, Dec. 8, 1703, *ibid.*, H 52. He was also suspended from the council, C.O. 28: 7, I 54. But he was restored in 1704, *ibid.*, K 7.

[5] Admiral Hovenden Walker to Earl of Nottingham, dated "*Boyne*, at Sea" (Barbadoes), Feb. 4, 1702/3, C.O. 28: 38, ff. 32-33. *Cf.* Gov. Bevill Granville to (same?), Barbadoes, Aug. 3, 1703, *ibid.*, ff. 44-45.

[6] John Helden, acting surveyor general at St. Christopher, to customs commissioners, Mar. 10, 1713, C.O. 388: 17, N 172. Depositions on the subject were given by Richard Kennedy, William Maxwell, Edward Rundell, John Sharpe, and John Hinton, customs officers at Barbadoes.

for French goods. When the first shipload of French goods was seized by John Sharpe, the customs searcher, it was released by Lowther's order.[7]

The illegality of any trade with the French rested upon articles 5 and 6 of the Treaty of Neutrality of 1686 between England and France. But English colonists at this period were commonly ignorant of the existence of this treaty—or had at least the appearance of being so. Thus Governor Lowther, in 1715, wrote to the Board of Trade that if the government desired the traffic with Martinique stopped, they would have to pass an act to that effect "for in time of peace, I know no Law against it—if They import no Prohibited Commodities."[8]

At this time, great quantities of French sugar, molasses, rum, ginger, cotton, and aloes from Martinique were being brought into Barbadoes and cleared from there to England as of the growth of Barbadoes. Finally, on March 21, 1715, the planting interest asserted itself in the Barbadoes legislature and obtained the passage of an act[9] to protect the planters against the further import of French products. The following prohibitive duties were laid upon the introduction of foreign goods.

Muscovado sugar . .	12s.	6d.	per cwt.
Improved sugar . . .	25s.		per cwt.
Molasses		18d.	per gallon
Rum		2d.	per gallon
Scraped ginger . . .	20s.		per cwt.
Scaled ginger . . .	12s.	6d.	per cwt.
Cotton		6d.	per lb.
Aloes		18d.	per lb.

7 Deposition of William Gordon, Feb. 12, 1719/20, C.O. 28: 15, T 244. See also George Newport to Board of Trade, Feb. 16, 1719/20, *ibid.*, T 246.

8 Gov. Robert Lowther to Board of Trade, Oct. 25, 1715, C.O. 28: 15, T 101. Du Quesne, governor of Martinique, to Lowther, Sept. 30, 1715, had complained against the English trading to Martinique, *ibid.*, T 104.

9 Hall, *Acts of Barbadoes*, no. 132 (C.O. 30: 1).

The act was confirmed October 17, 1717.[10] In soliciting the approval of the measure in England, the colonial agents admitted that sugar was cheaper among the French and Dutch than in Barbadoes. This they attributed to the wearing out of the soil in the latter island. Horses, asses, and slaves had been sold to the French and Dutch, and in return great quantities of their sugar had been exported from Barbadoes to England as the product of Barbadoes. Consequently, the British revenue had been defrauded of the double duty on foreign sugar and the price of sugar in England had been depressed. The object of the act was to put the English planter on a footing equal or superior to the French.[11]

This act was openly violated, as appears from a complaint to the king in 1720 from the speaker and members of the Barbadoes assembly. It was then stated that French goods in great quantities were daily imported from Martinique, and Barbadoes was being drained of money and provisions to the enriching of a few merchants. They concluded by urging the British government to take some means of stopping this trade.[12] Direct trade with Martinique continued for some years.[13] Governor Worsley, however, made several seizures and improved the efficiency of the customs service.[14] But these measures only served, apparently, to divert the

[10] Order in Council, Oct. 17, 1717, C.O. 28: 15, T 163.

[11] John Micklethwaite, John Koyd, and George Bampfield, to Board of Trade, received Sept. 25, 1717, C.O. 28: 15, T 127.

[12] Address to His Majesty from the Speaker and General Assembly of Barbadoes, Feb. 16, 1720, C.O. 28: 44, no. 1. Twenty signatures.

[13] Samuel Cox, pres. of council of Barbadoes to Board of Trade, Dec. 20, 1721. Cox declared Lascelles, customs collector in Barbadoes, had connived at the trade, C.O. 28: 17, V 91. See also Gov. Henry Worsley to Lord Carteret, Mar. 26, 1723, C.O. 28: 44, f. 14.

[14] Worsley to Board of Trade, Mar. 26, 1723, C.O. 28: 44, f. 14; C.O. 28: 17, V 109. *Cf.* John Mairly, searcher and waiter at Speights, to Worsley, Sept. 24, 1723, *ibid.*, V 121.

trade into a somewhat safer channel. For Worsley reported, in 1724, that the French were still supplied from Ireland and from these islands with beef and other provisions. "The vessels that carry on that Trade," he said, "clear out from hence either for St. Lucia, St. Eustacia, or St. Thomases but return with no clearances from those Islands, in that they are Free ports."[15]

It is interesting to note that Governor Worsley detected several unforeseen tendencies of the Barbadoes act of 1715 which were not generally admitted by West India planters. The first of these was that the act, instead of reducing the cost of food, really increased the price of North American provisions. For in the days of "free trade" with the French prior to 1715, Barbadoes being well stocked with sugar at moderate rates, the Northern colonists were satisfied to exchange their provisions for sugar. But since the exclusion of French sugar, its price in Barbadoes had been higher than in the French islands and consequently the North Americans, not being content to exchange their provisions for sugar, had insisted on payment in cash. The tendency, therefore, had been to drain the island of specie, increase the price of food and supplies, and force North Americans to buy their homeward cargoes of sugar from the French. Having grown accustomed to this practice and finding in the French islands a good market for their produce, many merchantmen from the North established direct connections with French houses in Martinique,

[15] Worsley to Board of Trade, Nov. 16, 1724, C.O. 28: 44, f. 116-117. Copy in C.O. 28: 18, W 38. See also Worsley to Board of Trade, July 16, 1723, C.O. 28: 17, V 116, where he declared that the trade went on through neutral ports and that it was impossible for him to break it up. Mar. 27, 1731, Worsley wrote to Newcastle that about a month before, the coast guard of Martinique went into "Carnache, a Port of St. Lucia, where Vessells generally resort to carry on a Clandestine trade with the French Inhabitants of Martinico, and meeting with a great many English Vessels, he Seized them all." C.O. 28: 45, f. 101.

Santo Domingo, and other French islands. By 1730, the English discovered that a large portion of British shipping, and Irish as well, had drifted away to the French. The result of this diminution of shipping was to raise the rates of freight paid by Barbadian planters on their exports. It seems probable that the amount of provisions and supplies diverted from Barbadoes was roughly equal in value to the value of foreign sugar actually excluded from Barbadoes as a consequence of the act of 1715. Finally, as Worsley observed, the duties paid the Crown, both in Barbadoes and in England were less after 1715 than before the passage of the act.[16] Similar opinions as to the disastrous results to Barbadian commerce from the act of 1715, though rare, are to be found among the writings of the time. A pamphleteer, in 1731, remarked that, during the era of virtually free trade, Barbadians got the sugar trade almost entirely into their own hands "or at least had made themselves the common channel of that Trade. But in 1715 they thought fit to lay a duty on the French and Dutch Sugars and molasses, to great to be supported: and by that means lost that trade

[16] Gov. Worsley to Board of Trade, July 7, 1730, C.O. 28: 21, Y 41. The passage referred to in the text is as follows: "I wish the People of this Island have not given some Occasion to the rise of this trade, by passing an Act in 1715 which was confirmed 17th of October 1717 for the laying a Duty of 12s 6d for every hundred weight of Muscovado Sugar, and £1 5s for every hundred weight of improved Sugar, which should be imported from Martinico, and other places not under his Majestys Subjection, and so in proportions upon Several other Commodities which were not of the Natural Product, growth, and Manufacture of some of his Majestys Colonies. The Occasion of the making of this Law, was the great Quantity of Sugar, Mellasses, Rum, Cotton, Ginger, and Alloes imported from those places, which it was apprehended would very much lessen the Value of the Manufacture of this Island. Yet it has been observed that the Current Cash in this Island was much more then, than it is now, for as there was then a greater quantity of Sugar here, they [Northern colonists] did not sell their provisions for money in order to go else where for a Loading and provisions were Cheaper and there was more Sugar to pay to Dutys to the Crown here as well as in England, besides the employing of more ships."

entirely, excepting what is carried on clandestinely, which is very common.'"[17] Another writer on the West Indies, in 1740, remarked that the act of 1715 "turned the Channel" of the sugar trade from Barbadoes, "an Oversight they will never recover.'"[18]

The trade to the French from Antigua, St. Christopher, Montserrat, and Nevis appears to have been much the same in character and extent as that from Barbadoes to Martinique. Also the method by which the planters sought to suppress it was the same and was copied from Barbadoes.

During the later years of the Spanish War, there were several reports of direct trade between the Leeward Islands and Martinique.[19] Just north of Antigua, the Virgin Islands were in a state of disorder and neglect, and offered a permanent field for illegal trade with the enemy.[20] It was a common practice for Barbadians, Leeward Islanders, and North Americans to clear for Anguilla or Spanish Town in the Virgin group, take certificates of entry there, and then reship their cargoes to St. Thomas. By this means the French and Spaniards were supplied with provisions and naval stores.[21] Thus through intermediate islands, and by a direct trade, large amounts of foreign sugar and other products came into Antigua and were reshipped to England as British

[17] Anon., *The Case of the Northern Colonies*, London, 1731, in C.O. 323: 9, M 29.

[18] Anon., *The Importance of Jamaica to Great Britain*, London, 1740, p. 58.

[19] B. T. Jour., June 12, 1710, C.O. 391: 22, p. 10. A letter from Perry, surveyor general of the Leeward Islands, *ibid.*, p. 53, read Aug. 25, 1710; Richard Oglethorpe to Lord Dartmouth, Antigua, May 19, 1710. The letters are in C.O. 152: 9.

[20] Petition of Capt. John Walton, lieut. gov. of Virgin Islands, to the queen, Jan. 1710, C.O. 323: 10, O 99.

[21] John Walton, late lieut. gov. of Virgin Islands, to Board of Trade, received and read Aug. 30, 1715, C.O. 152: 10, O 98.

goods. On November 17, 1715, Antigua passed an act[22] laying protective duties upon all such commodities imported. This was patterned after the Barbadoes act of eight months before. On June 19, 1716, this law was repealed and in its stead was passed an act absolutely forbidding the importation of French or other foreign sugar, rum, molasses, or cotton.[23] Governor Hamilton hoped that the Board of Trade would recommend this act for approval by the Crown.[24] In soliciting such recommendation from the board, the colonial agent, Nevine, described Antigua as being in a distressed condition. He said that, in the previous twelvemonth, the island had not made one half its usual amount of sugar, and that in the present year it would make only about one-fifteenth its usual complement, and the prospect for the next year was no better. All this was owing to an excess of dry weather. There was great scarcity of provisions for slaves and cattle; many negroes had been sold for lack of food. Some planters, he continued, had left the island and others were planning to go to the Northern Colonies. And finally, he claimed that Antigua was suffering from the low price of sugar and hoped to increase it by the exclusion of foreign goods.[25]

Of greater interest, however, is the criticism passed upon the policy to which the planters were now resorting. The conflict that accompanied the adoption of this protective policy was between the landed class on one side and British and colonial merchants and shippers on the other. It was in diminutive form an illustration of that

22 Acts of Antigua, C.O. 8: 4. Observations of the Customs Commissioners thereon, in letter from Sec. Carkess to Popple, Dec. 24, 1717, C.O. 152: 12, P 58.

23 Acts of Antigua, C.O. 8: 4; B. T. Jour., C.O. 391: 26, p. 234.

24 Hamilton to Board of Trade, July 12, 1716, C.O. 152: 11, O 193.

25 Memorial of Nevine to Board of Trade, received May 27, 1717, read July 1, 1717, C.O. 152: 12, P 122.

often recurring antagonism in modern English history between the agrarian and commercial classes. We may, therefore, consider rather fully the observation made on the situation at this time by Charles Dunbar, the surveyor general of customs for Barbadoes and the Leeward Islands. Dunbar had a seat in the assembly and heard the arguments for and against the act of 1716. His report to the commissioners of the customs, December 24, 1717, was as follows:

They who appeared for the Act were the Gentlemen Planters they who stood up ag^{t} it were either concerned in Trad themselves or well wishers to it. The Planters urged that the Importation of Sugar lowered the price of Sugar on this Island that it over stockd the Market in England & consequently sunk the price of it there also, that by an open Trade with the French Islands we supplyed them with Provisions & Negroes & by that means assisted our great Rivals in the Making of Sugar to fix & extend their Setlem^{ts} which in process of time would tend to a Manifest injury of all the British Sugar Colonies. Thus for the Planters. The Traders replyd that the Credit of this Island had been long ruined by keeping up here the prices of Sugars above their real value, that it could avail no Man to sell off his Sugars at a fourth part more than they were worth, when whatever was purchased with them was rais'd by the Trader 25 p Cent on that very Account. That the French comonly governed them selves nearer to the Market in Europe with respect to the difference of Exch^{a} which certainly ought to be the Standard of ours and therefore the price of Sugar was sunk no lower than it should be and the Importation of other Sugars could be no injury to any one pson, but on the contrary a free Trade was a very great advantage to Britain in General & to this Island in particular, for first it consumes many of our British Manufacturers, Secondly it encourages Trade & Navigation to English Subjects, & employs a Number of Our Men & Ships Thirdly it makes this Island a good Mart for Negroes, Provisions & other Goods necessary for the Support of the Collonies & Consequently we have not only the first Choice of any of these but that too on the easiest terms, 4th the Concourse of Ships which this draws to Us makes the freight low of our Sugars home an Advantage of it self sufficient to weigh down much greater inconveniencys than that of lowering the price of our Sugars here, which in truth is rather an imaginary than real injury.

That as to over Stocking the Market in England & sinking the price there of Sugars It must be considered that it is not the Consumption at home but the demand from abroad that keeps up the Market, Britain alone consumes but a small part of the produce of these Colonies and the rise of Sugars always is upon the encouragem^t^ there is for the Exportation of it & while Britain continues the chief Mart of Europe for that great & Staple Comodity whereon the Foreign Markets have often their dependence there can sure be no danger of over Stocking the Market which hitherto has risen in proportion the better it was stocked.

That as to supplying our great Rival in the making of Sugar with Negroes, provisions &c & thereby assisting them to carry on their setlmts to the detriment of the British Sugar Colonys, it must be confessed this objection would have more weight If it were once proved that the French could not any other way be supplyd but from Us But We know the Contrary & the truth of it is We supply them now only in part, but if we are out of humour and will forwardly throw that part out of our hands, the Dutch stand ready to take it into theirs who comonly know their own Interest & are wise enought to close with it and no doubt they would bless & greedily grasp the occasion that opend their way to a branch of Trade they have long secretly envied Us in & would be glad to dispossess Us of—The result of all this is very Natural, instead of British, *Dutch* Shipping would be employed, Dutch Manufactures vended, and Holland so famous for the Art of engrossing would in process of time rival, if not out vie Britain in being the Mart of Europe for that great & Staple Comodity Sugar But notwithstanding these & many other Argumts of less Moment the Bill pass'd the House & went home for the Royal Assent.[26]

[26] Charles Dunbar to Commissioners of the Customs, Barbadoes, Dec. 24, 1717, copy, C.O. 152: 12, P 96.

Dunbar also pointed out that the act would defraud the British revenue by a clause which declared that the four and a half per cent on all exportations from Antigua should no longer be paid to the Crown, and that it violated the 5th and 6th articles of the treaty of 1686 by penalizing any French ship that should ever touch at Antigua, ignoring the treaty provision whereby Frenchmen might put into British ports in case of storm and refit in case of distress. Influenced, it would seem, by the interests of English manufacturers, merchant shippers, and the revenue, the Board of Trade recommended to the Privy Council the disallowance of the law.[27] It was accordingly disallowed by an order in Council of May 26, 1719.[28]

Trade with the French from Antigua continued and was threatened only by the endeavor of Governor Hamilton to enforce the 5th and 6th articles of the treaty of 1686.[29] It was in vain that the commissioners of customs warned their officers in the colonies to watch for and report the reshipment of French sugar to England.[30] Trade with the French was "altogether by Stealth" while that with the Danes and Dutch was "publick and open."[31] "Upon their coming to anchor here," wrote Governor Hamilton of the smugglers, "and notice given to their correspondents, they immediately hire the Suger droging Vessels (light coasting craft for lightering), these they employ to unload their Sloops, and Shift the Suger into English Cask, bring them into Harbour, as if they brought them from other parts of the Islands, and take out Cockets for them at the Custom House, and Ship them as our own produce for Great Britain, tis true they

[27] B. T. Jour., May 7, 1717, C.O. 391: 27, p. 231. The attorney general had, however, approved the act, May 7, 1718, C.O. 152: 12, P 122.

[28] C.O. 152: 12, P 188.

[29] Hamilton to Board of Trade, Oct. 7, 1717, C.O. 152: 12, P 60.

[30] Sec. Carkess to Popple, Dec. 24, 1717, C.O. 152: 12, P 58.

[31] Hamilton to Board of Trade, July 20, 1719, C.O. 152: 12, P 207.

pay the four and a half per Cent here, but then his Majesty is defrauded of the alien Duty at home, and of great part of the enumerated duty upon Sugars Shipt to North America.'' Hamilton said it was impossible for the customs service to cope with the problem: ''For, the Custom house boat can no Sooner put to Sea but they have notice from their friends, and thereupon they immediately weigh anchor and to Sea, and keep out till some Signals are made that all is secure.''[32]

Once again, in 1721, Antigua passed an act laying prohibitive duties on the import of foreign sugar, molasses, rum, cotton, and ginger.[33] The objects of the law were to cheapen the cost of merchandise and provisions, raise the price of sugar, and embarrass the French. It was maintained that the act would not lessen the amount of shipping or revenue, since it would promote a great development of the island.[34] Dunbar, however, again condemned the policy of restricting commerce as he had done five years before. His arguments against the Antigua act constitute in fact an able defense of free trade in the West Indies and may be summarized as follows: (1) It would decrease the consumption of English supplies and manufactures, and deprive Antigua of a larger and better assortment of imports. (2) French sugar lowered the price of Antigua sugar, but it was unreasonable, he said, that the value of sugar should be governed by ''the prices of our Species here'' instead of by the British market. This attempt of Dunbar, by theoretical reasoning, to discredit the popular belief that French sugar would permanently depress English prices was the weakest point in his argument. (3) The French

[32] Hamilton to Board of Trade, June 30, 1721, C.O. 152: 14, R 4.

[33] Acts of Antigua, C.O. 8: 6; B. T. Jour., C.O. 391: 31, f. 78.

[34] Hamilton and the Council of Antigua to Charles Dunbar, June 26, 1721, C.O. 152: 14, R 5.

trade increased the colony's shipping and England's sea power, especially at times when English ships would have had to return in ballast or pursue a less profitable northern trade. The act would damage shipping and should be rejected on this ground alone "since the Navy of England is the Bullwork of Britain, and Merchants Ships Nurserys for the Navy." (4) The act would decrease the revenue by diminishing the importation of sugar into England. The French sugar reshipped from Antigua increased the four and a half per cent even if the English duties on foreign sugar were evaded. A duty one-fifth or one-sixth the present prohibitive one on foreign sugar entering Antigua might, he believed, yield a good revenue to the island. (5) The French trade tended to prevent French sugar from going direct to Holland, Hamburg, and other markets, and thus increased the Dutch and German demand on England for sugar. (6) The act, concluded Dunbar, would not embarrass the French since they would still be supplied by the Dutch, Irish, and North Americans.[35] In England it was also apprehended by the Customs Commissioners that the act would force merchants to carry French sugar direct to Europe instead of by way of the British colonies and England. This would reduce an important source of Crown revenue.[36] Whether the Antigua act of 1721 was confirmed or disallowed we cannot say, but in any case, the direct trade with Martinique and Guadeloupe continued to be a source of great annoyance to the government of the Leeward Islands.[37]

In Jamaica, at the beginning of the Spanish War, trade with the French was reported to be a matter of daily

[35] Charles Dunbar to Hamilton, 1721, C.O. 152: 14, R 5; Dunbar's letter of June 30, 1721, C.O. 152: 14, R 22. Compare his attitude in 1730, p. 217.

[36] Carkess to Popple, Aug. 23, 1721, C.O. 152: 14, R 22.

[37] Gov. Hart to Board of Trade, July 12, 1721, C.O. 152: 14, R 101.

occurrence. An act of the island in 1702 forbade the export of arms, ammunition, and provisions to the enemy.[38] In 1710, Governor Handasyd reported that he had brought actions under this act against several persons for clandestine trade. None of the prosecutions, however, had any effect, because juries always brought in verdicts of ignoramus, or to that effect. One sloop was condemned in the vice-admiralty court, but this decision was later reversed in England.[39] The returns for the slaves, provisions, and dry goods sent to the French were largely made in indigo brought from Santo Domingo and reshipped to England as of British growth.[40] French sugar from Santo Domingo also reached England through the same channel.[41] It was stated that several leading members of the assembly were concerned in this trade.[42] In 1715, the assembly and council, after some friction, agreed upon an act to prevent this fraudulent trade with Santo Domingo,[43] but the law proved ineffectual.[44] It was reported in 1717, that Jamaicans were furnishing the French with all the materials for sugar works. "Very lately," wrote a Jamaican to Sir Nicholas Lawes, "a Frenchman came from Hispaniolo [Santo

[38] Acts of Jamaica, C.O. 139: 9, no. 62.

[39] Gov. Thomas Handasyd to Board of Trade, June 4, 1710, C.O. 137: 9, M 13.

[40] Assembly Minutes of Jamaica, Nov. 11, 1712, C.O. 140: 12, f. 242. There was some talk of legalizing the trade for indigo at this time. See Gov. Hamilton to Board of Trade, May 15, 1712, for the David Creagh case, C.O. 137: 10, N 14, 16. It was stated that many Frenchmen were residing in Jamaica and draining the island of coin to buy indigo and other French goods. Jamaica Council Minutes, Mar. 22, 1714/5; C.O. 140: 13, f. 89. See also Council Minutes, Mar. 13, 1715/6, C.O. 140: 13, f. 410.

[41] Jamaica Council Minutes, Jan. 17, 1715/6, case of the sloop *Sea Flower*, C.O. 137: 13, f. 368.

[42] Memorial of Jamaica Council to Board of Trade, Mar. 13, 1715/6, C.O. 137: 11, O 48.

[43] Council Minutes, Dec. 7 and 16, 1715, C.O. 140: 13, ff. 159, 182.

[44] Memorial of Council to Board of Trade, Mar. 13, 1715/6, C.O. 137: 11, O 48.

Domingo], and bought all the Lead and Mill Work in the Island, so that several People were forced to Stop Work, till more came in the London Ships. The Trade is now grown Barefaced. They bring in Sugars as well as Indigo. There are now at North Side Three Ships from London, under pretence of loading there, which is in truth not able to Load one ship; but we are assured they go home freighted with French Sugar and Indigo. Besides the French have raised our Bitts i.e. 7½d to 10d per Bitt; by which we shall soon be Stript of all our Currant Money: for Gods Sake endeavour to put a stop to it.''[45] The feeling on the subject between the planters and the traders grew tense, and in 1721 the assembly directed that the governor "would be pleased to nominate and point out those people who countenance the French Trade.''[46] Another act was passed in 1721 for preventing commerce with Santo Domingo, but the Board of Trade reported against it, and the law was disallowed by order in Council of August 9, 1721.[47] The question remained a cause of friction between planters and traders.[48] Another act, in 1726, imposed heavy penalties for importing French sugar, indigo, and other products, and reshipping them to England.[49] Governor Hunter reported, in 1730, that Jamaica's chief foreign trade was with Santo Domingo where negroes and specie were sent in exchange for indigo and sugar. "Some People are of Opinion," said Hunter, "that this is not a Beneficial trade to this Island, others have different Sentiments."

45 Gov. Sir Nicholas Lawes to Board of Trade, London, Nov. 4, 1717. The passage is quoted by Lawes from a letter written to him from Jamaica, C.O. 137: 12, O 162.

46 *Journal of the Assembly of Jamaica*, II, 351.

47 Representation of the Board of Trade, Aug. 9, 1721; order in Council, Aug. 9, 1721, C.O. 5: 4, f. 27; C.O. 137: 14, O 29.

48 Memorial of Merchants trading to Jamaica to Board of Trade, May 31, 1724, C.O. 388: 24, R 145.

49 Acts of Jamaica, C.O. 139: 11, f. 42. Passed Mar. 9, 1725/6.

To conclude, the situation in Jamaica was similar to that found in other British sugar colonies. Merchants who were engaged in importing provisions, English manufactures, and slaves, were selling their surplus supplies to the French planters in Santo Domingo. In return, French sugar in large quantities passed through Jamaica and entered England as the product of that island. Repeated attempts by the planters to break up this trade, in order to secure cheaper plantation supplies and dearer sugar, altogether failed.

The policy of the English government toward the subject of trade with the foreign sugar colonies was for many years unsettled. A certain degree of protection was granted the British planters in 1657, when a preferential tariff admitted British sugar into England at one-half the duties paid on foreign sugars.[50] But when, in the early eighteenth century, great amounts of French sugar were slipping through the protective barrier and enjoying the advantages of British products,—even then the government hesitated long before taking any step to afford further protection to the planters. In the meantime, planters sought by their own protective acts to defend themselves. But these laws were difficult to enforce, and, in so far as they were carried out, largely defeated their own ends by driving supplies and shipping away from the British West Indies to the French. It was with great reluctance and a full knowledge of the dual interests involved, that the government finally, in 1733, fell in with the movement for protecting the sugar industry on an imperial scale. This attitude of the home

50 By an Act of June 26, 1657, a preferential rate was accorded to English sugar, *Acts and Ordinances of the Interregnum, 1642-1660*, ed. C. H. Firth and R. S. Rait, London, 1911, II, 1128. The planters had insisted on this, inasmuch as they were denied all trade with foreigners. Brit. Mus. Add. MSS. 11411, ff. 9-10. *Cf.* Beer, *Origins of the British Colonial System, 1578-1660*, pp. 408-409, and note.

authorities had already been foreshadowed partly by their disallowance of colonial legislation which sought to prohibit the French trade, and partly by the refusal of the Board of Trade to recommend for legislation any of the numerous proposals that reached Whitehall during the previous twenty years for its prohibition. The position of the government may be further determined from a study of their orders regarding intercourse with the French.

The Board of Trade, in 1716, referred the question of the legality of trade with the foreign West Indies to the consideration of the Commissioners of the Customs. The latter replied "That the goods of French, Dutch, Danish or other Foreign plantations are not by any Law past here prohibited to be Imported into his Ma^{ts} (Dominions) provided they are Imported in English Ships."[51] But upon examining the 5th and 6th articles of the treaty of 1686 with France, the Board of Trade found that both nations had agreed to the policy of mutual prohibition. Thereupon the board sent to Secretary Methuen to inquire if the treaty was still in force.[52] Methuen answered that the treaty was in force, and that Governor Lowther and the other governors should be instructed by the Board of Trade not to suffer the English to trade with the French or the French with the British settlements.[53] On May 16, 1717, therefore, the Board of Trade issued a circular letter to the colonial governors, reciting the substance of the 5th and 6th articles of the treaty of 1686 and commanding them to "take particular care for the Future that the forementioned Treaty be observed and put in Execution, and that no illegal Trade be carried

51 Carkess to Popple, C.O. 5: 866, V 119, received Nov. 20, read Nov. 21, 1716.

52 B. T. Jour., Feb. 6, 1716/7, C.O. 391: 26, ff. 188-189.

53 Methuen to Board of Trade, Apr. 8, 1717, C.O. 28: 15, T 116.

on between his Majty's Island of Barbadoes (or other English Colonies) under your Government and the French Settlements in America, by any of his Majesty's Ships of War attending Barbadoes, or by other British Ships; As likewise that none of the French Subjects be allowed to trade from their said Settlements to Barbadoes."[54] It is worthy of note that, at this time, the board interpreted the treaty of 1686 as giving the English power to prohibit their own subjects from carrying on trade with the French. In compliance with the instruction drafted by the board, the colonial governors issued proclamations prohibiting all such "illegal" trade. Both Governor Shute[55] of Massachusetts and Governor Hunter[56] of New York issued such proclamations. Hunter remarked, however, in regard to his order that "what effect it may have in deterring men from it I cannot tell, . . . there being no Acts of Parliament forbidding that trade, or inflicting penalties for such, all I can do is to vex them, which has always provok'd the Spleen of some of the trading sort, in the meantime whatever is in my power to discourage it shall be executed." Governor Walter Hamilton of the Leeward Islands wrote that throughout all the islands in his group he should prohibit the trade, but expressed the belief that it was "almost impossible" for the officers to carry out such an order.[57]

Efforts to suppress the trade with the French islands from Ireland and New England, by authority of the

[54] Board of Trade to Gov. Robert Lowther of Barbadoes, C.O. 324: 10, ff. 113-114. Similar letters were sent to the other governors, B. T. Jour., May 16, 1717, C.O. 391: 26, f. 259.

[55] Shute to Board of Trade, Nov. 9, 1717, C.O. 5: 866, V 161. He sent the order of the Board of Trade to Rhode Island and Connecticut.

[56] Hunter to Board of Trade, C.O. 5: 1051, Bb 115. *Cf.* also Hunter to Board of Trade, Nov. 3, 1718, C.O. 5: 1051, Bb 179.

[57] Walter Hamilton to Board of Trade, Nevis, Oct. 7, 1717, C.O. 152: 12, P 60.

treaty of 1686, proved quite disappointing. From Barbadoes, where the treaty had been reinforced by legislation, Northern and Irish traders who had used the island as an entrepôt now turned away to engage in direct trade with the French.[58] Knowing of this, the Board of Trade wrote, in 1722, to Lord Carteret that they were "apprehensive the Laws at present in Force for regulating the Trade and Navigation are defective in this particular, We thought it necessary to lay the same, before your Lordship, that a proper Remedy may be thought of to prevent this growing evil," and desired that he might command the lord lieutenant of Ireland and the governor of New England to "prevent the Progress thereof as far as in them lies."[59] It should be noted that what the Board of Trade called "pernicious" and a "growing evil" was not the sale of *English* manufactures, supplies, and slaves for French sugar which came into England as British, but the trade in *Irish* food stuffs, and *North American* provisions, lumber, and live stock for French sugar and molasses.

The grounds upon which the Board of Trade recommended the suppression of the North American and Irish trade to the French islands appear to have been political rather than economic.[60] It seemed advisable for a time, at least, to run the risk of sacrificing the industrial interests of North America, Ireland, and British manufactures for the sake of weakening the power of France in the West Indies. For at this time the board was in possession of authoritative opinions on the damage that

58 Gov. Worsley to Board of Trade, June 13, 1724, C.O. 28: 18, W 28.

59 Board of Trade to Lord Carteret, Aug. 15, 1722, C.O. 324: 11, ff. 1-2.

60 *Cf.* the wording of the instructions to Gov. Burnet of New York: "By all possible methods to endeavour to hinder all trade and correspondence with the French whose strength in the West Indies gives very just Apprehensions of the Mischief that may ensue, if the utmost care be not taken to prevent them." C.O. 5: 1053, Cc 145.

would be inflicted on New England in suppressing the trade between North America and the French West Indies. We also know that the board, in their representation[61] of September 8, 1721, had admitted the necessity of such trade on economic grounds, although it was detrimental to the British West Indies. But to offset the damage likely to result from its abolition, they recommended the promotion of naval stores in North America.

Neither the Northern Colonies nor Ireland could be induced to act upon the suggestion that the treaty of 1686 be enforced by colonial legislation. Finally, in 1728, the construing of trade to the French by English subjects as illegal by virtue of that treaty was found to be unwarrantable. The attorney general and solicitor general, in that year, pronounced the opinion: "That it was not the intent of this Treaty to provide that either of the contracting powers should seize and confiscate the Ships and Goods of their own Subjects for contravening the said Articles; and if such intention had appeared We are humbly of opinion that it could not have had its' effect with respect to his Maj's Subjects, unless the said Articles had been confirmed either by Act of Parliament of Great Britain, or by Acts of Assembly within the Respective Plantations."[62] They suggested that the instructions to the governors in this matter be so revised as to make clear the distinction that they had power under the treaty to stop Frenchmen holding intercourse with English colonies, but not to prevent British subjects trading to the French. In accordance with this opinion

61 C.O. 324: 10, ff. 296 *et seq.*

62 Report of Attorney General Yorke and Solicitor General Talbot to Board of Trade, June 3, 1728, C.O. 323: 8, L 94. The Board of Trade had requested their opinion on the legality of the instructions to the governors on the matter in 1717. Correspondence relative to the dispute about the interpretation of the treaty is to be found also in Brit. Mus. Add. MSS. 22671, ff. 120-140.

the committee of the Privy Council declared that the governors had been acting upon a mistaken sense of the treaty in proceeding to condemn ships and cargoes of British subjects for trading to the French plantations; the Treaty "could only entitle His Majesty's Governours to condemn French Ships trading to our Plantations, there being no law to justify the Condemnation of the ships belonging to his Maj^{tys} Subjects for such Trade." The committee ordered the Board of Trade to draw up a revised instruction to prevent the like mistake for the future. The Board was also to consider what laws might reasonably be passed in the colonies for restraining British subjects from importing such French products as might interfere with British trade.[63]

We have now reached the point where only new legislation by the imperial parliament could meet the supposed evils of international trade in the new world. In 1730, the West Indians started a propaganda in that direction, which culminated in the Molasses Act of 1733.

[63] Order of a committee of the Privy Council to Board of Trade, June 6, 1728, C.O. 323: 8, L 95.

CHAPTER XI

THE MOLASSES ACT, 1730-1733

The attempt of the West Indians to discourage English trade with the foreign sugar plantations by colonial acts excluding the products of those places from the British West Indies had, as we have seen, completely failed. Indeed the British West Indians had only hurt themselves by driving many North Americans away from them to trade exclusively with the French. Three tendencies unfavorable to the British West Indies grew out of this situation: freight rates increased, Northern supplies became more expensive, and the total loss of the North American market for English sugar, molasses, and rum became imminent. Besides, the French plantations were benefiting by all these advantages which the English were losing. French sugar was fast driving the English product from the markets of Europe, and now it had invaded the British colonies. Moreover, an unknown but presumably great amount was being carried by Northern vessels direct to England, or by way of American ports, and entered there as of British growth. Thus the royal revenues were being cheated to an untold extent by the non-payment of export duty in the West Indies, of enumeration dues on intercolonial trade in sugar and molasses, and of the duty on foreign sugar entering England.

That there was real economic depression in Barbadoes in 1730 there can be no doubt. Apart from the evils arising from the North American trade to the French, Barbadoes was suffering from soil exhaustion. The occupation of all the land for sugar culture had seriously diminished the raising of provisions and meat. The timber both here and in the Leeward Islands was practically all cut off. Barbadoes was extraordinarily dependent, therefore, upon outside sources for supplies. While the decline of Barbadoes was a topic for comment during the preceding half century,[1] the evidences for it in the decade from 1725 to 1735 are more impressive than ever before. The decrease in the exports of Barbadoes and the Leeward Islands is reflected in the marked falling off in revenue from the four and a half per cent export duty.[2]

[1] See Gov. Kendall to Sec. of State, Apr. 4, 1691, C.O. 28: 37, ff. 137-138; same to same, July 12, 1693: "the whole island is a Spectacle of Poverty and Misery," *ibid.*, f. 196; agents and others to the king, Jan. 31, 1700/1, on the decline of shipping to Barbadoes, C.O. 29: 7, ff. 265-266; William Bridges and Robert Heysham, agents, to Board of Trade, Oct. 7, 1702, on the emigration of small freeholders to North America "where provisions are more plenty," C.O. 28: 6, H 9; petition of Barbadoes planters to the queen, 1702, C.O. 28: 38, f. 25, to the same effect; Sir Bevill Granville to Board of Trade, July 10, 1706, C.O. 29: 10, f. 93; Raynes, Bate, and Stewart to Royal African Co., July 28, 1708, on decline in the value of Barbadoes, Treas. 70: 8, f. 72; Col. Sharpe before Board of Trade, July 22, 1715, B.T. Jour., C.O. 391: 25, ff. 187-188; Sharpe's address to council and assembly of Barbadoes, May 4, 1715, on the declining fertility of soil, C.O. 28: 14, T 44. See Chart XI opposite p. 244 for balance of trade for Barbadoes, etc.

[2] 1715-24. An Account of the Produce of the Duty of 4½ per Cent in Barbadoes and the Leeward Islands, for nine Years since 1715 (shillings and pence omitted):

		£
From Christmas 1715 to Christmas	1716	9721
	—1717	10736
	—1718	13411
Net Money paid to the Rec. Gen'l	—1719	8679
in London by the Sales of the	—1720	3916
King's Sugars & other Goods sent	—1721	8237
home to England and by Bills of	—1722	2900
Exchange remitted hither within	—1723	2200
the time above mentioned	—1724	4000
		63800
Out of which has been paid for freight, customs, & other incidental expences		17646
Net Money paid into the Exchequer in the nine years 1715-24		46153.

(C.O. 28: 18, W 68.) See also Chart XIII opposite p. 302.

On the other hand, the quantities of sugar imported into England and recorded as the growth of Barbadoes and the Leeward Islands do not show a corresponding decrease.[3] Probably this indicates that a considerable

Compare also the decline in revenue from exports at Barbadoes between the periods 1723-1728 and 1729-1734:

Average Produce per year of the 4½% in Barbadoes on Sugar, etc.

Commodities	The six years 1723-1728 inc. £	The six years 1729-1734 inc. £	Difference less £
Rum & Wool . . .	2164	2036	128
Musc. Sug.	8009	3796	4213
First White	31	7	24
Second White . . .	50	44	6
Third White . . .	116	13	103
Fourth White . . .	—	—	—
Coarse Clayed . . .	1412	980	431
Ravel Cotton . . .	33	53	—
Vine Cotton	91	77	14
Scraped Ginger . . .	208	178	30
Scaled Ginger . . .	5	1	4
Aloes	19	14	5
Totals	12138	7189	4960
			Gained 20

C.O. 28: 25, Aa 60.

[3] An Account of what Quantitys of Sugar has been Imported from the British Sugar Islands respectively into England

From	From Christmas 1726 to Christmas 1727			1728			1729			1730		
	cwt.	qrs.	lb.	cwt.	qrs.	lb.	cwt.	qrs.	lb.	cwt.	qrs.	lb.
Antigua	96112	0	25	187260	1	2	205532	2	25	182,277	2	17
Barbadoes	164013	2	10	238664	0	18	159843	2	22	250,075	2	25
Jamaica	176230	1	23	271665	1	10	354686	3	10	319,456	1	25
Mountserat	32689	3	19	50635	3	7	36316	0	25	44,751	1	8
Nevis.	30233	2	22	50293	3	9	71540	1	12	66,059	1	17
St. Chris.	141366	1	10	165642	2	7	158728	0	26	162,407	0	6
W. Ind. in Gen'l	2485	1	0	4397	2	22	3430	0	3	1,742	1	23
Total	643131	1	25	968499	2	19	990078	0	11	1,020,770	0	9

Custom House
London
8 February 1731

John Oxenford
Ass. I. Genl.
C.O. 28: 22, Y 118.

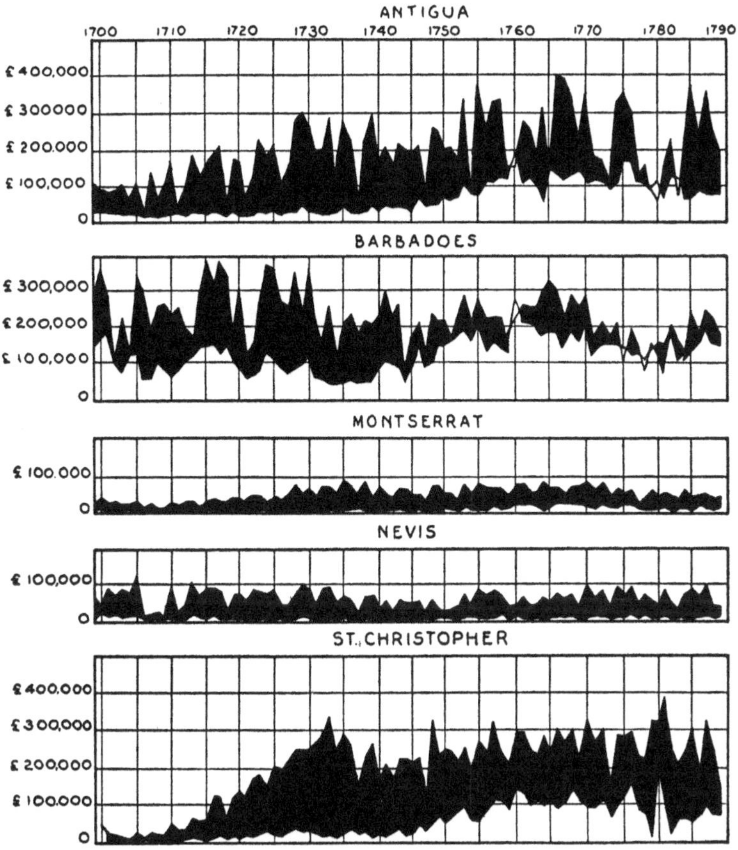

CHART XI. THE BALANCE OF TRADE BETWEEN BARBADOES, THE LEEWARD ISLANDS AND ENGLAND, 1699-1789

The above chart represents the balance of trade, in the eighteenth century, between Barbadoes, the principal Leeward Islands and England. Black areas represent balances in favor of the colony; enclosed white areas, which are very few in number, represent balances in favor of England. Of the black area, the upper margin represents the value of all exports from the colony to England and the lower margin the value of all imports into the colony from England. Of white areas, where the balance is against the colony, it is the upper margin that represents imports into the colony from England and the lower that represents exports from the colony to England. Based upon tables compiled from Custom House Accounts, Ledgers of Imports and Exports.

amount of French sugar was carried directly to Great Britain and falsely entered as English sugar. The excessively heavy duties on the export trade of Barbadoes and the Leeward Islands must have been an additional temptation to many a shipper to smuggle in the cheaper French sugar. For besides the £10,000 yielded by the four and a half per cent export duty on sugar in Barbadoes, about £25,000 a year was paid in duty upon the importation of Barbadian sugar into England.[4]

It was universally agreed that Barbadoes was now far less fertile than the French plantations. Thirty acres in the French islands, it was stated in 1717, could be worked with thirty to forty negroes and a few horses and cattle, which in Barbadoes would require one hundred and fifty negroes, fifty or sixty cattle, and twelve horses.[5] The lands of Barbadoes and the Leeward Islands were almost entirely cultivated and, except for Jamaica, practically no virgin soil remained in the British West Indies.[6]

4 Petition of Duke of Montague to the king, Jan. 23, 1727/8, C.O. 28: 19, X 16. Barbadoes was no doubt too heavily taxed at all times. Among the colonial expenses was one of £8000 which had been paid annually to Governor Worsley since 1722. There was probably much truth in the statement of the assembly that "the Poor inhabitants of this Island . . . are already but too Sensibly impoverished and reduced to great Extremities, not only by heavy Taxes, but the Calamitys of a lost, and decayed Trade." Address to Gov. Worsley, Nov. 25, 1730, C.O. 31: 19, f. 9.

5 Agents of Barbadoes to Board of Trade, received Sept. 25, 1717, C.O. 28: 15, T 127. *Cf.* Thomas Ekines to Hans Sloane, London, May 2, 1724, on exhausted soils, Brit. Mus. Sloane MSS. 4047, f. 170. Governor Worsley wrote in 1723 that one negro in Martinique could cultivate as much as two in Barbadoes. C.O. 28: 17, V 119.

6 Gov. Worsley to Board of Trade, Oct. 18, 1724, C.O. 28: 18, W 51; Gov. Hart to Board of Trade, July 12, 1724, C.O. 152: 14, R 101; Francis Freelove to William Wood, July 1, 1729, C.O. 28: 21, Y 10. The lower price of French sugar was owing in the main to their superior soil. "It is to this Cause only that the Fall of Sugars is owing; the French have the Advantage of us in their having better Ground, as well as better Management." Captain Fayrer Hall's *Evidence before a Committee of the House of Commons*, April 1731, C.O. 5: 4, f. 273.

Herein lay the fundamental cause for the lower price of French products.[7]

Probably long before the middle of the eighteenth century, part of the exhausted sugar land in Barbadoes had been abandoned or turned into pasture.[8] The decline of the island's productivity may be measured by comparing its estimated yield of 22,769 hogsheads of sugar in 1736 with an annual average of 13,948 hogsheads during 1740-1748, and 9554 hogsheads in the years 1784 to 1786.[9] The deceptive feature about most official statistics of exports from Barbadoes and the Leeward Islands is that they include so much sugar that was really of foreign origin but was smuggled into England as of the growth of Barbadoes. The decline of Barbadoes was far greater than is indicated by English customs accounts.[10]

[7] "The principal Causes that concur in creating this Difference in the Price between English and Foreign Sugar, are the great Charges attending our Navigation, the high Duties imposed upon our Sugars at Importation, the Importation of French Sugars into Ireland, and above all the great Expense our Planters are at in cultivating this Commodity, some of our Sugar Islands being almost worn out, especially Barbados, where many more hands and much more manure are requisite than in the fresh lands lately planted by the French in Hispaniola and other Parts of the West Indies." Representation of Board of Trade to House of Lords, Jan. 14, 1734/5, C.O. 5: 5, f. 114. See also John Ashley, "The Settlement and Insolvency Bill," 1737, C.O. 28: 25, Aa 60.

[8] The governor of Barbadoes in 1747 wrote: "The whole Island is cultivated by the Inhabitants in some manner or other, except such parts as are worn out and Impoverished, that the Labour of Manuring the same would not be recompensed, of wch Sort there are several thousand Acres, part whereof the Owners of the Sugar Work Plantations Use as Pasture for the large Quantitys of Cattle they are obliged to keep for the Manure even of their best Lands and for carrying their Effects to Markett." Thomas Robinson to Board of Trade, Feb. 20, 1746/7, C.O. 28: 57, Bb 57.

[9] Edwards, *West Indies*, I, 352.

[10] The following apprehension in the assembly's address to Gov. Worsley in 1730 (Nov. 25) was to a considerable degree justified by facts: "This Fatall truth we are too sensible of, that unless thro His Majesty's most gracious Interposition to save this . . . colony, and by his most Glorious

That the decade from 1729 to 1739 was one of real distress for Barbadoes there can be little question. Aside from the controversial pamphlets of the period, which manifestly exaggerate every evil, there is much evidence that Barbadoes was undergoing a period of hard times. Prices for sugar were the lowest reached in the century.[11] And to deepen the gloom of the picture a violent and devastating hurricane swept over Barbadoes[12] in August

Influence, our Trade is revived, or our Taxes Lessened most of the Inhabitants, as many have Already done, must desert their habitations, quit the valuable utensils of their Sugar Works, and resort to some other of His Majesty's Colonys that are happily in a more flourishing condition.'' C.O. 31: 19, f. 12.

11 ''We do indeed labour under Disadvantages on Account of the low Prices which our Commodities bear at home.'' Samuel Barwick, president of council, in an address to council and assembly of Barbadoes, Dec. 7, 1731, C.O. 31: 19, f. 63.

12 On August 13, 1731, a violent tempest and hurricane began and lasted all that and the succeeding day, ''with the utmost fury, to the Inexpressable terror and imense damage of the Inhabitants, who have had not only a great number of their Corn Fields, Plantain Walks, Fruit and Timber Trees blown down, broken or torn up by the Roots, and their Canes damaged, but their dwelling Houses, Wind mills, Boyling houses, and other their best and most Substantial buildings, some of them wholly demolisht, and others overset, rent, uncovered or otherwise greatly damnified; And so general has the Calamity been, that there is scarce a person throughout the whole Island, but who has received a considerable loss by this dreadfull Storm, the Consequences of which are still more greevous, for that there is not in this Island (nor has been here for some Years since that pernicious Trade between the Northern British Colonies and the Foreign Sugar Colonies began) Lumber sufficient to repair a tenth part of the Buildings damaged by this Tempest.

''This scarcity of Lumber is one of the mischievous effects of that Trade, and great is the Number of poor Inhabitants who now have no place to lay their heads in, and ly exposed to all the Injuries of the approaching rainy Season for want of those Northern Supplies which our Neighbours the French are plentifully furnished with. So great is our present desolation that many of the poorer Inhabitants, unable to rebuild their ruined Houses, will be driven to quit the Island; And thus our Strength decays, and at the same time the Exhorbitant power of the French at our very doors, threaten us with instant destruction in case of War; For their Islands are full of Men and Arms, whilst the Inhabitants of this Island grow every day thinner, and want almost everything necessary for their Defence.''

of 1731, to be followed in 1733 by a withering drought.[13] We may be sure that Barbadians did not fail to portray these calamities for all that the facts would warrant. Such was the background of economic distress out of which emerged the movement for parliamentary aid.

The Assembly of Barbadoes, in their address to Governor Worsley, November 11, 1729, attributed the cause of the distressed state of the island to the deflection of the Northern and Irish trade from the British to the French and Dutch settlements. Worsley was urged to represent this to the British government and to permit the assembly to employ agents at public expense to solicit redress for the injury.[14] Worsley was not disposed to intercede with the home authorities in behalf of the planters. This governor's chief interest in the colony appears to have been to obtain a larger personal remuneration than Barbadoes could afford to pay. But his coöperation was not needed for, on July 7, 1730, Worsley wrote to Newcastle and to the Board of Trade: "Mr. Ashley Deputy Auditor General here has set on foot a Petition to his Majesty (which is Sign'd by himself, and many other Gentlemen) praying (as I am informed) that his Majesty would prevent the Importation of any Sugar, Rum, and Molasses

Representation of Barbadoes to Board of Trade, Aug. 27, 1731, C.O. 5: 4, f. 287.

[13] "The Terrible Prospect of having no Crops this Year by the Excessive Drought, The Number of People continually running off (Notwithstanding I take all the Care I can to prevent it) and the Miserable Condition and Poverty of the Island in General" made it impossible to bear the expense of fortifications. Gov. Lord Howe to Newcastle, Jan. 6, 1733/4, C.O. 28: 45, f. 159. Much of the correspondence of this period deals with economic distress. *Cf.*, *e.g.*, James Dottin, president of council of Barbadoes, to Newcastle, Aug. 2, 1735: "I can't help . . . Acquainting you of the very bad State this Island at present is in, the principal Planters daily declining and deserting it; Credit almost lost among us for want of a Currency, and the fear of losing their debts by the great Incumbrances the Inhabitants are under." C.O. 28: 45, f. 188.

[14] Council Minutes, Nov. 18, 1729, C.O. 31: 18.

from any of the French and Dutch Colonies into Ireland, or the Northern Colonies till those Commodities had paid the Duty in Great Britain." In the original petition the Barbadians also prayed that the sugar colonies might have direct trade to Spanish ports in Europe on payment of a one per cent export duty. The propaganda included in the beginning, therefore, not only a suppression of Northern and Irish trade with foreign sugar plantations but also a partial removal of sugar from the "enumerated list" and a lowering of export duties. Worsley agreed with the planters that the trade complained of was a detriment to Barbadoes. Speaking of the origin of the movement for reform he said: "This affair has been carried on by the particular Gentlemen concerned in it; and indeed if I had been Applyed to, I could not have concern'd my self in it, in that by His Majesty's 96th Instruction to me, which I published here, such a Trade is not contrary to the Sense of the Treaties, and that there is no Law that prohibits it, unless as to this Island the Law [of Barbadoes of 1715] above mentioned laying a Duty upon Sugars brought hither from Martinico &c. by the greatness of the Duty may not amount to a prohibition of that Trade. Many of the Gentlemen who have sign'd the Petition to his Majesty have contributed pretty largely by Bills of Exchange they have given Mr. Ashley, and Mr. Osborne for bearing the Expence in carrying on their point." How the promoters extended the organization of their movement to England is contained in Worsley's further remark that "Mr. Ashley has recommended Mr. Peter Le Heup to be one of their Agents, and has with three, or four more wrote to Mr. Forster to be the other, they give them each at first £500. Sterling by Mr. Osborne, and Mr. Ashleys Bills: the money gather'd upon this Occasion is betwixt £2 or

£3000.'"[15] Peter Le Heup and Samuel Forster were by act of assembly appointed colonial agents and granted salaries of £250 a year apiece. The council refused to approve this act, but these persons appear, nevertheless, to have served in furthering the planters' movement in England.[16] Le Heup and Forster at first employed John Sharpe, a London solicitor and brother of Governor Sharpe of Maryland and of Joshua Sharpe, Treasury solicitor, to intercede for the planters, but they soon dismissed him upon orders from Barbadoes; whereupon Ireland and the Northern Colonies immediately retained him to defend their interests.[17]

The proposals of the West Indians were formally presented to the British government in the autumn of 1730. Two petitions to the king in behalf of Barbadoes were

[15] Gov. Henry Worsley to Newcastle, July 7, 1730, C.O. 28: 45, ff. 65-67. Same to Board of Trade, July 7, 1730, C.O. 28: 21, Y 41. Worsley informed Newcastle that many vessels loaded with sugar from St. Lucia for Ireland; many also cleared out of Barbadoes for Newfoundland and the Isle of Man when really they went to the French West Indies. The French governors permitted the trade. "Sugar at Martinico, tho' not so fine as ours here, yet perhaps as good for the Sugar Bakers is much cheaper, then the Sugars of Barbados, and Mellasses is there worth about 4d p Gallon, and here 'tis 9d & 10d." If the French could not dispose of molasses to the English they would throw it away.

[16] "The Assembly has passed a Bill in which they choose Samuel Forster, and Peter Le Heup Esq^{rs} which by the Council was unanimously rejected, And in the same Bill they had appointed a Salary of £250 Sterling to each of them, and they have already received £500 each, and a considerable more has been subscribed for, in order to support their Petition in relation to the Sugar Trade, which is founded upon the prohibition of a trade that is not only allowed of, but were encouraged, by my 96th Instruction; yet at the same time they would make the world believe, they are almost reduced to want, tho' I am Assured, that there was £6000 Spent lately in two of the Parishes where the Elections were contested." Gov. Worsley to Board of Trade, Nov. 27, 1730, C.O. 28: 31, Y 89. See also Council Minutes, Nov. 26, 1730, C.O. 31: 19, f. 4.

[17] It became known that Sharpe was an agent of Worsley who was not a friend of the planters' movement. Worsley to Board of Trade, June 27, 1731, C.O. 28: 22, Y 103; Worsley to Newcastle, June 27 1730, C.O. 28: 45, f. 103.

received by the Privy Council and referred on November 23 to the Board of Trade for its consideration. The plantations committee of the Privy Council appointed January 21, 1731, as the date for the hearing of counsel on these two petitions and on three counter petitions from the Northern Colonies, and the Board of Trade was requested to be present at the same time. The three counter petitions were also referred to the Board of Trade,[18] and on December 1, 1730, all the petitions were read at a meeting of the board.[19] Of the two in behalf of the sugar colonies one was from the planters, traders, and other inhabitants of Barbadoes and was signed by one hundred and fifteen persons; the other was from merchants, planters, and others interested in and trading to the British sugar colonies and contained eighty-seven signatures, some of which were of English slave traders.[20] Both petitions complained of the pernicious trade from the Northern Colonies and Ireland to the French sugar colonies, which, the petitioners said, would ruin the British sugar islands and advance the French. They argued that the French had at least three advantages over the Barbadians in the marketing of their produce. First, the duties and expenses born by the French were less than those of the English planters. Secondly, British sugar could not be shipped to Europe except at a charge of twenty per cent more than the French, owing to ''enumeration'' and high duties. Thirdly, British sugar cost upwards of ten per cent more to North American consumers than French sugar. These facts, they said, would explain why the Northern Colonies ''are most

18 Vernon to Popple, Nov. 23, 1730, C.O. 28: 21, Y 50.

19 B. T. Jour., Dec. 1, 1730, C.O. 391: 39.

20 The original petitions are in C.O. 28: 21, Y 51, 52. The canvass for signatures to the Barbadoes petition had been conducted by John Ashley and Osborne. In England, the support of absentee planters and West India traders for the English petition was secured by Le Heup and Forster.

Supplied with Foreign Sugar,'' and why Ireland imported sugar directly from the French islands. The petitioners prayed, therefore, that foreign produced sugar, rum, and molasses might be required to pass through Great Britain and pay the foreign duties, or, that the British sugar planters might be accorded the same advantages the French have, that is, direct trade to Europe and lower duties. At possibly the same time, or a little later, a similar petition to the king, containing fifty-eight signatures, was received by the board in behalf of the sugar planters from the mayor, aldermen, merchants, and ship owners of Liverpool.[21]

The petitions in behalf of the Northern Colonies, which were also read to the board, December 1, disclosed the existence of an organization which, though not as extensive as the West Indian, was formidable. Richard Partridge, agent for Rhode Island and New Jersey, peti-

[21] The original petition, undated, is in C.O. 28: 21, Y 58. The men who signed, of whom many were slave traders, were

Chas. Goore
Tho. Steers
Sam. Powell
Hen. Trafford
Tho. Seel
Jas. Holding
John Clarkson
Edw. Parr
Wm. Penkett
Pet. Rainford
Edw. Trafford
Dan Danvers
Wm. Pole
Wm. Carr
Jas. Percivall
Jos. Manekly
Roger Brooke
Jas. Ansdell
Augustine Woodward
Robt. Dixon
James Sandiford
Tim^{o}. Gamon
Thos. Parker
Thos. Chafferr
Ra. Williamson
John Lyon
Wm. Crockit [?]
Wm. Evered
Edwd. Markland
John Parr
John Headman
Jos. Clegg
John Entwisle
John Brooks
Tho. Crockshutt
R^{d}. Windsor
Tho. Hughes
Edw. Kenyon
John Hughes
Hen. Clough
Rob. Armitage
Jos. Lomar
John Atherton
Geo. Tyrer, Mayor
Wm. Webster
Tho. Coore
Forster Cunleffe
Sam. Ogden
John Danvers
Wm. Marsden
John Pemberton
Bryon Blundell
John Cobham
John Earle
Tho. Seel, Jr.
Hen. Cottam
Edw. Ratchdale
John Hughes.

tioned, November 2, 1730, to be heard by counsel in defense of those colonies,[22] and Ferdinand John Paris of London prayed that the inhabitants of and merchants and traders to Pennsylvania might be heard in defense of a trade whose suppression would ruin thousands in North America, and especially in Pennsylvania.[23] The third petition, signed by forty-two London merchants and others trading to and interested in those colonies,[24] prayed that counsel might be heard for Massachusetts, New Hampshire, New York, and South Carolina. A petition to the king from Dublin merchants praying that their counsel might be heard against the West India petition was also referred to the Board of Trade and read February 2, 1731.[25]

After two postponements, the Committee of the Privy Council began a hearing on all the petitions February 2, 1731. But the counsel for the Northern Colonies, on plea that the petitions of the sugar interests had only just

22 C.O. 28: 21, Y 54. For Partridge's services see *Correspondence of the Colonial Governors of Rhode Island*, ed. by Miss G. S. Kimball, 1902, I, 19-89, *passim*.

23 C.O. 28: 21, Y 55.

24 C.O. 28: 21, Y 53. This petition was signed by

Benning Wentworth	Walter Newberry	Charleton Thruppe
Tho. Gainsborough	Richd. Allurd [?]	Jas. Church
Sam'l. Wragg	Wm. Terry	Tho. Eames
John Wilkinson	John Boll	Tho. Thomas
John Gerd. Moller	Tho. Plumstead	John Miles
Sam. Storke	Sam'l. Russell	Robert Scott
Phill. French	Hen. Forster	Wm. Hore
John Gilbert	Fra. Wilks	Rd. Brown
J. Bull	Sam'l. Waldo	John Wentworth
John Lloyd	Tho. Sandford	W. Shepperd
John Radburne	John Harris	Bernd. Eales
Tho. Bayly	W. Bury	Sam'l. Baker
Tho. Church	John Forster	Rd. Janeway
Richd. Devon	Ralph Gulston	John Mico.

25 Referred to Board of Trade from committee of Privy Council, Jan. 29, 1730/1, C.O. 28: 21, Y 58.

been sent to America, prayed for further postponement. Whereupon, the hearing was postponed until April 26, 1731. But on March 30, 1731, the West India party petitioned the committee that they might withdraw the three petitions—from Barbadoes, the English sugar traders, and the Liverpool merchants, as they had now made application to parliament for relief. An order in Council of April 8, 1731, granted the withdrawal of the three petitions.[26] Exasperating delays and a loss of faith in the usefulness of orders in Council and "additional instructions" to colonial governors determined the sugar colonies to turn to parliament as the only sure agency for reform in colonial affairs.[27]

The first application to parliament was made by the Barbadians, February 23, 1731. Their petition to the House of Commons complained of the trade which "enriched the colonies of other nations, so it was injurious to the trade of this kingdom, and greatly impoverished the British Sugar-Colonies."[28] On March 22, the committee to whom the petition had been referred made its report which, for its thoroughness in consideration of a colonial problem of the eighteenth century, is an extraordinary parliamentary document.[29] The house seemed

[26] C.O. 28: 21, Y 57, Y 80; *Acts of the Privy Council, Col.*, III, 222.

[27] On the colonial tendency in this respect see Prof. Andrews in *American Historical Review*, XX, 767, and note 20.

[28] *Parl. Hist.*, VIII, 856.

[29] *Commons Journal*, March 22, 1730/1, XXII, 685-689. Mr. Sisson in a testimony before the committee said that molasses could be bought in the English West Indies for 12d. per gallon. This was very much higher than in the French islands where Worsley said it could be bought for about 4d. per gallon. Sisson said the English planters were in great need of lumber, horses, and provisions, which were high there. "Some of our planters," he declared, "have already removed to New York, and others are going." Martinique already had more men, he believed, than all the British West Indies excepting Jamaica. This evidence was confirmed by three sea captains, Richard Nanfan, Robert Taylor, and James Nisbett. John Lewis, a sugar trader, testified that he had bought French sugar at

persuaded that the British planters were undergoing a genuine hardship owing to a scarcity of Northern supplies and a vanishing American market for their produce. Accordingly, it was ordered that Sir John Rushout, Mr. Bromley of Cambridgeshire, and Mr. Sandys should prepare a bill for the encouragement of the sugar colonies. The bill was reported March 26, and met every desire of the West India interest. It prohibited the importation of foreign sugar, molasses, or rum into Great Britain, Ireland, or any British dominion

12s. 6d. per cwt., which in Jamaica would have cost 25s. per cwt. Northern staves, he said, had increased at Jamaica from £6 or £7 to £20 per thousand. He knew of Northern vessels that had sold their cargoes in Jamaica for money which they had carried to Hispaniola to buy molasses. William Frazier testified that in November and December 1729 he saw about 30 New England vessels at Martinique and St. Lucia trading horses and lumber for molasses. New Englanders informed him that if they brought in 60 horses alive they paid nothing for their permission to trade. Frazier had paid the Northerners money at Barbadoes for cargoes, and they had taken it to Martinique to buy molasses. Horses and lumber at Barbadoes had increased 25 or 30 per cent in price in the last seven years. Barbadian rum had fallen in the same proportion. Patrick Blake, a planter of St. Christopher, gave similar evidence. Francis Blower and John Tomlinson had also witnessed the extensive trade from the French to New England. Richard Harris said that the fall in price of English sugar was due to the enormous increase in French production, which was made possible largely by the New England trade. Jeremiah Dunbar, Surveyor of H. M. Woods in America, stated that the French were getting the best of Northern timber, that while the king's forests were being exhausted the British West Indians were getting no benefit from them, but only the French. According to Dunbar, the Northern trade to the French was developing crude manufactures for export in New England, which in time would make that region industrially independent of the mother country. Also many ships were built in New England for the French. Mr. Forster mentioned the 4½ per cent export duty in Barbadoes and the Leeward Islands as a severe handicap to their trade; the export duty in the French islands was only 1 per cent. This, he said, made English molasses and sugar dearer than the French. Besides, there was the "enumeration duty" of 1s. 6d. per cwt. on brown sugar exported from one English colony to another which discouraged Northerners from purchasing in the British West Indies. These two duties constituted a burden of 12 per cent on the export of brown sugar from the British sugar colonies to New England.

and also forbade the exportation of horses or lumber to any foreign sugar colony.[30]

A momentous controversy now began in the Commons' committee and was echoed outside in a war of pamphleteers between the North American and West India interests. The Northern contention, that New England and the Middle Colonies must have access to a larger tropical market than the British West Indies could afford, was, as far as facts went, unanswerable. The productive resources of the English temperate zone colonies were far beyond the consumptive power of the British West Indies. To confine Northern produce to so narrow a market as the British sugar colonies alone would be ruinous to Northern dealers in lumber, provi-

[30] *Commons Journal,* XXI, 700. A printed text of the act is in C.O. 5: 4, ff. 257-259. March 30, the Commons were informed of the value of the exports from England to the sugar islands from 1714 to 1726, *ibid.,* 703. March 31, the act received the second reading, *ibid.,* 689. April 2, a petition from Francis Wilks, agent for Massachusetts and Connecticut, was read against the bill. April 6, the committee reported that they had heard counsel for and against the bill, *ibid.,* 718. April 9, the house were informed of the value of English exports to Maryland, Virginia, and the Carolinas, *ibid.,* 722. The same day, the committee in charge of the bill reported having heard further counsel on both sides. April 10, a petition against the bill was read from Samuel Wragg, agent for South Carolina, *ibid.,* 725. April 12, the bill was again reported to the house and certain amendments adopted, *ibid.,* 728. April 14, a petition against the bill was read from John Penn, Thomas Penn, and Richard Penn, in behalf of Pennsylvania. The petition was laid upon the table, *ibid.,* 732. The same day the bill was passed by the Commons, *ibid.* Among the evidence taken by the Commons' committee in charge of the bill that of Captain Fayrer Hall was printed and is in C.O. 5: 4, ff. 272-277. Hall opposed the bill of 1731 with the customary arguments of New Englanders. See also his article of the same character in *The Post Boy,* March 1, 1730/1, and his pamphlet entitled *Remarks upon a Book, entitled, the Present State of the Bill now depending in Parliament, concerning the British Sugar-Colonies.* Printed for J. Peele, 1731. In 1732, Hall was won over to the support of the West India interest. Richard Partridge to Wanton, Mar. 3, 1732/3, *Corr. of Col. Governors of Rhode Island,* I, 27. In *The Post Boy* of March 6, 1731/2, Hall published another article in support of the British planters.

sions, and live stock. Furthermore, the inability of the British West Indies to furnish the Northern Colonies with sufficient molasses to supply their rum distilleries would be equally disastrous. For the manufacture of rum was not only an important industry in the North, but slave traders, Indian traders, and fishermen were dependent upon an abundant supply of cheap rum. The proposed scheme, by depressing the prices of the staples of the North and raising the prices of its tropical imports, would seriously diminish its earning power. This ultimately would hurt England by decreasing the amount that North Americans could purchase of British manufactures.

To offset this argument, the English were asked to believe that Northerners obtained a considerable supply of European manufactures through the French West Indies.[31] It was also asserted, though without facts to support the statement, that the British West Indies could consume more supplies than North America could furnish them, and that, should the bill pass, New Englanders would not be confined to a glutted market. Physical geography, however, might have suggested to those acquainted with it a strong presumption to the contrary. Also, the assertion of the British sugar planters that they would be able to supply the American colonies with molasses in an amount and at a price that would not inconvenience Northern industries carried no conviction to practical merchants. While the West Indians, in their attempt to prove that the proposed reform would not harm the Northern Colonies, are weak and unconvincing, they were successful in persuading parliament of their own depression and of the certainty that the sugar bill, if enforced, would provide a sure cure

[31] Jeremiah Dunbar's evidence in Commons' committee, March 22, 1730/1. *Commons Journal*, XXI, 688.

for their distress and would check the prosperity of their French rivals.

The Sugar Bill of 1731 passed the Commons April 14. In the House of Lords, between April 14 and April 29, elaborate preparations were made for a review of the matter on the occasion of the second reading of the bill, April 29.[32] On the whole it would seem that the Lords gave more attention to the Northern side of the controversy. But the remainder of the session proved too brief for decisive action upon the bill, and on May 7 the session ended.

Up to this point the campaign for the sugar colonies

[32] *Lords Journal,* XXIII, 676-695 *passim.* April 15, the Lords requested the Board of Trade to submit Cadwallader Colden's report on the navigability of the St. Lawrence, a portion of the board's report of 1721 on trade to the French sugar colonies, the Massachusetts act of 1694 for the better government of the Indians, customs accounts of trade with the West Indies from 1714 to 1726 and with the Northern Colonies in the same period, the Barbadoes act of 1715, the order in Council confirming the same, and the 96th instruction to Governor Worsley. The customs accounts were delivered to the Lords April 22 and the other documents April 27. Too brief a time remained in this session for their proper consideration. April 21, the bill was read a first time. April 22, petitions against the bill were heard from (1) London merchants trading to New York, (2) London merchants trading to Virginia and Maryland, (3) Richard Partridge for New Jersey and Rhode Island, (4) Francis Wilks for Massachusetts and Connecticut, (5) Samuel Wragg for South Carolina, and (6) William, John, Thomas, and Richard Penn for Pennsylvania. Consideration of these petitions was deferred till April 29. Henry Caswell, a Queen Street merchant, and Walter Newberry, a Cornhill merchant, were ordered (Apr. 24) to attend at the second reading. Counsel for the bill as well as Jeremiah Dunbar and John Tomlinson were also ordered (Apr. 27) to attend on the same occasion. April 28, a petition from merchants to the Northern Colonies was heard and it was ordered that the Northern Colonies be heard by counsel at the second reading. John Elliott and Peter Le Heup were ordered to be present also. The treaty of 1686 between England and France was also submitted by Newcastle to the Lords in response to their request (Apr. 21). April 29, the second reading occurred, counsel against and for the bill were heard and some recent proceedings from the journal of the Massachusetts assembly were read. Further consideration of the bill was postponed. The session ended May 7, 1731.

had been conducted entirely by the Barbadian planters. It became apparent that the support of all the British sugar colonies would be needed to overcome the opposition or lethargy of the House of Lords. Accordingly, on August 27, 1731, the Barbadians made a second representation to the British government for relief.[33] Their statement of the situation was now confirmed by similar representations from Antigua,[34] St. Christopher,[35] and

[33] "To the Right Honble the Lords Commissioners for Trade & Plantations. The Humble Representation of the General Assembly of the Island of Barbados." Agreed to Aug. 27, 1731, C.O. 5: 4, ff. 283-289. Another copy is in C.O. 28: 22, Y 110. It was received by the Board of Trade Oct. 25, 1731, and read Oct. 28. The content is very similar to the petition of 1730. Barbadoes recommends that the import of French produce into British dominions be prohibited, *or* that the trade between the Northern Colonies and the French be abolished, *or* that duties be reduced in the English West Indies and sugar be removed from the enumerated list.

[34] "To the Honble the Lords Commissioners for Trade and Plantations. The humble Representation of the Lieutenant Governor Council and Assembly of Antigua." Received from Yeamans, agent for the Leeward Islands, Nov. 17, 1731. The document states that, since the peace of Utrecht, Antigua has "scarcely increased one Man; . . . Mountserrat and Nevis have much decreased." Whereas the population of Martinique has increased tenfold and their trade and sugar works in the same proportion. The French were equally prosperous at Guadeloupe, Hispaniola, St. Lucia, and Mariegalante. The trade from the Northern British colonies was at the root of this progress. Surinam's progress was also due to this "pernicious trade." With a better market and cheaper supplies they could produce considerably more sugar, etc. The representation was signed by John Burke, Tho. Stephens, John Frye, Jr., John Tomlinson, Edw. Chester, Wm. MacKinen, Steph. Blizard, Tho. Mathews, Tho. Watkins, Jacob Morgan, Jas. Parke, John King, Jacob Hubon, Rich. Oliver, Benj. King, Tho. Kerby, Geo. Thomas, Fra. Carlisle, Edw. Byam, Nathaniel Crump, John Hayes, and Geo. Lucas. Antigua and the Leeward Islands, according to Gov. Mathews (Aug. 31, 1734), had suffered by a blasting of their sugar crops which had increased during the years 1729 to 1734. C.O. 152: 20, V 46.

[35] "The Humble Representation of the President, Council, and Assembly of St. Christophers to the House of Lords." Sept. 24, 1731, C.O. 5: 4, ff. 218-221. An identical document of the same date was addressed "To the Right Honble the Lords Commissrs for Trade & Plantations," and received Dec. 2, 1731. C.O. 152: 19, T 115. The main argument is that the suppression of the trade from the Northern Colonies to the French will put an end to their mastery of the sugar trade. The Northerners also drained the

Nevis.[36] After long hesitation Jamaica also joined in the movement. The representations, at this stage of the controversy, instead of proceeding from private persons, were drawn up by the colonial governments in the sugar islands. In November 1731, the Board of Trade brought the subject of the representations to the attention of Newcastle, and desired him to propose that the consideration of the sugar trade be recommended to parliament at its next meeting.[37] In the winter of 1731-1732, the Secretary, Newcastle, and the Board of Trade were again bombarded with representations from Northern Colonies[38] and with answers from the West Indies.[39]

coin of the British West Indies. Also they obtained foreign manufactures from the French, which decreased the consumption of British manufactures. The signers were J. Spooner, speaker of assembly, John Griffes, clerk of the assembly, Jos. Esbridge, Cha. Payne, Chas. Pym, Wavel Smith, John Douglas, Abraham Payne, and Jos. Phipps, members of the council.

36 "To the Right Honourable the Lords Commissioners of Trade & Plantations. The Humble Representation of the Honble Michael Smith Esqr Commander in Chiefe of all his Majesty's Leeward Charribee Islands, and the Councill and Assembly of His Majesty's Island of Nevis." C.O. 5: 4, ff. 301-303, received Feb. 8, 1731/2, C.O. 152: 19, T 123. The "pernicious," "detrimental," and "destructive" trade was the "cause of the Rise and Great Improvement of the ffrench Islands." Lumber, horses, and fish were scarce and very expensive in Nevis. But the French and Dutch were well supplied. The French in the Caribbean were becoming very formidable to the English in military respects. The signers were John Dasent, speaker, Mich. Smith, Cha. Bridgewater, Ja. Symonds, Mich. Williams, Jas. Browne, Rich. Abbott, C. Brodbett, Daniel Smith, Tho. Pym, Sam. Clarke, Jos. Herbert, Geo. Webbe, Roger Pemberton, Robt. Pemberton, Jas. Earle, Tho. Stewart, and John Woodley.

37 Board of Trade to Newcastle, Nov. 9, 1731, C.O. 28: 40, f. 21; C.O. 5: 4, f. 24. Signed by P. Dominique, Ja. Bondenell, T. Pelham, and M. Bladen. The matter was again referred to Newcastle in the board's letter of Nov. 17, 1731, C.O. 5: 4, f. 25. Newcastle on Nov. 24, 1731, directed the board to report to him as soon as possible on the practicability of the planters' proposals. C.O. 28: 22, f. 111.

38 In this period, the following documents in defense of Northern interests were received. "To the Right Honble the Lords Commissioners for Trade and Plantations. Some Short Observations on, & Reasons in Answer to the three Representations from the Assembly of Barbadoes & the Islands of Antigua & St. Christophers, . . . for & on behalf of several of the

The Board of Trade seemed to be hopelessly perplexed by the magnitude and complexity of the issue. In reply to Newcastle's repeated request for their report on the

Northern Colonies, vizt. The Massachusetts Bay, Rhode Island and Providence Plantations, New Jersey, Connecticut, New York.'' C.O. 5: 4, ff. 222-228. Another copy in C.O. 28: 22, Y 112. This was received and read Dec. 20, 1731. On Dec. 20, 1731, the board received from Richard Partridge *The Case of the British Northern Colonies,* four printed folio pages. There are copies in C.O. 5: 4, ff. 260-261, 262-263, 264-265, and a manuscript copy in ff. 266-267. Another printed copy dated Dec. 10, 1731, is in C.O. 323: 9, M 29, and another in C.O. 28: 22, Y 113. This is a bitter attack on the extravagance and prodigal methods of British sugar planters. ''To the King's most Excellent Majesty. The Humble Representation of the President Council and General Assembly of the Colony of New York.'' Agreed to Oct. 29, 1731, received by the Board of Trade Feb. 14, 1731/2, C.O. 5: 1055, Dd 161. Another copy is in C.O. 5: 4, ff. 215-217. On Oct. 29, 1731, also, Rip Van Dam, president of the New York council, wrote a letter to the Board of Trade supplementary to the above, which is in Newcastle Papers, Brit. Mus. Add. MSS. 33028, f. 263. ''To the Right Honour^ble^ The Lords Commission^rs^ For Trade & Plantations. Reasons humbly Offer'd on behalf of the Province of Pensilvania in America, against the Restraint on One Side, & Monopoly on the other prayed by the Assemblys of the particular Islands Barbados, Antigua, & St. Christophers.'' Six and a half folio pages, C.O. 5: 4, ff. 297-300. The opposition of Virginia to the West India scheme is contained in a letter from the lieutenant governor, William Gooch, to the Board of Trade, Sept. 8, 1731, C.O. 5: 4, ff. 208-214. See also the ''Extract of a Representation from the Council of Virginia to the Board of Trade, in Answer to the Representation from the Sugar Islands.'' C.O. 5: 4, ff. 204-207.

[39] In defense of the Sugar Islands Peter Le Heup sent the Board of Trade, Dec. 29, 1731, ''A Reply humbly offered on behalf of the Island of Barbadoes to the Observations on their Petition and to the Address from the Northern Colonies,'' three folios, C.O. 28: 22, Y 115. Another copy is in C.O. 5: 4, ff. 241-245. The board wrote to Newcastle, Jan. 4, 1731/2, relative to the controversy, C.O. 5: 4, f. 35. From the Customs Commissioners the board tried to obtain facts on the trade between the West Indies and North America, Board of Trade to Carkess, Feb. 1, 1731/2. B. T. Jour., C.O. 391: 41, f. 42.

The following material was also received in defense of the sugar islands: ''The Case of the British Sugar Colonies,'' six and a half pages, C.O. 5:4, ff. 268-269. ''A Reply humbly Offered in behalf of the Leeward Islands and Jamaica to a Paper Intituled some Short Observations on and Reasons in answer to the three Representations from the Assembly of Barbadoes and the Islands of Antigua and St. Christopher complaining of the Trade

subject[40] the board answered, Feb. 10, 1732, that they had not made up their minds on a "matter of so much Nicety and Importance."[41] They sent him all the papers they had received on the subject, "which consist of many Allegations, but of Allegations only, and not of Proofs, Which has brought this Matter hitherto no further than to an Issue upon the Facts in Dispute between the opposite Parties."

For the second time, the House of Commons passed the Sugar Bill, March 15, 1732.[42] Next day the act was reintroduced in the House of Lords. During March and April, both sides to the controversy were again heard and copies of all the material the Board of Trade had received on the subject were submitted to the Lords. May 2, the Lords postponed action for a month but, on June 1, adjourned without further considering the bill.[43] The

Between the Northern Colonies and the French and Dutch plantations for and on Behalf of several of the Northern Colonies," twenty pages, received from Yeamans, agent for the Leeward Islands, Jan. 19, 1731/2, C.O. 152: 19, T 122. Another copy is in C.O. 5: 4, ff. 229-240. "Answers *to all* the Objections *made to the* Bill for supporting the Sugar Colonies," two and a half pages, C.O. 5: 4, ff. 270-271. "Observations on the Argument of the Northern Colonies that Sugar Cotton and Indigo are not permitted to be exported from the French and Dutch Islands," C.O. 5: 4, ff. 249-250. "Remarks on Objections to the clauses in the Sugar Bill as to the Onus Probandi," C.O. 5: 4, ff. 251-252. The writer attempts to justify the clause requiring the owner of a seized vessel to prove his innocence.

40 Newcastle to Board of Trade, Feb. 4, 1731/2, C.O. 28: 22, Y 117.

41 Board of Trade to Newcastle, Feb. 10, 1731/2, C.O. 5: 4, ff. 278-282. Among the enclosures is a representation from Nevis which had just been received. The board hoped to be excused from making a full report because the issue was about to be threshed out in parliament. The commissioners who signed this letter were Martin Bladen, Edward Aslee, James Bindenell, and Robert Bridgeman.

42 Jan. 26, 1731/2, leave was granted to introduce the bill. In February, counsel for and against the act was again heard. The vote on March 15 was 110 yeas to 37 nays. The Board of Trade were required to submit to the house copies of all papers it had received on the controversy. Newcastle to Board of Trade, Jan. 29, 1731/2, C.O. 28: Y 116.

43 *Lords Journal,* March 16, 1731/2, May 1732. See also *An Anony-*

mercantile power which had twice mastered the House of Commons was for the second time thwarted in the House of Lords.

For the third time the Commons passed the Sugar Bill on March 21, 1733.[44] In its last form the act contained one change, namely, British colonists were not absolutely forbidden to carry freight to the foreign West Indies, but their return cargoes of sugar, molasses, and rum were subject to prohibitive import duties. April 3, 1733, the bill was introduced in the Lords, petitions from nine Northern colonial agents were heard, and the bill was finally passed May 17, 1733.[45]

In the management of the British empire the passage of the so-called Molasses Act of 1733 marked the first triumph of the West India interest in England. Aware of their power, the planters continued to improve their organization and, until the emancipation of slavery just a century later, constituted an important factor in English politics. Lord Howe, who claimed to have promoted the planters' bill in England, became governor of Bar-

mous Letter to a Member of the House of Lords, 1732, two pages printed, C.O. 5: 4, f. 248. This was probably written by Fayrer Hall and is a defense of the bill by one who had been converted to the West India side.

44 *Commons Journal,* Feb. 22-Mar. 21, 1732/3.

45 *Lords Journal,* Apr. 3-May 17, 1733. The act is 6 Geo. II, c. 13, and is commonly known as the Molasses Act. It was to be in force for five years but was continued by 26 Geo. II, c. 32, until 1764, when it was made perpetual by the Sugar Act (4 Geo. III, c. 15) but with a reduction by one-half of the duty on molasses and sugar.

The Molasses Act imposed duties in all British colonies at importation on foreign sugar and paneles 5s. per cwt., on rum 9d. per gallon, and on molasses 6d. per gallon. No French sugar, molasses, or rum was to be shipped into Ireland. To assist the planters in recovering the European market a drawback of the entire import duty was allowed on British sugar re-exported from Great Britain within a year. To promote the export from Great Britain of English refined sugar an additional bounty of 2s. per cwt. was granted, making the total bounty on refined sugar for export 6s. per cwt. See also *Parl. Hist.,* VIII, 918, 902-1002, 1195-1200, 1261-1266.

badoes in the spring of 1733.[46] John Ashley, the leader of the West India movement, on February 26, 1732, was made a member of the Barbadoes council.[47] The Assembly of Barbadoes (Aug. 21, 1733) and the council (Sept. 4) adopted addresses of thanks to the king for the passage of the act in their behalf.[48]

The position of Jamaica, with reference to the movement which we have just reviewed, deserves especial attention. Jamaica was the "frontier" settlement in the British West Indies, its extensive areas were but sparsely settled, its natural resources were far from being exhausted, and in 1730 general prosperity prevailed throughout the island.[49] "The countrey in general as to the planting Interest is in a flourishing Condition, the Seasons lately having been better than for many years past," wrote Governor Hunter in 1728.[50] Again in 1731, he declared: "The Island is Richer in its produce but poorer in people than ever."[51] In the absence of that distress which existed in the older islands, Jamaica felt slight inclination to promote the movement for the better regulation of West India trade. The assembly of

46 The Barbadoes council declared to Lord Howe April 17, 1733: "But we now flatter Our Selves, there is still room to save us from ruin, since several Resolutions have already been agreed to by the British Parliament in Our Favour, which we are well assured, is in great measure owing to your Excellency's earnest Endeavours." Council Minutes, C. O. 31: 19. See Howe's letter to the Duke of Rutland, Jan. 28, 1733, in which he urged that the House of Lords pass the planters' bill. *Hist. MSS. Com. Rep.* XII, Pt. 5, p. 194.

47 Barbadoes Council Minutes, C.O. 31: 19, Apr. 18, 1732.

48 C.O. 28: 45, f. 150; C.O. 31: 19.

49 Gov. Hamilton, in 1712, referred to "indefinite vast Tracts of Land uninhabited" in all the parishes of Jamaica. C.O. 137: 10, N 14. Hurricanes, to be sure, had devastated Jamaica in 1722 and 1726. That of 1722 is described in C.O. 137: 14, Q 43, 44, 47, 48, 50. The effects of that of 1726 are described in C.O. 137: 16, R 64; 21, S 7. A great deal of destruction and suffering attended both. See also Chart XII, opposite.

50 Hunter to Board of Trade, Nov. 9, 1728, C.O. 137: 17, S 41.

51 Same to same, Oct. 8, 1731, C.O. 137: 19, S 145.

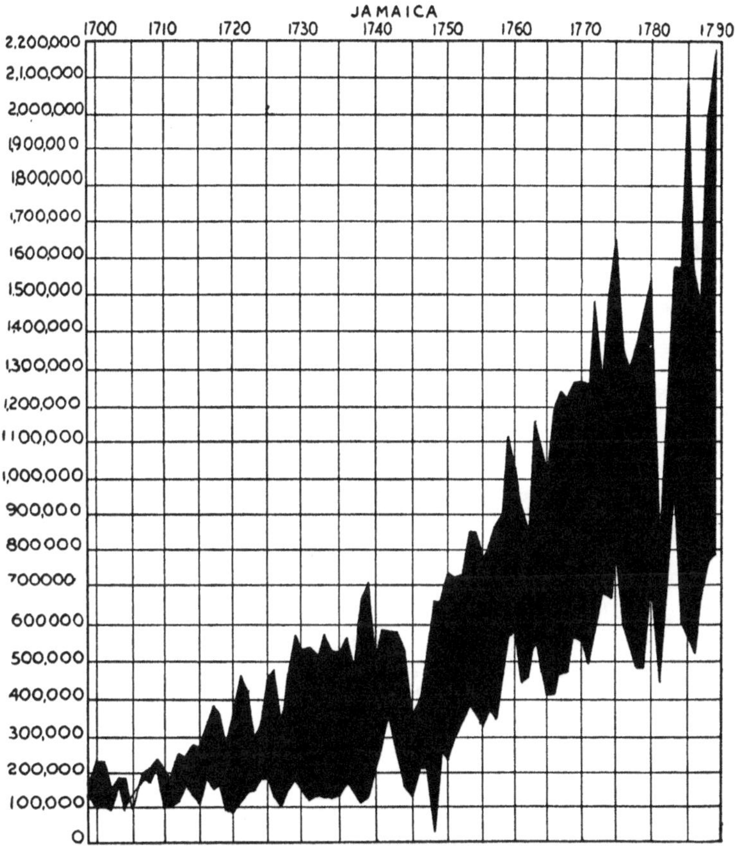

Chart XII. The Balance of Trade in Pounds Sterling Between Jamaica and England, 1699-1789

The above chart represents the balance of trade, in the eighteenth century, between Jamaica and England. Black areas represent balances in favor of Jamaica; the enclosed white area, in 1705, represents a balance in favor of England. Of the black area, the upper margin represents the value of all exports from Jamaica to England and the lower margin the value of all imports into Jamaica from England. Of the white area of 1705, where the balance is against Jamaica, it is the upper margin that represents imports into Jamaica from England and the lower that represents exports from Jamaica to England. Based upon tables compiled from Custom House Accounts, Ledgers of Imports and Exports.

1730 displayed no interest in the Barbadian propaganda. But a new assembly was elected in the spring of 1731 and the governor observed that "there are more Merchants in it than in the last."[52]

This assembly was anxious to commit Jamaica to the scheme to suppress commerce between British North America and the foreign West Indies. But the council and assembly could not agree on a common petition. Therefore, on November 27, 1731, each body sent an address to the king in support of some measure to stop trade between the Northerners and the French and Dutch planters. Two hundred pounds was appropriated to be used in England by Humphrey Morris to promote the object. Baron Scroope, Richard Harris, John Beecher, and Peter Day, all in Great Britain, were also chosen to solicit support for the cause.[53] Jamaica's contention was that a prohibition of Northern trade to the French would increase the sale of Jamaica molasses, encourage the settlement of her vast unoccupied areas, and stop the drain of specie occasioned by New Englanders taking coin to Santo Domingo to buy return cargoes.[54]

52 Same to same, Apr. 21, 1731, C.O. 137: 19, S 128.

53 *Journal of Assembly*, III, 39-45, Nov. 17-27, 1731. The addresses of the council and assembly are in C.O. 137: 19, S 152, 153, received by Board of Trade Feb. 14, and read Mar. 23, 1731/2.

54 "If the Island were Improved, there is Land uncultivated sufficient to make Sugar to serve all Europe." Gov. Hunter to Board of Trade, Dec. 24, 1730, C.O. 137: 19, S 124. In 1741, Gov. Trelawny estimated that two-thirds of Jamaica was uncultivated. To Board of Trade, Nov. 21, 1741, C.O. 137: 23, W 48. In 1752, the council and assembly declared there was room for at least 150 new sugar works producing an average of 1200 hhds. a year each beside molasses, and that old estates could produce at least a third more if sufficient slaves, lumber, live stock, and provisions could be had at low rates. Address to the king, Nov. 20, 1752, C.O. 137: 25, X 115. Soil exhaustion as an argument was not available for Jamaicans to any great extent in 1730. Twenty-five years later, however, the governor did allude to the "well wore" soil of St. Thomas in the Vale. Charles Knowles to Board of Trade, Dec. 31, 1754, C.O. 137: 28, Y 43.

Early in 1732, Richard Harris informed the Jamaicans that, until then, the propaganda had been carried on entirely by the other islands "with great murmuring that Jamaica had been so backward." As no funds from Jamaica had reached England, Harris was advancing his own money for the cause. But the assembly resolved to reimburse him and the other gentlemen what they should spend in the service of the island. November 15, 1732, the assembly voted that £200 should, if necessary, be borrowed at twelve per cent and be paid at once to Harris. Also a loan of £300 at thirty-seven and a half per cent was authorized for lobbying in England. The agents were instructed by the council and assembly to urge, not only the prohibition of trade with the French, but liberty to carry sugar directly to foreign markets and a reduction of import and excise duties on rum in Great Britain.[55]

While Jamaica's contribution to the movement of 1730-1733 was small, belated, and perhaps half-hearted, it was nevertheless important in unifying the sentiment of the planting interest and securing the consent of the House of Lords to the Molasses Act. On October 18, 1733, the council and assembly returned thanks to His Majesty for the passage of so encouraging a measure.

NOTE: Pamphlet Literature on the Controversy of 1730-1733. The text of the above chapter is based upon official letters and representations. But the controversy gave rise to a multitude of printed anonymous pamphlets probably written, in many instances, by the agents of the various colonies to influence parliament. The following is a partial list of such material. *Some observations upon Jamaica as to its natural history, improvement in trade, etc. By An English Merchant,* [London, 1731 ?, possibly written in 1726]. *A Letter to the Right Reverend the Lord Bishop of London* [1730], *with A Short Essay concerning the*

[55] *Journal of Assembly,* III, 106, Nov. 15, 1732.

conversion of the Negro-Slaves in our Sugar Colonies [written by Rev. Robert Robertson of Nevis, June 1727]. *The Importance of the British plantations in America to this Kingdom with the state of their trade, and methods for improving it; as also a description of the several colonies there,* Pp. 144, London, J. Peele, 1731, a clear and unbiased description of the trade connections to various parts of the empire by a merchant who had visited nearly all the trade centers, non-argumentative and supplied with valuable commercial estimates. The authorship was claimed by Captain Fayrer Hall in his testimony before the Commons' committee in April 1731. *Whitehall Evening Post,* April 10, 1731, containing a dissertation on the sugar question. *The Free Briton,* April 15, 1731, containing *A Letter on the Bill now depending in Parliament. The Daily Post Boy,* March 6, 1731/2, containing a letter by Captain Fayrer Hall on the controversy; *The Importance of the sugar colonies to Great Britain stated, and some objections against the Sugar Colony Bill answered. In a Letter to a member of the House of Commons,* Pp. 40, London, 1731, very hostile to the Northern colonists. *A short answer to an elaborate pamphlet, entitled, The importance of the sugar plantations etc.,* London, 1731, a reply to the preceding book by a Northerner. *The present state of the British sugar colonies consider'd. In a letter from a gentleman of Barbadoes to his friend in London,* Pp. 28, London, J. Wilford, 1731. The author, apparently a planter, describes the declining state of the English as compared with the French West Indies and urges that sugar be removed from the enumerated list for Ireland and southern Europe and advises the exclusion of foreign sugar, etc., from the British empire. *The British Empire in America Considered,* 1732, in a second letter from the same author. *Remarks upon a Book, entitulated, The Present State of the Sugar Colonies Consider'd. Wherein some consequences and effects of restraining our trade are examined,* Pp. 34, London, 1731. The author, a native Englishman engaged in West India trade, opposes a prohibition of trade to the French, and advocates free trade in sugar from the islands to southern Europe. *The Case of the Provinces of Massachusetts Bay and New Hampshire and the Colonies of Rhode Island with*

Providence Plantations, and Connecticut in New England, and the Province of New Jersey, with respect to the Bill now depending in the Honourable House of Commons, intituled a Bill for the better securing and encouraging the Trade of His Majesty's Sugar Colonies in America [1731]. *Captain Fayrer Hall's Evidence, before a Committee of the House of Commons, in April 1731, concerning the Sugar Colony Bill,* Pp. 12, in C.O. 5:4, ff. 272-278, endorsed February 10, 1731/2. Pp. 10-12 contain "*A Letter to a Noble Lord concerning the Advantage to the French Sugar-Colonies from the Northern Colony Trade,*" the same letter as appeared in the *Daily Post Boy,* March 6, 1731/2. *The Case of the Northern Colonies,* 1731. *The Case of the British Northern Colonies,* Pp. 3, [1731(?)]. Ascribed to John Sharpe, opposes the planters' bill, describes the luxury of West Indians, declares they consume less manufactures than Northerners do, and that the bill would diminish England's exports to North America. *Observations on the Case of the Northern Colonies* [1731]. *The Case of His Majesty's Sugar Plantations* [1731 ?] in Lord Sumner's *Collection of Scarce and Valuable Tracts,* 1748. *Arguments against the Bill for the better securing and encouraging the Trade of His Majesty's Sugar Colonies in America* [1731 or 1732]. *The Controversy between the Northern Colonies and the Sugar Islands respectively Considered* [1732]. *Some Considerations humbly offer'd upon the bill now depending in the House of Lords, relating to the trade between the Northern Colonies and the sugar islands. In a letter to a noble peer,* 1732, Pp. 19. *A True state of the case between the British Northern colonies, and the sugar islands in America, impartially considered. With respect to the Bill now depending in the Right Honourable the House of Lords relating to the Sugar Trade,* 1732. Pp. 46. *The Case of their Majesties Sugar Plantations* [1732 ?]. *A Comparison between the British Sugar colonies and New England, as they relate to the interest of Great Britain. With some observations upon the state of the case of New England. To which is added A letter to a member of Parliament,* Pp. 43, London, J. Roberts, 1732. *The Case of the British sugar colonies* [1732 ?]. *The dispute between the Northern colonies and the sugar islands, set in a clear view* [1732 ?]. *A Detection of the State and Situation of the Present Sugar Planters of Barbadoes and the Leeward Islands* [1732]. The author was Robertson, a planter of Nevis. He gives an able presentation of the declining profits from sugar culture in the long settled group of islands. *A supplement to the Detection of the state and situation of the present sugar planters of Barbadoes and the Leeward Islands: showing among other matters, that the surest way for England to command the Sugar market abroad, is to contract rather than enlarge her sugar colonies. In a letter from an inhabitant of one of his Majesty's Leeward Carribbee islands, to a member of the House of Commons in England. To which is added a letter from a traveller in the Carribbees to his friend in London,* Pp. 92, London, 1733, also by Robertson. *An Enquiry into the methods that are said to be now proposed in England, to retrieve the Sugar Trade. By the author of the Detection of the state and situation of the sugar planters of Barbadoes and the Leeward Island,* Pp. 31, London, J. Wilford, 1733. *A Short Account of the Hurricane* [1733], also by Robertson. *Proposals offered for the sugar planters redress, and for reviving the British Sugar commerce in a further letter from a gentleman of Barbadoes, to his friend in London,* Pp. 35, London, J. Wilford, 1733. *Some further Considerations of the Consequences of the Bill now depending in the House of Lords, relating to the Dispute of the Trade of the British Colonies in America, in a Letter to a Noble Lord, by a Person familliarly acquainted with the English, French, and Dutch Settlements on the Continent, also on the West India Islands.* [Signed] *Philo-Americus,* folio Pp. 4, April 6, 1733. The author, an experienced trader, opposed the planters' bill and declared that industrious planters in Barbadoes could still become rich. John Ashley, *The sugar trade, with the encumbrances thereon laid open. By a Barbados planter,* Pp. 22, London, 1734. Mainly a plea for direct trade in sugar to Europe. The following also are valuable for contemporary references to the controversy. *Representation of the Commissioners for Trade and Plantations to the Right Honourable the Lords Temporal and Spiritual in Parliament assembled relating to the State of the British Islands in America, with*

Regard to their Trade, their Strength, and their Fortifications, January 14, 1734, folio Pp. 7, London, 1734. *Caribbeana,* London, 1741. An edition of *Caribbeana,* in 2 volumes, edited by Vere Langford Oliver, has been published by Mitchell, Hughes, and Clark, London, 1909.

CHAPTER XII

ILLICIT COMMERCE, 1733-1748

Had the British government possessed the disposition and power to enforce the Molasses Act, New England and the Middle Colonies of North America would undoubtedly have suffered a severe economic depression. To new communities with abundant resources and power to produce enormous supplies of fish, lumber, livestock, meat, and provisions, access to outside markets is of vital importance. The markets of Great Britain were, except for certain kinds of timber, practically closed to these staple commodities of North America. England and Scotland were adequately supplied with the products of Northern agriculture, while Ireland competed with New England in the output of horses, meat, and provisions. Indeed, as the Northern Colonies developed, it became increasingly difficult to find within the empire adequate markets for their expanding industries. The Roman Catholic countries of southern Europe had long offered an attractive outlet for the better varieties of fish. The Indian trade, the fisheries, and West Africa took a large share of the rum from Newport and Boston distilleries. But by far the largest vent for Northern industry existed amid the tropical communities, under many flags, stretching from Nassau in the Bahamas to Dutch Guiana in South America. Here was a field which, taken altogether, was quite as populous as British North America. Its demands upon Northern lumbermen, stock-raisers, and farmers, furnished a powerful incentive for the clearing and settlement of the continent. Access to this great tropical market for Northern traders had never been legally free except to the Dutch and British colonies. But by hook or by crook Northern sea-captains had gained admission everywhere, and made their supplies indispensable to Frenchmen, Dutchmen, Danes, and Spaniards, as well as to the English. In fact, by 1730, the British West Indies were generally regarded as only part of a legitimate and necessary sphere for exploitation by North American merchants.[1] Northerners were fully aware of what the West Indies as a whole meant to them and they were not to be misled into supposing that the British West Indies by themselves would ever again play the rôle they had in the seventeenth century. Mercantilism, so far as it implied the suppression of international trade in the western world, found no favor outside the official class in Boston, New York, and Philadelphia. The ideal of British planters, as revealed in the Molasses Act, came therefore as a challenge to the future progress of the whole region from Portland to Baltimore. From the sugar planters' point of view, which happened to coincide with that of the mercantilist, the new arrangement involved the treble advantage of reducing the cost of living, raising the price of sugar and molasses, and checking the export of specie. But to the Northerner the project was selfish and reactionary. To have complied with the law would have meant for him not only a serious limitation of business and profits, but a denial of the future development of America. It was the serious attempt by the British government in 1760 to enforce this law, and again in 1764 by lower duties to interfere with trade to the foreign West Indies, that ushered in the revolutionary movement. In view of this aspect of the

1 The historian seeking for the growth of "internationalism" might find indications of it in the gradual tendency toward the legalization by Holland, France, and Spain of international trade in the American Mediterranean. According to British *legislation*, however, the tendency was in the opposite direction—toward nationalism.

subject a history of West India commerce in the generation following the Molasses Act becomes of more than local importance.

From the first proposal of the Molasses Act until the opening of the revolutionary period, it was constantly asserted by Northern colonists that enforcement of the law would seriously curtail the sale of British manufactures. First, because the Northerners would be limited in their purchasing power, and secondly, because a stop would be put to the peddling of English wares throughout the settlements in the Caribbean. In this connection the home authorities were warned that, when America was no longer able to pay for manufactures from England, she would be forced into manufacturing for herself. Such action would of course tend to violate the whole colonial policy of Great Britain. Many officials and merchants in England were undoubtedly impressed by these arguments and their influence deprived the Molasses Act of the necessary public support for its enforcement.

As early as 1734, the apprehension of Northern people as to what might happen to their industry if the law were enforced was expressed by Governor Cosby of New York. He wrote to the Board of Trade: "The main bent of our ffarmers is to raise wheat, and they are like to remain in that way untill the price of it becomes so low, that Necessity puts them on Some other way of Cultivation; which in process of time is like to happen, because the Sugar Islands cannot encrease in the proportion which the Northern Colonies do, And whether some other encouragement may bring them over Sooner I cannot affirm."[2] But the farmers were indeed likely to remain

[2] Gov. William Cosby to Board of Trade, Dec. 6, 1734, C.O. 5: 1057, E 58, 59. Mercantilists were not displeased, however, with the prospect of decline for colonies that were likely to compete with the mother country.

in the way of raising wheat, for Northern traders had no intention of giving up their market in the foreign West Indies.

Even in the British sugar colonies there were some misgivings about the wisdom of the step that had been taken. If the law could not be enforced, might not the islands be worse off than before its adoption? For, in former days, Barbadoes and the other English islands were at least ports of call for Northern ships and entrepôts for foreign trade. In 1735, James Dottin, president of the council of Barbadoes, wrote to the Board of Trade: "It is found by experience that in contempt of the late Act of Parliament, very great quantitys of these [foreign sugar, molasses, and rum] are Still carryed to the Northward, it is alleged that the Same had much better be purchased in this Island where the Cash given for them wou'd then Center, rather then it Should be carryed elsewhere, which prevents us reaping any advantage from it, and still Occasions a less demand for our produce."[3] Evidently there were some British West Indians ready to coöperate with the Northerners in violating the Molasses Act.

The actual revenue collected in the Northern Colonies under the Molasses Act may be seen in the tables printed below.[4] The returns fell off considerably after the first

[3] James Dottin to Board of Trade, Nov. 29, 1735, C.O. 28: 24, Aa 39.

[4] Account of the quantity of foreign rum, molasses and sugar imported into the British Northern Colonies with the duty paid thereon (shillings and pence omitted):

Christmas to Christmas	Rum or Spirits Gals.	Duty	Molasses or Syrups Gals.	Duty	Sugar Cwt.	Duty
1733-4	5439½	£204	3183	£ 79	33	£ 8
1734-5	996¼	37	90	2	597	149
1735-6	...	..	187	5	888	222
1736-7	170	6	2708	68	512	138
	6605¾	£247	6168	£154	2030	£507

Treas. 64: 273.

Account of the gross duty collected on molasses in the Northern Colonies under the Molasses Act (shillings and pence omitted):

1734	£79	1740	£ 23	1746	£ 57	1752	£475	1758	£ 519
1735	2	1741	51	1747	51	1753	577	1759	447
1736	5	1742	233	1748	56	1754	975	1760	1170
1737	68	1743	286	1749	202	1755	530	1761	1190
1738	25	1744	173	1750	1236	1756	236	1762	718
1739	..	1745	72	1751	510	1757	95	1763	1541
								1764	2100
								Total,	£13,702

The total duty paid on molasses in the Northern Colonies under the Sugar Act of 1764 up to Feb. 1767 was £24,133. Treas. 64: 274.

An account of *all* the duties collected under the Molasses Act from 1733 to 1750 (shillings and pence omitted):

Christmas to Christmas			On Merchandize	On Prizes
1733-4	£330	1741-2	£ 722	£ 140
1734-5	151	1742-3	462	41
1735-6	293	1743-4	235	67
1736-7	220	1744-5	98	3082
1737-8	69	1745-6	355	125
1738-9	109	1746-7	461	1259
1739-40	25	1747-8	693	2762
1740-41	101	1748-9	1279	140
		Total,	£5603	£7616

Custom House Accounts, C.O. 5: 38, Appendix, no. 4.

An Account of all the duties collected under the Molasses Act in each of the Northern Colonies from 1733 to 1750 (shillings and pence omitted):

	On Merchandize	On Prizes
Bahama Islands	£ 777	£1880
South Carolina	672	3073
North Carolina	...	529
Virginia	61	587
Maryland	64	...
Pennsylvania	600	145
New Jersey	46	...
New York	2003	...
Connecticut	...	99
Rhode Island "N.B. No Accots received since Michs 1744"	...	...
Massachusetts	1043	313
Nova Scotia ...."Nil"............	...	...
Bermuda	337	1020
C.O. 5: 38, Appendix, no. 5.	£5603	£7616

year; they increased somewhat during the years 1742-1744 as the result of naval activity, and again in 1750 due to a stricter administration prompted by protesting planters; and, finally, under Pitt's orders, the act was rigidly enforced and the largest revenues yielded by it were during the years 1760-1764.

The imperial machinery for the enforcement of trade laws in America was weak and inadequate. There were, in 1733, three surveyors-general of the customs in the British American colonies. Charles Dunbar had charge of Bermuda and all the West Indies except Jamaica; George Phenney surveyed Jamaica, the Bahamas, the Carolinas, Virginia, Maryland, and Pennsylvania; and John Peagram was in charge of the Jerseys, New York, Connecticut, Rhode Island, Massachusetts, New Hampshire, and Newfoundland.[5] Colonial governors were instructed at this time to admit the surveyors-general of the customs as extraordinary councillors in the respective governments and to coöperate with them to secure the collection of revenues.[6] The salaries of customs officers could not have put them beyond the temptations of bribes. At the port of Boston the collector received £100 a year, the comptroller £70, and the surveyor £100. At Piscataway the "collector and surveyor" was paid £100. Salem's collector and surveyor received £40, Rhode Island's £100, New London's £80, New York's collector and comptroller £55 each; New Jersey had two collectors at £40 each and one at £30; Pennsylvania had three collectors and a comptroller who received respectively £160 plus £80 for a boat, £90, £90, and £80.[7] The

[5] Carkess to Board of Trade, Aug. 7, 1733, C.O. 323: 9, M 44.

[6] C.O. 323: 9, M 45.

[7] Chamberlayne, *Notitia Dignitae Angliae* (1755), pp. 96-99. As compared with the few officers listed for the Northern Colonies, Chamberlayne states that Newcastle (England) had one collector and a clerk receiving £160, and 55 other officers receiving £1160.

total charge of managing the revenue in the colonies in the eleven years preceding the Molasses Act (1722-1732) was £55,129, or less than £5000 per year, of which only £2798 went for salaries.[8]

The British naval patrol stationed in America was likewise altogether too small to prevent smuggling. In 1735, there were but seventeen ships of the royal navy stationed in North America and the West Indies.[9] This number, although considerably increased during the following thirty years, was never sufficient to cope with the illusive Yankee smugglers along so extensive a coast.

Evasions of the Molasses Act became, almost immediately after its passage, a matter of common knowledge. Governor Mathew of the Leeward Islands, in 1736, attributed this to "the negligent behaviour" of Northern customs officers. As a last resource acts were passed by

[8] Treas. 38: 364. The total cost of the customs service in England during the same period was £2,077,558 or about £190,000 a year.

[9]

Admiralty Office 8 December, 1735		List of His Majesty's Ships Stationed in America, of what Strength and in what Ports			
Rate	Ships	Captains	Men	Guns	Disposition
4	Dunkirk	Degby Dent	400	60	
	Argyle	Geo. Slater	300	50	
	Antilope	Timy Bridge	240	40	
5	Kinsale	Jno. Forrester	250	40	Jamaica
6	Shurnese	Miles Stapylton	140	20	
				c. s.	
Sloop	Drake	Tho. Fox	100	4 10	
4	Newcastle	Ellis Brand	300	50	
5	Pearl	Jno. Toller	250	40	Leeward Islands
				c. s.	
Sloop	Spence	Jno. Russel	100	6 10	
5	Diamond	Rd Herbert	250	40	Barbadoes
6	Fox	Hen. Reddish	140	20	
Sloop	Shark	Rd Symonds	100	8.12	Bahama Islands
	Hawk	Jas. Gascoigne	70	6.10	Georgia
6	Rose	Chas. Wyndham	140	20	S. Carolina
	Sea Horse	James Compton	140	20	Virginia
	Tartar	Math. Morris	140	20	N. York
	Scarborough	Tho. Durrell	140	20	New England

C.O. 323: 10, N 13.

West India legislatures themselves, authorizing the seizure by colonial patrols of foreign ships returning with cargoes from British colonies. Such acts were passed by Montserrat and St. Christopher in 1736. The *Fleuron* of St. Malo was accordingly seized near Montserrat, July 22, 1736, and, in September, the French sloop *Dolphin* was seized within one league of Montserrat by Captain Welch of the *Mulberry*, commissioned by Montserrat. The court of vice-admiralty condemned the *Dolphin*, one-third being awarded to the Crown and two-thirds to the owners of the *Mulberry*. In the following winter, two Dutch ships from St. Eustatius were likewise captured.[10]

Jamaica was a center for much clandestine trade with the foreign sugar islands. Masters of vessels from the Northern Colonies were often indulged to enter and clear their vessels at one and the same time, and they frequently cleared out their ships as having on board the commodities of Jamaica when in fact they had no such produce aboard or even design to take any but, instead, carried empty casks for holding and money for purchasing molasses, sugar, and other produce of the French islands. But by means of their cockets obtained in Jamaica they easily entered the Northern Colonies with such sugar, etc., as the produce of Jamaica. A Jamaican act of 1738 provided, therefore, that a customs officer, before clearing a ship, should have the master appear before him and show certificates of sale of all the goods he had on board of the produce of Jamaica, such certificates to be attested by a magistrate of the parish where the goods were purchased.[11] This law was generally re-

[10] Correspondence on these acts and cases is in C.O. 152: 22, 52, 57, 61, 62, 63, 64, 65, 66, 67, 78, 84, 85, 86.

[11] Acts of Jamaica, Feb. 1737/8, C.O. 139: 15, no. 1. Re-enactments are in *ibid.*, nos. 10, 22, 43; *ibid.*, 16, nos. 1, 9, 18; *ibid.*, 17, nos. 2, 21; acts of 1752, no. 18.

newed annually during the period under review. The practice of entering foreign sugar in Great Britain and North America as the product of Jamaica did not cease for, in 1752, an act of the island, declaring that Jamaicans themselves were engaged in such trade with the French, Spanish, Dutch, and Danes,[12] provided that, after January 1, 1753, no sugar, rum, or molasses should be imported into Jamaica from the French, Spanish, Dutch, or Danes, under penalty of forfeiture of ship and cargo and fines of £100 on the master and £100 on the owner of the goods.[13] In 1756, it was further provided that Northern colonial ships should not, as theretofore, enter and clear out the same day.[14]

Some effort was made by the French in the West Indies to confine commerce to national limits. In the spring of 1737, apparently in retaliation for an act passed in Montserrat, the British ship *Scipio* and other vessels from Europe and North America bound for Jamaica were seized by the French in the Windward Passage between Santo Domingo and Cuba. A French edict of 1727 had directed the seizure of British vessels sailing within one league of the coast of the French West Indies. But the most favorable route for reaching Jamaica was through the Windward Passage, and British ships were in the habit of anchoring in Donna Maria Bay and Tiburon Bay for wood and water. English merchants contended that these bays should be open ports, and urged England to reciprocate by opening certain ports in the British

12 Gov. Trelawny wrote to the Board of Trade Nov. 21, 1741: "The French afford Sugar, Molasses, Rum, Coffee &c[a] cheaper than a Jamaica Planter can do, which invites His Majesty's Subjects of the Northern Colonies to buy of them. It is believed that a considerable quantity of their produce is every year clandestinely shipped from this Island to Great Britain." C.O. 137: 23, W 48.

13 Acts of Jamaica, 1752, C.O. 139: 17, no. 18.

14 *Ibid.*, C.O. 139: 18, no. 4.

sugar islands. Negotiations between Earl Waldegrave, the British minister, and Maurepas were undertaken to exempt innocent traders from arbitrary arrests. France finally agreed to soften the execution of the edict of 1727: foreign ships within a league of French territory would be seized only when the nature of their cargoes or other evidence proved that they came to engage in commerce. In cases of distress foreign vessels might remain in French waters only a necessary time. In conformity with the old Treaty of Neutrality of 1686, England and France again agreed to restrain each other's subjects from direct international trade in America.[15]

Direct trade between the British and French colonies in America after this time attracted less notice. But commodities were still exchanged in increasing amounts between the two peoples. This was accomplished through the neutral ports of the West Indies. Colonial agents of English sugar planters in 1737 wrote: "The Trade which is Carryed on Between the French and the English at the Islands of each Nation respectively is of very little Consequence, compared with that which is carried on at St. Lucia, and at the neutral Ports of St. Eustatia, St. Thomas, and Curacoa, and therefore it will be very immaterial to prevent the one, without preventing the other.'"[16] The Board of Trade was concerned with this phase of

[15] Colonial agents of the Sugar Colonies to Board of Trade, Apr. 1, 1737, C.O. 152: 22, W 99. Newcastle to Board of Trade, Mar. 21, 1736/7, enclosing correspondence of Maurepas to Earl Waldegrave, Mar. 1736/7, C.O. 323: 10, N 20. Waldegrave to Newcastle, May 22, 1737, C.O. 388: 36, Y 84. Newcastle to Board of Trade, July 8, 1737, *ibid.*, 37, Y 96. See also *ibid.*, 37, Y 97. John Sharpe to Newcastle, Nov. 13, 1738. A plea for opening Donna Maria and Tiburon Bays, C.O. 137: 48, ff. 46-48. See also Representation of Bristol Merchants, C.O. 152: 23, X 17. The rupture with Spain suspended negotiations for the opening of free ports. C.O. 323: 17, R 22, 28, 29.

[16] John Sharpe, John Yeamans, Richard Cooper, colonial agents, to Board of Trade, June 6, 1737, C.O. 388: 37, Y 88.

the subject from time to time during 1737. Wilks, the agent for Massachusetts, urged, however, that no further restraints be put upon commerce between the Northern Colonies and the neutral islands.[17]

The commerce of New England was little changed by the Molasses Act. Governor Belcher reported in 1737: "The Sea Coast of the Province is so extensive & has so many Commodious harbours, that the small number of Custom House Officers are often complaining they are not able to do much for preventing illegal Trade. Nor does the Sugar Act take any great Effect; great Quantities of foreign Molasses are still brought into this Province, and much of it by way of Rhode Island."[18] New York's imports from foreign sugar colonies were said to be small, but she had a considerable export trade in provisions, lumber, and horses to St. Thomas, Curaçoa, and Surinam.[19] Governor Mathew of the Leeward Islands declared it was well known that hardly fewer than one hundred sloops from the northward traded at St. Eustatius yearly for French rum and molasses. The business was managed through two agents there—Sagernan, a Frenchman, and Steward, an Englishman. Mathew was alarmed at the great amount of French sugar that was serving "to Glutt Our British

17 C.O. 391: 46, ff. 132-133, June 10, 1737. Col. George Thomas informed the Board on July 14, 1737: "The Island of St. Eustatia, belonging to the Dutch, will remain as it is at present, the chief Mart for the Commodities of both English and French; and as that is a free Port and the Dutch not proposed to be included in the Treaty, an illicit Trade will be carried on there without any risque or additional expense." The only effective suppression of trade between British and French would be the confiscation anywhere of foreign goods found on each other's ships. C.O. 388: 37, Y 97. *Cf.* also Thomas, Coope, Yeamans, Spooner, and Sharpe to Board of Trade, Apr. 1, 1737. C.O. 152: 22, W 99.

18 Gov. Belcher to Board of Trade, Mar. 2, 1736/7, C.O. 5: 879, Cc 38.

19 Archibald Kennedy, collector of customs to Board of Trade, Jan. 10, 1737/8, C.O. 5: 1059, Gg 19.

Markets, and ruin the Sale of Our Own Sugars." "And the French," he declared, "who have beaten us out of all Foreign Marketts, must also be lett into our own British Markets. Is it hard to Forsee what miserable plight The English Sugar Planter is Daily reducing to?"[20] Through the activity of the Montserrat coast guards, Governor Mathew had effectively intercepted the transport of sugar from Martinique and Guadeloupe to St. Eustatius for the Rhode Islanders. The latter, nevertheless, secured great quantities of French sugar from Santo Domingo through St. Eustatius. "The Sta Eustatia Bay is coverd with them Rhode Islanders, and persons of best Distinction there, without the Least Reserve, averred . . . that no fewer than Six Thousand Cask Shifted into English Hogsheads, were Disignd through these Islands for the British Marketts."[21] The annual importation into Great Britain of French sugar disguised as English produce was estimated, in 1740, as at least 15,000 hogsheads.[22]

Another center for illicit trade between Northerners

[20] Gov. Mathew to Board of Trade, May 26, 1737, C.O. 152: 23, X 5. Mathew was engaged at the time in a vigorous enforcement of the Montserrat Act and mentions the seizure of two French sloops, the *Dolphin* and *Catherine,* and of a ship called the *Two Sisters.*

[21] *Ibid.,* June 14, 1737, C.O. 152: 23, X 8. Mathew was informed of the St. Eustatius trade by a Mr. Arbuthnot who had just returned from there. The following description of St. Eustatius was made by General Fleming to the Board of Trade in 1740: "There is but a small part of this Island fitt for Canes, and they don't make above 500,000 lb. Sugar, and 200 Hogsheads of Rum. There is one Town of pretty great trade Containing about 150 Houses having very large Shops of Dry Goods from Holland, and there comes 2 or 3 large Ships from Guinea every year with Slaves, most of them are sold to the French, for Sugar, and Molasses, and great part of the Sugar is put into English Casks and Shipt for English Sugar; The New York, Rhoad Island, and New England Traders, purchase the Melasses from the French, and Duch, with what Money they have pickt up in the British Colony's for Lumber Horses &c and thus the Sugar Act is eluded and ineffectual as to the chief Intent of it." C.O. 152: 23, X 78.

[22] Perrin, *Present State of British and French Sugar Colonies,* p. 18.

and the French was in the British Virgin Islands—the northernmost group of the English Antilles. The settlers there, notorious for immorality, sold the produce of their plantations to the French and Dutch, in return for all kinds of European goods. A part of these goods was then sold to North Americans, who carried them home in violation of the navigation laws.[23]

To a conscientious British official the state of trade in the Caribbean must have appeared in great disorder. The Commissioners of the Customs informed the Board of Trade that, so frequent was opposition met in enforcing the Molasses Act, they desired the governors to be instructed to coöperate further in the execution of the law.[24] Accordingly, an additional article was added to the royal instructions to colonial governors requesting them to aid the collectors and other officers of the customs in collecting the duties and making seizures of goods clandestinely imported.[25] In March 1739, a petition to the House of Commons, drawn by John Sharpe in behalf of the planters, asked for such alterations and additions to legislation as would effectually stop clandestine commerce.[26] A bill for rendering the Molasses Act more effectual pended for a time in parliament but did not pass.[27] What was needed to realize the planters' idea was not more law but administrative machinery.

Admiralty courts encountered constant opposition to

[23] Robert Dinwiddie's report on the Virgin Islands (1739), C.O. 152: 23, X 77.

[24] Carkess to Board of Trade, Apr. 15, 1737, C.O. 323: 10, N 21.

[25] C.O. 323: 10, N 21.

[26] Although addressed to the Commons this petition was presented at the Board of Trade, March 25, 1739, by Sharpe, several planters attending. C.O. 152: 23, X 52.

[27] "The Inhabitants are acquainted by their Agent [Sharpe] that a Bill is now pending in Parliament, for rendering more effectual the Law," etc. Gov. Byng of Barbadoes to Board of Trade, Jan. 7, 1739/40, C.O. 28: 25, Aa 80.

the enforcement of their decisions against smugglers. A typical instance was the Fowles case in New York late in August, 1739. The collector, Archibald Kennedy, brought suit in the court of vice-admiralty against Thomas Fowles for importing illicitly in the sloop *Mary and Margaret* foreign gunpowder and molasses from St. Eustatius. Fowles put in the usual plea of no jurisdiction for the vice-admiralty court, since seizure was made in New York bay. The bay, it was contended, was within one or more counties and the case, therefore, should be determined in a common law court with a jury. The vice-admiralty court proceeded to prove the libel. But before the day of hearing Fowles applied to the supreme court for a writ of prohibition, commanding the vice-admiralty court to cease from further prosecution of the suit. The advocate general and counsel for the collector pled for admiralty jurisdiction. The judges of the supreme court, however, granted the prohibition. Kennedy's comment on this case reveals the very serious predicament that all American collectors were in during the colonial period. "If the prohibition in this Cause was well Issued," he wrote, "no Breach of the 15th Car: 2d Cap. 7 [the Staple Act of 1663] can be tryed in the Admiralty (but must be tryed at Common Law by a Jury, who perhaps are equally concerned in carrying on an Illicit Trade and its hardly to be expected that they will find each other guilty) For if the Importation into the province makes the Breach of the Act, and no part of the province but is within one of the Counties, And whatsoever is done within a County Cannot be tryed in the Admiralty, but must be tryed by the Common Law, Consequently no Breach of that Act can be tryed any way but by a Jury." The only resort for the collector lay in a possible appeal to the king and council. The lieutenant governor, George Clarke, in reporting the case to the Board of Trade, ventured the opinion that "if some Method be not fallen upon whereby Illicit Trade may be better prevented, I doubt it will be to little purpose to bring any Cause of that kind to tryall by a Jury, and the officers of the Customs will from thence be discouraged from exerting themselves in the discharge of their Duty."[28]

Before the end of 1739, Great Britain went to war with Spain, over the alleged barbarities of Spanish coast guards, though fundamentally to force larger concessions for British trade with Spanish America. With the opening of the War of the Austrian Succession, the following year, England entered into a military alliance against France, and hostilities continued until the peace of Aix-la-Chapelle in 1748. From a commercial point of view, this period is strangely characterized by British Americans supplying their enemies, the French in the West Indies and their fleets, with all the essentials of life in exchange for sugar, molasses, and rum. In fact, from the wider notice which this trade attracted it might seem that its volume was greater in war than in peace, though that was probably not the case. As in time of peace, the Dutch and other neutral islands served as important entrepôts.

To isolate commercially the French in America became, as the war advanced, England's primary aim. The royal navy proved incapable, however, of achieving this end—a failure due more to New Englanders than to the enemy. "There is one thing which if we can do," wrote Horace Walpole in 1740, "would disable both France and Spain to maintain long their maritime force, which is, if we could hinder them from being furnished with salt provisions from Ireland, and with fresh provisions from our Northern Colonies; for their fleets are now victualled

[28] Lieut. Gov. Clarke to Board of Trade, Dec. 15, 1739, enclosing a report by the customs collector on the Fowles Case, C.O. 5: 1059, Gg 45.

by this means.'"[29] Admiral Vernon, writing from his ship in Port Royal, Jamaica, took the same view: "As I well know, the best way to distress the Enemy here with their large Fleets, is to keep them from having supplies of Provisions from Our Northern Colonies, or Ireland." He had written a circular letter to all the governors of the Northern Colonies requesting them not to permit any provisions to go from them unless security should be given for their being landed in British colonies only, and not to allow any to go to the French or Dutch.[30]

Throughout the war the English West Indies suffered severely for the want of such provisions as were going to the French. As a relief to them, an order in Council of February 19, 1741, relaxed the British embargo on provisions by allowing the export to the sugar islands of biscuit, beans, and oats from Great Britain. The beans and oats were needed for horses. Security treble the value of the cargo was required to ensure their being landed only in the British West Indies.[31]

Massachusetts by act, April 10, 1741, prohibited all provisions being exported except, under bond, to English possessions.[32] A special report from Governor Shirley, in 1743, revealed extensive violations of the Staple Act throughout New England and showed that vice-admiralty courts were encountering effective opposition from smugglers of European and foreign West India goods.[33]

29 Horace Walpole to Robert Trevor, Oct. 3, 1740, *Hist. MSS. Com. Rep.* XIV, Pt. 9, p. 57.

30 Adm. Vernon to Newcastle, Dec. 12, 1740, C.O. 137: 48, f. 142.

31 C.O. 323: 11, N 82.

32 C.O. 5: 883, Ee 51. Fish might be carried to any part of Europe or Africa at peace with England.

33 Gov. William Shirley to Board of Trade, Feb. 26, 1742/3, enclosing a report on illicit trade and inadequate admiralty jurisdiction from William Bollan, advocate general and later colonial agent. C.O. 5: 883, Ee 87. See also Robert Auchmuty, judge of vice-admiralty in New England, to Board

Trade with the enemy, however, according to a Massachusetts act of June 1744, was declared an act of treason, punishable by death; provision was made, nevertheless for "flags of truce."[34] At this moment, Governor Shirley averred that Louisburg, Martinique, and other French settlements were drawing the chief part of their provisions from the British colonies.[35] Early in 1745, embargoes were laid on all shipping in the four New England colonies, and, according to Shirley, strictly observed in three of them, in order to prevent intelligence and supplies being carried to the enemy.[36]

Perhaps the most statesmanlike suggestion made, in this period, for improving the administrative machinery and removing the chief incentive to smuggling was that proposed by the judge of vice-admiralty at Boston, Robert Auchmuty. He recommended that the duties levied on landing foreign sugar, rum, and molasses, which were prohibitive, should be so lowered as not greatly to exceed the contingent charges of smuggling, and, with the commodities bulked, he was confident that merchants would not find it profitable to run the risks of smuggling. With the increased revenue more "preventive officers" might be provided, who now were really wanted, especially in Rhode Island and Connecticut, and the salaries

of Trade, May 31, 1743, *ibid.*, Ee 89; Board of Trade to Newcastle, May 11, 1743, C.O. 5: 5, f. 184; Shirley to Board of Trade, Feb. 6, 1747/8, C.O. 5: 885, Gg 3. An interesting correspondence describing illicit trade carried on in 1745 by Edmund and Josiah Quincy with Thomas and Adrian Hope in Holland, in collusion with customs officers at Newcastle and Kirkwall, may be found in C.O. 323: 13, O 113.

34 *Acts and Resolves*, 1744, Ch. VI, p. 120, C.O. 5: 885, Gg. The goods specified were arms, ordnance, powder, shot, lead, pitch, tar, hemp, masts, cordage, iron, steel, brass, pewter, saltpeter, provisions, clothing, or any other supplies.

35 Shirley to Board of Trade, June 16, 1744, C.O. 5: 884, Ff 27.

36 *Ibid.*, Mar. 27, 1745, C.O. 5: 885, Ff 79.

of existing officers paid instead of being charged to the Crown.[37]

In Rhode Island and Connecticut, illicit trade with the French was, according to Auchmuty, much more extensive than in Massachusetts. Such trade was facilitated, not alone by their more advantageous location, but by the absence of such officers in both those colonies as were present in Massachusetts. In Rhode Island, the collector was only a deputy; his principal who rented the district was never present. In both colonies, comptrollers were wanting.[38] Such officers were indispensable, and their pay should have been received out of the seizures made. Masters of many ships that arrived in Rhode Island from foreign plantations refused to take the usual oath upon entering, but instead voluntarily paid the penalty of £100 in bills of credit which were worth about one-fifth that amount in sterling money, and were thereupon admitted to entry. Others escaped both requirements. Some sent mates to swear and enter. The oath was frequently tendered in a tavern or coffee house and the examination became a mere form. The whole system cried out for more officials, comptrollers, and scrupulously managed custom-houses.[39]

Flags of truce from Rhode Island disguised an extensive trade with the French West Indies in the later years of the war. The brigantine *Victory*, Charles Bardin, Master, and Joseph Whipple of Newport, owner, sailed from Rhode Island, January 20, 1747, as

[37] Auchmuty to Board of Trade, London, May 31, 1741. Customs *collectors,* he said, for want of duties to collect in New England, were virtually but *search officers* and were with no propriety termed *collectors.* C.O. 5: 883, Ee 97.

[38] Auchmuty to Board of Trade, London, May 31, 1741, C.O. 5: 883, Ee 97.

[39] *Ibid.,* Nov. 23, 1743, C.O. 5: 884, Ff 12.

a flag of truce to Cap François in Santo Domingo, and under pretense of carrying five French prisoners had three hundred quintals of refuse codfish, twenty barrels of shad, and two hundred bunches of onions. She arrived at Cap François in twenty-three days, was there several weeks, and returned to Newport with one hundred and seventy-four casks of molasses and other private ventures. The French governor made no objection to the trade. The expense of the prisoners was borne by the shipowner. The outbound cargo was cleared without any custom-house permit. The trade at the Cape was also without written permit or clearance papers. At Newport, the molasses was landed without payment of any duty.[40] Over sixty such flags of truce, according to Governor Shirley's informers, in the eighteen months prior to February 1748, had sailed from Rhode Island to Cap François, Léogane, and other points in Santo Domingo and Martinique laden with provisions of all kinds by which the French were enabled to fit out great numbers of privateers. In consequence, agents of British victualling contractors declared the possibility of supplying English naval and military forces in America greatly endangered. It was said that the sending of a single French prisoner in one of these flags of truce was thought sufficient to give countenance for carrying on this trade.[41]

Connecticut was generally classed with Rhode Island by observers of illicit trade, but there is little data for specific cases. In May 1748, the assembly there ordered the publication of a royal proclamation against trade

[40] C.O. 323: 13, O 113. It was said that, at this period, even if flags of truce paid the duties they would get foreign sugar, etc., cheaper than they could be had in the British West Indies.

[41] Gov. Shirley to Board of Trade, Feb. 6, 1747/8, C.O. 5: 886, Gg 3. Shirley said he believed no flag of truce, engaged in trade, had sailed from Massachusetts. See also Shirley to Bedford, July 2, 1748, C.O. 323: 12, O 4.

with the enemy by the sheriffs in the counties of Hartford, New Haven, New London, Fairfield, and Windham.[42] Something more than publication was probably needed along the Connecticut shore.

From New York, flags of truce occasionally sailed with provisions to Santo Domingo and returned with sugar and molasses. The sloop *Endeavour,* owned by Alderman John Painter of New York, Joseph Ball of Newport, and Lawrence Paine, sailed January 31, 1747, as a flag of truce to Cap François with 70 quintals of dry fish, 20 barrels of pickled fish, 500 lbs. of cheese, 1000 hoops, 800 staves, 6 barrels of beef, 4 barrels of flour, 600 lbs. of bread, 8 or 9 barrels of beer, and the same of cider, beside private ventures. She took in at Cap François a cargo of Jamaican prize sugar, French molasses, and loaf sugar, besides the private ventures. About the same time, the sloop *Mary of Bermudas,* as a flag of truce, accomplished a similar commercial voyage to Léogane. In this case the vessel cleared for Jamaica and traded under a license from the French governor.[43]

In Pennsylvania and Maryland, during this period, there is little evidence of illicit trade, though the lieutenant governors were told of frequent violations of the Molasses Act and were urged to prosecute smugglers.[44]

Of quite as damaging a character to England and to the British sugar planters was the trade of Ireland with the French West Indies directly and via St. Eustatius. In 1745, Irish ships were carrying provisions from Cork to Martinique where trade went on under French licenses.

42 *Connecticut Colonial Records,* IX, 360.

43 C.O. 323: 13, O 113. During the years 1744-1746, 14 French prizes and 1 Spanish, all heavily laden with foreign sugar, were condemned by the vice-admiralty court in New York and sold. C.O. 5: 1061, Gg 215, 216.

44 Instructions to Thomas Bladen, Lieut. Gov. of Maryland, May 6, 1742, C.O. 5: 1294, f. 189; and to James Hamilton, Lieut. Gov. of Pennsylvania, July 29, 1748, *ibid.,* f. 319.

Some Irish vessels exposed themselves to capture by the French, following which, friendly exchanges were effected.[45] The trade at St. Eustatius seems to have been enormous. Heyligger, the Dutch governor, had commissioned over fifty French vessels which were daily carrying provisions to the French. Dutch fly boats also cleared from Cork with provisions ostensibly for Surinam but actually for Martinique.[46] Under admiralty orders of March 14, 1745, to seize neutral ships laden with provisions for the French, numerous seizures of Dutch vessels were made by the British navy. Judge King of the vice-admiralty court at Antigua held that it would be impossible, however, to condemn Dutch vessels unless it were officially announced that the French island to which they were bound was in a condition of blockade.[47] In January 1746, the *Princessa* and *Drednought* seized a Dutch schooner laden with fish bound to Martinique, which by false passes pretended to be bound to Essequibo, a Dutch settlement on the mainland. Vice-Admiral Townshend detected many Dutch vessels carrying on this trade, all countenanced by Heyligger who was "more like a pedlar than the States General officer."[48] Heyligger complained of the injustice of such seizures, but Townshend justified them on the ground that ships'

[45] Gov. Thomas Robinson of Barbadoes to Board of Trade, Mar. 24, 1744/5. Robinson cites several instances and the evidence in the case of the snow *Recovery* of Cork in the vice-admiralty court at Barbadoes. C.O. 28: 26, Bb 29.

[46] Gov. William Mathew to Vice-Admiral Townshend, St. Christopher, Apr. 23, 1745. Adm. 1: 305; Townshend to the Admiralty, Nov. 8, 1745, *ibid.* The Dutch snow *De Jonge Benjamin,* with provisions from Cork, and suspected of being bound to the French on the supposition that she carried French and Spanish passes, was seized by Captain O'Hara and brought into Antigua and libeled in the vice-admiralty court, but was acquitted. O'Hara to Commodore Lee, Apr. 5, 1747; Lee to Corbett, Apr. 6, 1747. Adm. 1: 305.

[47] Commodore Fitzroy to Thomas Corbett, July 9, 1745. Adm. 1: 305.

[48] Townshend to Corbett, Desiada, W. I., Jan. 12, 1745/6. Adm. 1: 305.

papers were often falsified and British captains must detect their virtual destinations and act accordingly.[49] Even neutral ships, laden with provisions brought into British ports but acquitted, were still detained unless they disposed of their cargoes to English subjects.[50] The daily supplies which Heyligger sent the French islands made provisions more plentiful and cheaper there than in the English islands. Also all sorts of military stores, ammunition, and cordage, for fitting out privateers were sent to them. Of the many neutral ships prosecuted, those condemned always lodged appeals to be tried again in England. This made officers very cautious of having anything to do with them for fear of future trouble. As for those acquitted and asked by the naval commanders to unload and sell their provisions, a difficulty arose, because, such goods being foreign, the custom-houses could not enter them.[51]

The status of provisions needed more careful definition if these difficulties were to be overcome and England achieve her object of isolating the enemy. This status

49 Townshend to Heyligger, Jan. 28, 1745/6. Adm. 1: 305. Townshend refers to Heyligger as "complaining of very great hardships and illegal practices, committed by the ships of the squadron under my command, in seizing their vessels, carrying them into port, and detaining them, which vessels were detected in carrying on an illicit trade and commerce with the Enemy, by furnishing them with Fish, and other Provisions, althô they knew I had environed Martinico." To Corbett, Feb. 8, 1745/6. *Ibid.*

50 "You are hereby required & directed in case you meet with any neutral ships laden with provisions for the supply of those [French] Islands to seize and bring them into port, in order to their being proceeded against in the Admiralty Courts, And thô the said ships should be there acquitted (in case of my Absence) You are not to suffer them to proceed with their lading of provisions to sea, or any other ships to go out of his Majesty's Islands with Provisions for the enemy which you are to hinder by all means possible, leaving a liberty to the commanders of the neutral ships to dispose of their provisions to His Majesty's subjects, and then go to sea." Commodore Lee to Capt. O'Hara, Barbadoes, Feb. 28, 1745/6, Adm. 1: 305.

51 Lee to Corbett, May 26, 1746, Adm. 1: 305.

was settled by a decision from the High Court of Admiralty, in 1746, that "Provisions are and always have been esteemed contraband."[52] Neutrals carrying provisions to the French were thereafter seized under the law of contraband, and the special orders of the Admiralty were on May 13, 1746, revoked.[53] The Dutch were said to have posted placards prohibiting trade with the French, notwithstanding which they continued to use every opportunity to supply their needs.[54] In the last year of the war, the French were chiefly supplied from the Danish islands of St. Thomas and Santa Cruz. Several Danish vessels were stopped and one of them condemned.[55]

The British West Indians themselves were not entirely innocent in the matter of trade with the enemy. Barbadians carried slaves to the French and in return clandestinely imported French soap, candles, wine, gold and silver brocades, laces, silks, calicoes, dimities, cambrics, and pistereens.[56] The Leeward Islands constantly supplied the French via St. Eustatius with provisions and anything they needed. Old Governor Mathew had fought a long fight to enforce the trade laws but, in the absence of honest coöperation from customs officers, he felt that his struggle had proved vain.[57] Trade between the Lee-

52 F. T. Pratt, *Law of Contraband of War*, London (1856), p. 93, noted by Beer, *British Colonial Policy, 1754-1765*, p. 93, n. 3. Beer observes that Holland did not accept this definition of contraband, which was opposed to her treaties with England.

53 Lee to Corbett, July 20, 1746, Adm. 1: 305.

54 Admiral Osborn to Corbett, Antigua, June 23, 1748, Adm. 1: 306.

55 *Ibid.*

56 Governor Robinson to Board of Trade, Nov. 27, 1742, C.O. 5: 5, ff. 186-190. See also Robinson to Board of Trade, Feb. 20, 1746/7, C.O. 28: 27, Bb 57.

57 "It is as visible as day light at Noon that the French Islands have allways been Supplyd from these Islands with Provisions or any thing they wanted, by their being sent hence first to Sta Eustatia. I have laboured incessantly to prevent it but to no Purpose. I applyd time after time to Mr.

ward Islands and Martinique was also facilitated by flags of truce. Commodore Lee reported capturing one from Martinique loaded with Madeira wine and cocoa. It appeared that this venture was carried on by a Mr. Rowland of St. Christopher in partnership with Bannister, collector of customs at Antigua. The vessel and cargo were condemned in the vice-admiralty court, and the collector, strange to record, lodged an appeal against the decision. Lee said that this was but one of many such cases.[58]

The effect of the war of 1744-1748 upon the British West Indies was twofold: great distress on account of the scarcity of all kinds of supplies, and a considerable loss of market in North America for sugar, etc., due to heavy importations of French produce there.

Often when supplies from the Northern Colonies were not clandestinely diverted to the French, they were seized by French privateers. During one month, in January and February, 1746, thirty-six English vessels bound for Antigua were captured. This not only deprived the island of absolutely indispensable supplies and threat-

Dunbar who was at the head of the Custom house to oblige all Vessels bound for Sta Eustatia to clear out Regularly in the Custom house here, under penalty of forfeiture, I urgd 'twas going as much out of the Government as going to any more distant foreign Port. But t'would not do, and grounds of suspicion were not wanting that he himself was Concerned in this Trade, as he made no Secret of the Strict Correspondence between him & Mr. Heyligger Governour of Sta Eustatia, through whose hands nine Tenths of that Trade was allways carryd on. I have not seen Mr. Dunbars Successor yet, he continues at Barbadoes. I may happen to be as little regarded by him, unless I can say something to him from Your Lordships on it. I grow very old and sickly.'' Gov. Mathew to Board of Trade, St. Christopher, Oct. 16, 1747, C.O. 152: 25, Y 174. See also Mathew to Board of Trade, Antigua, Oct. 15, 1748, C.O. 152: 26, Z 36.

58 Commodore Lee to Corbett, St. Christopher, May 26, 1746, Adm. 1: 305. In retaliation for his activity against illicit traders a few merchants of Antigua induced the council and assembly to draw up an address to the king charging Commodore Lee with neglect of duty in protecting British trade! Lee to Corbett, Aug. 2, 1746, *ibid.*

ened famine, but made the island's produce a drug on the market.[59] The experience of commerce with the French West Indies also advertised more widely the fact that those islands could undersell the British in tropical staples.[60] In Jamaica, also, distress was widespread at the close of the war.[61]

British North America, in 1748, was commercially less dependent upon Great Britain and its colonies than at any previous time. Its export trade to foreign markets was firmly rooted, and from those markets it derived, in violation of the navigation laws and the Molasses Act, the bulk of its tropical produce and European manufactures. Prosperity and the standard of living in New England and the Middle Colonies were intimately connected with the existence of this trade. Commerce with England was obstructed by the small demand in Great Britain for the staples of New England and by a scarcity of currency in the latter with which to buy British goods.

59 Vice-Admiral Townshend to Lucas, Mar. 28, 1746; Lucas to Townshend, Mar. 26, 1746. Adm. 1: 305. "We are distressed to the greatest Degree, all most to famine by the French taking all our Trade for Provisions. They and Plantation Necessarys have been & still are at most Exhorbitant Prices, from their Scarcity. No Planter can Subsist much longer under the Pressure. . . . And now Our Calamitys are Encreased by a Drawth (in this Island especially) That our Cattle begin to dye for want of Forage and Water & Consequently little or no Crop for next Year." Gov. Mathew to Board of Trade, Antigua, Sept. 19, 1746, C.O. 152: 25, Y 162. See also Admiral Osborn to Corbett, Antigua, May 24, 1748, Adm. 1: 305.

60 "The Effects the French Settlements have upon this Island [Barbadoes] are many & injurious, they hinder the consumption of our Sugar & Rum in the Northern Colonys by underselling us considerably to the Dutch at St. Eustatia . . . Their soil is much more fertile than ours; by means whereof they make greater quantitys so they do it with much less Expence & Trouble, their Dutys are trifling, & their Navigation with these Commodities are free: Upon all which Accounts they can Outsell us at Foreign Marketts." Gov. Robinson to Board of Trade, Barbadoes, Feb. 20, 1746/7, C.O. 28: 27, Bb 57.

61 Gov. Trelawny to Board of Trade, June 8, 1749, C.O. 137: 25, X 37.

The Northern Colonies were getting more and more out of harmony with the mercantilist ideal of empire, or with what in more recent times would be called imperial federation.[62]

[62] The following interesting observations on North American trade were submitted by Henry McCulloch to the Duke of Bedford, July 28, 1748: "That your Memorialist also observed in his Progress through America, that as the Colonies there Increase in Riches, so they increase in Luxury; And that the Silks, Hollands Linnen Cambricks, Threads, Stuffs, Tea, Sugar, Molasses &c Imported into the Northern Colonies from Holland, France, and the Levant, and from the Dutch and French plantations, are much more in value, than from Great Britain and its Colonies, And although all Encouragement ought to be given to the Exportation of Fish from New England, and Newfoundland, and also to the Exportation of Corn, Flower, and Bread from New York and Pensilvania, and even to the shipping of lumber, Horses, and Provisions, to the Dutch, and French Colonies, yet it is Conceived that in Return for those Commondities a very great Illicit Trade is now Carried on, And that it will be very much for the service and Benefit of His Majesty's Subjects in Great Britain, to have the state of the said Colonies taken into Consideration, and to provide such Remedies as may be thought most Expedient to prevent, for the future, an Undue Importation of Foreign Merchandizes into the said Plantations.

"That One great Reason of the Trade of His Majesties Northern Colonies being Carried on in a wrong channel proceeds from the British Merchants not being fully secured in their Properties, by the frequent alterations made in the Current bills of the said Provinces, And by the Arbitrary Exertion of Power in the Governours thereof, All which naturally lessen the Credit of the said Colonies, And from that lessening of Credit the merchants in America are induced to furnish themselves with Goods from those places, where they dispose of the Merchandise usually Exported by them to Foreign Markets." C.O. 5: 5, f. 246.

CHAPTER XIII

ILLICIT COMMERCE, 1748-1763

In the period of truce between the close of the War of the Austrian Succession and the opening of the Seven Years' War, illicit commerce continued along the lines that had proved so profitable.[1] The period is characterized chiefly by attempts of the navy to suppress the evil, by memorials from planters against it, and by investigations of the situation by the Board of Trade and the House of Commons. Under Commodore Holburne's command a number of illicit traders were seized among the Caribbee Islands, but the presence of an unscrupulous vice-admiralty judge at Barbadoes, during the years 1750-1751, did much to relax the prosecution of smugglers.[2]

The most notable instance of illicit trade in this period was the case of the *Enterprise* in 1749.[3] The affair was brought to the attention of the British government in a representation to the king from the governor, council,

[1] Gov. Henry Grenville to Board of Trade, Barbadoes, Feb. 8, 1748/9, C.O. 28: 29, Cc 28; William Bollan to Board of Trade, Boston, Oct. 24, 1749, C.O. 323: 12, C 61; Gov. Mathew to Board of Trade, Antigua, Nov. 17, 1748, reporting seizures by the French of vessels carrying flour and bread from Antigua to the French islands, C.O. 152: 26, Z 40; Gov. George Clinton to Board of Trade, New York, May 23, 1749, C.O. 5: 1062, Hh 48; Commodore Holburne to the Admiralty, Barbadoes, Sept. 25, 1751. Adm. 1: 306.

[2] Commodore Holburne to the Admiralty, Barbadoes, Feb. 3, 1750, May 31, 1751, and Aug. 24, 1751, Adm. 1: 306.

[3] C.O. 5: 6, ff. 3-25; C.O. 137: 25, X 49, 50, 51, 52, 53. See Appendix X, p. 401, for documents on the *Enterprize* case. Intercourse with Santo Domingo from the Northern Colonies was carried on by an easier navigation, than that to Jamaica, the Jamaicans point out, because it was geographically nearer North America. C.O. 5: 6, f. 9.

and assembly of Jamaica, and was agitated as an indication of a widespread evil in the colonial customs service that required reform by parliament. The sloop *Enterprize,* owned by Jonathan Nichols, deputy governor of Rhode Island, Richard Mumford, master, had cleared in August 1749 for Jamaica. Sailing directly for Léogane (Santo Domingo), and being laden there with sugar and molasses, she was captured by H. M. S. *Cornwall* in Tiburon Bay with important papers and correspondence. A letter to Nichols from Lawrence Cholet, a French merchant at Léogane, revealed interesting details of a regular intercourse between Rhode Island and Santo Domingo, in violation of all the trade laws and in collusion with French authorities and Rhode Island customs officers. All this was corroborated by the testimony of Richard Mumford, who declared that the *Enterprize* was but one of many vessels engaged in this trade.[4] But the British West Indies were scarcely better off than the French for honest customs administration, for at Tortola, Spanish Town, and Anguilla, in the Virgin group, collectors were granting false clearances for Danish sugar from St. Thomas and Santa Cruz to be carried to Bristol and London as of British growth.[5]

The British planters in 1750 were deeply alarmed about the security of their monopoly of the British and colonial sugar markets. Memorials were sent to the Board of Trade by the governments of Jamaica, Antigua, St.

[4] The great advantage of French molasses over English in the manufacture of rum is indicated by the fact that in New England, in 1750, West India rum sold for about 2s. 6d. per gallon, while the price of New England rum was from 1s. 4d. to 1s. 6d. per gallon. Capt. John Thomlinson to Thomas Hill, secretary of the Board of Trade, London, Nov. 15, 1750, C.O. 323: 13, Q 78.

[5] William Mathew to Board of Trade, London, Mar. 8, 1750/1, C.O. 323: 13, O 91. These collectors were under the surveyor-general for Barbadoes and the Leeward Islands; they received no salaries and were very mercenary. The Virgin Islands produced little but cotton.

Christopher, and by an association of planters and merchants in England, asking for an absolute prohibition of intercourse with foreign sugar settlements in America.[6] In the winter of 1750-1751 the Board of Trade devoted a number of meetings to canvassing the whole subject of illicit trade in America.[7] Many interesting testimonies on West India commerce from 1733 to 1750 were made by naval officers, merchants, and traders. John Sharpe managed the side for the planters, while William Bollan, agent for Massachusetts, the other agents for the Northern Colonies and their counsel, Ferdinand John Paris, attended in behalf of the continental colonies. Of these, only Bollan and Paris spoke, very briefly, declaring that a suppression of trade, legal or illegal, to the foreign West Indies, would subject the settlement of North America to a most serious limitation. The planters' association finally modified their proposal for a prohibition of foreign intercourse by making the following recommendations: 1, that the Northern Colonies be prohibited by act of parliament from importing sugar, rum, or molasses from any foreign sugar settlement; 2, that ships of the royal navy be empowered to visit and search any merchant vessel and to seize and prosecute in the vice-admiralty courts vessels laden with foreign sugar, rum, or molasses; liberal rewards being given to naval officers and crews; 3, that four sloops of war be stationed to patrol the New England coast, that the ships on the Leeward Island station prevent illegal trade at St. Eustatius,

[6] Lieut. Gov. Fleming's address to council and assembly of St. Christopher, Nov. 10, 1750, C.O. 152: 27, Aa 6; memorial from St. Christopher to Board of Trade, Dec. 20, 1750, *ibid.*, Aa 17; memorials from Antigua to Board of Trade, received Jan. 23, 1750/1, C.O. 152: 27, Aa 3; memorial of sugar planters, merchants, and others to Board of Trade, received from John Sharpe, Oct. 18, 1750, C.O. 323: 12, O 59; another copy of the same is in C.O. 5: 38.

[7] See Appendix XI, pp. 414-430 for the official report of the investigation.

Santa Cruz, and St. Thomas, that those of Barbadoes prevent the same at Surinam, Berbice, and Essequibo, that those at Jamaica guard against illicit trade to Santo Domingo; and that these patrols be ordered to cruise as constantly as possible and exercise every power to detect illegal trade; 4, that customs officers, by a stricter system of rewards and penalties, be stimulated to honest and efficient service; 5, that West India customs officers be strictly enjoined not to give cockets to Northern vessels without strict search of their cargoes first being made; and 6, that governors be required to do their utmost to enforce the law.[8] The Board of Trade would not recommend any modification of the Molasses Act. In March, upon petition from the planters for relief, a brief inquiry into the illicit sugar trade was started by the House of Commons, but came to nothing.[9] The planters intended to renew their application to the Commons at its next session,[10] but the matter was allowed to drop. The net result of the agitation was a circular letter to the governors directing them to endeavor to have their colonies pass laws to prevent the illicit importation of foreign goods.[11] The Northern Colonies took no action, but St.

[8] Proposals received through John Sharpe, Jan. 10, 1750/1, C.O. 323: 13, O 87; B. T. Jour., C.O. 391: 58, f. 5. Paris thought these recommendations very detrimental to the Northern Colonies. January 15, 1750/1, the board considered the sugar trade for the last time during this winter. *Ibid.*, ff. 14-15.

[9] House of Commons' request for papers on the illicit trade from the Board of Trade, Mar. 15, 1750/1, C.O. 323: 13, O 93.

[10] Richard Partridge, agent for Rhode Island, Connecticut, New York, and Pennsylvania, to Gov. Green of Rhode Island, July 3, 1751. Kimball, *Cor. of Col. Gov.*, II, 133-134. The Board of Trade gave Sharpe and the Northern agents a final hearing on the controversy, Nov. 5, 1751. C.O. 391: 58. Dec. 19, 1752, the board received from Sharpe a memorandum on "Cases of Flags of Truce," C.O. 323: 13, O 113.

[11] Representation of Board of Trade, Jan. 27, 1762, Brit. Mus. Add. MSS. 35916, f. 71. Two papers bearing on the controversy are: "State of an Illegal and Clandestine Trade carried on by the British Northern Colonies

Christopher and Nevis passed acts, in 1752, for the better enforcement of the Molasses Act. These required every shipper of sugar, molasses, and rum to produce affidavits that the same were produced in those islands before customs officers could grant clearances or the governor grant "let passes."[12] Jamaica also, in 1752, passed an act prohibiting the importation of foreign sugar, rum, and molasses. The act, however, was disallowed, February 15, 1762, because it was in violation of the Molasses Act.[13]

It is clear that, by the middle of the eighteenth century, the British government could not be persuaded by the planting interest to take any further step to enrich the West Indies to the detriment of the Northern Colonies. Public sentiment in fact, as far as it existed at that period, was becoming critical of the monopolizing tendencies of West Indians, and was disposed to blame them in part for the underlying causes of illicit trade. A thoughtful observer, Postlethwayt, asked why, if North Americans could not trade to as good advantage in the English as in the French islands, more land had not been broken up in the British colonies, so that sugar, rum, and molasses would be fully as cheap as in the French islands? This would probably have remedied the commercial malady. Certain it was that the French in the West Indies had sold their sugar from thirty to forty per cent cheaper than the English.[14]

in America" (ca. 1750), Brit. Mus. Add. MSS. 33030, ff. 401-402; and *An Inquiry Concerning the Trade, Commerce, and Policy of Jamaica*, 1757 (but written in 1751), p. 18.

12 Lieut. Gov. Gilbert Fleming to Board of Trade, June 19, 1752, C.O. 152: 127, Aa 57, 58. Gov. Mathew to Board of Trade, Nov. 30, 1752, C.O. 152: 28, Bb 3.

13 C.O. 139: 18, no. 43; renewal of act in 1756 in *Acts of Jamaica*, ed. 1787, II, 4-6; *Acts of the Privy Council, Col.*, IV, 517-518; Brit. Mus. Add. MSS. 35916, f. 71.

14 "If our northern colonies could not trade with our own sugar islands

Among the causes that contributed to lessen trade between the British West Indies and North America, mention should be made of the export duties, collectable on produce leaving the British sugar islands for other British colonies. These were the "enumeration dues" of five shillings per hundred weight on white sugar and one shilling six pence on brown. Also, in Barbadoes and the Leeward Islands, an export duty of four and a half per cent on all dead commodities was collected. As early as 1689, Littleton had called the "enumeration dues" a "grievous clog" on commerce with the Northern plantations.[15] Again in 1751 a friend of Jamaica declared that these duties were a detriment to Northern trade, and that their abolition would remove a large temptation to smuggling.[16] The government of Jamaica, in an address to the king, November 20, 1752, coupled with its petition for a parliamentary prohibition of French imports a prayer for the repeal of the one shilling six pence per hundred weight export duty on brown sugar.[17] At this time, the amount collected upon white and brown sugar

to so good advantage as they have done with the French, why have not proper laws been enacted to break up more land in our island colonies, that sugars, rum, and melasses might be rendered, full as cheap as those in the French colonies? For this we have proved to be the sovereign specific for most of our commercial maladies as well in America as Europe. Certain it is that the French in the West Indies, have sold their sugars from 30 to 40 per cent. cheaper than the English have done. Whether our northern colonies, or our West-India traders, have been the most to blame with relation to what we have been speaking of, I shall not here enquire. That they both have been highly culpable is not to be doubted.'' Malachi Postlethwayt, *Great Britain's Commercial Interest Explained*, London, 1757, I, 494. In Postlethwayt's estimation one of the deepest injuries of illicit trade was its detriment to the increase of British ships and sailors. *Ibid.*, I, 487.

[15] Littleton, *Groans of the Plantations*, p. 4.

[16] *Inquiry concerning the Trade, Commerce, and Policy of Jamaica*, 1757, p. 18. This pamphlet was written, according to the printer's note, in 1751.

[17] C.O. 137: 25, X 115.

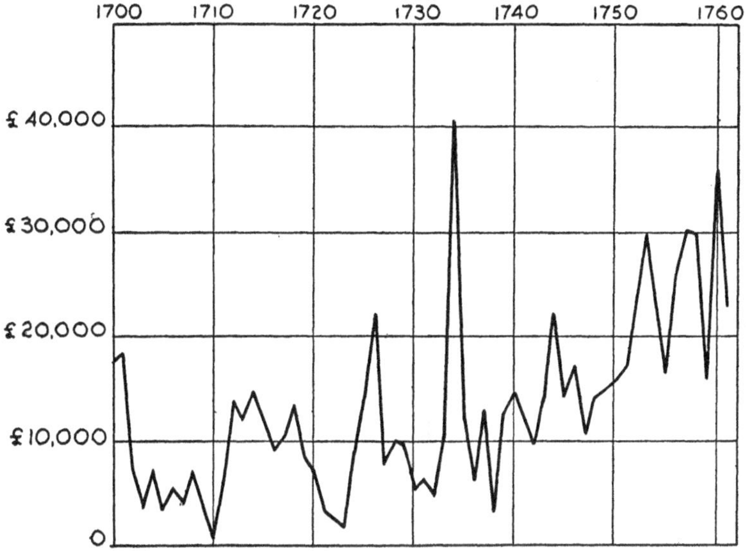

CHART XIII. AMOUNT OF THE REVENUE PAID INTO THE ENGLISH EXCHEQUER FROM THE FOUR AND A HALF PER CENT DUTY LEVIED ON ALL DEAD COMMODITIES EXPORTED FROM BARBADOES AND THE LEEWARD ISLANDS, 1700-1761

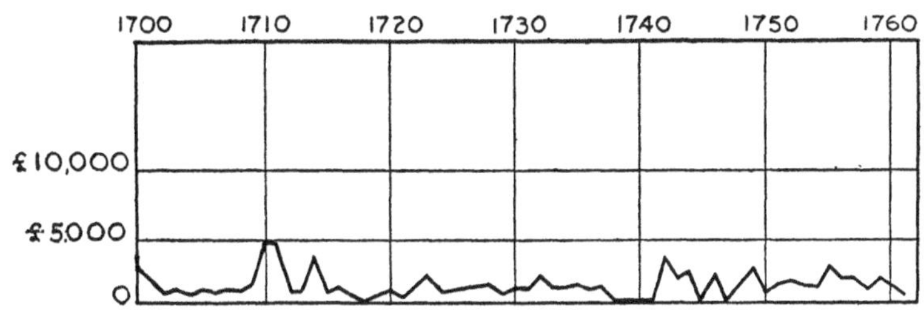

CHART XIV. AMOUNT OF THE REVENUE PAID INTO THE ENGLISH EXCHEQUER FROM THE DUTIES LEVIED ON ENUMERATED COMMODITIES EXPORTED FROM ONE BRITISH COLONY TO ANOTHER BY 25 CHARLES II, C. 7, 1700-1761

The above charts are based on Custom House Accounts in Treas. 38: 340.

exported from the West Indies to the Northern Colonies amounted to a little over £2000 a year.[18] But the duties

[18] The net produce of the duty of 1s. 6d. per cwt. on brown and 5s. on white sugar collected (by 25 Car. II, c. 7) in the West Indies on its export to North America for the years 1745-1751 was as follows (shillings and pence omitted, hence apparent discrepancies in addition):

Year	Jamaica	Barbadoes	Antigua	St. Christopher	Nevis	Montserrat	Total
1745	£ 389	£ 507	£ 211	£136	£ 48	£ 15	£1307
1746	373	529	191	70	7	14	1184
1747	311	417	50	75	8	8	870
1748	344	675	203	136	15	25	1398
1749	205	507	121	66	9	8	915
1750	736	955	182	131	19	17	2040
1751	847	918	248	132	16	16	2179
	£3205	£4508	£1206	£746	£122	£103	£9891

The following table gives the quantities of goods exported from Jamaica to North America on which "enumeration dues" were paid in the same period:

Year	Sugar White at 5s. per cwt.	Sugar Muscovado at 1s. 6d. per cwt.	Ginger at 1s. per cwt.	Dyeing Wood at 6d. per cwt.	Indigo at 2d. per lb.	Cocoa at 1d. per lb.	Cotton at ½d. per lb.
1745		10891 cwt.	85 cwt.				2669 lbs.
1746		10453 "	58 "				2873 "
1747		8715 "	78 "				——
1748		9628 "	23 "	15 cwt.		6061 lbs.	7354 "
1749		5724 "	80 "			257 "	2384 "
1750	156 cwt.	20074 "	109 "	220 "	127 lbs.	1084 "	3625 "
1751	297 "	22739 "	—— "		298 "	——	15164 "
	453 cwt.	88224 cwt.	433 cwt.	235 cwt.	425 lbs.	7402 lbs.	34069 lbs.

The amount of the duties collected in Jamaica on "enumerated goods" exported to North America in the same period was as follows (shillings and pence omitted):

Year	On Sugar	On Other Goods	Total
1745	£ 817	£ 6—	£ 822+
1746	784	10	794
1747	654	3	657
1748	722	44+	767
1749	429	7	436
1750	1544	23	1567
1751	1780	39	1819
	£6730	£132	£6863

Treas. 64: 274.

were probably largely evaded, as the radical increase in amounts paid during the years of commercial investigation, 1750-1751, seems to indicate. If the plantation duties had been scrupulously collected, they would have constituted undoubtedly too great a burden on colonial commerce. In any case, the influence of export duties in the British West Indies seems to have been to stimulate the Northerners to purchase return cargoes in the foreign islands.

An effect of illicit trade with the French, which was the subject of criticism at all times but especially at this period, was its tendency to drain the British West Indies of their coin. This was felt particularly after the interruption of trade with the Spanish Main had diminished the inflow of specie. Governor Knowles of Jamaica attempted, by an inspection of the custom-house books in 1752, to measure the extent to which North American traders were draining that island of cash. He observed that, from January 1, 1735, to October 19, 1752, the number of vessels cleared from Jamaica for the Northern Colonies was 2503, of which 763 had cleared in ballast only and more than half the remainder had cleared with but one to five hogsheads of sugar or molasses. Those leaving with little or no cargoes must, as a rule, have carried cash. Again, the value of imports into Jamaica from North America, in 1751, amounted to £112,825. But the highest value of the produce of Jamaica exported in any one year to North America amounted to no more than £29,222. The difference between these amounts, excepting commissions and port charges, estimated at £9060, must have been carried off in money, so that the island was drained annually of a balance of £78,902, all of which money, he believed, was carried to the French. Also, the clearances in ballast, according to Knowles, facilitated entries in the Northern Colonies. In reality, ships sailing in ballast purchased French produce, which was clandestinely landed in the numerous inlets of the American coast, and then entered port in ballast with Jamaica clearances to correspond. It was incredible that ships clearing from the West Indies in ballast in winter, as many did, could have made their way through the stormy Atlantic without cargoes.[19] Two years later Knowles stated that the amount of specie carried from Jamaica to Santo Domingo was £50,000 sterling a year.[20] Another estimate of the coin drained from Jamaica, made in 1751, put the amount at £64,977, but it was charged that Jamaicans themselves were in part responsible for the outflow.[21] This permanent balance of trade against Jamaica prompted many suggestions that the island undertake the breeding of its own horses, and the production of shingles, headings, corn, certain provisions, and even lamp oil, and that other foods and salt meats might better come from England.[22] The British West Indies constituted but one of many slave communities in modern history that became alarmed at its dependence in a multitude of ways upon more highly organized societies, and its aspirations for industrial independence are paralleled by those of our own Southern commercial conventions in the two decades preceding the Civil War.

The failure of the planters' association to secure an absolute prohibition of Anglo-French trade in America

19 Gov. Charles Knowles to Board of Trade, Nov. 18, 1752, C.O. 137: 25, X 112. He states that the American trade ''has drained the Island of Specie to such a degree now, as to become a matter of the last importance.'' These figures were also given in the Jamaica colony's address to the king, Nov. 20, 1752. C.O. 137: 25, X 115.

20 To Board of Trade, Dec. 31, 1754, C.O. 137: 28, Y 55.

21 *An Inquiry concerning the Trade, Commerce, and Policy of Jamaica*, p. 16.

22 *Ibid.*, 17-18. The suggestion is made in Long, Edwards, and elsewhere. With slave labor, however, it was unprofitable to diversify industry enough to lessen appreciably the dependence of the West Indies on outside supplies.

must have tended, notwithstanding admonitions of the Board of Trade to the contrary, to confirm North Americans in their feeling that the Molasses Act could still be safely evaded. In the decade from 1750 to 1760, illicit trade increased rather than diminished, even during the French War, and its suppression in 1760 was based on military necessity rather than in response to the desires of sugar planters.

Jamaica, as we have seen, was an important point of departure for vessels engaged in trade with Santo Domingo and thence departing to North America. With a coast line extending over three hundred miles, Jamaica possessed but one port of entry, Kingston. The settlement of the island, however, caused a number of outlying bays and inlets to become exporting points for plantation products. Vessels that had unloaded cargoes in Kingston and then reloaded with sugar at some distant point, such as Montego Bay or Port Antonio on the north shore, found it a great inconvenience to return to Kingston to obtain clearance papers. It was a practice of long standing, therefore, for a ship to enter at Kingston and then, in a few hours, receive clearance papers specifying a fictitious clearance of so much sugar, etc. The vessel's cargo could then be landed at some outport and such produce taken in as the master desired. Consequently customs officers in England or America never knew from the clearances just how much of the ship's cargo came from Jamaica and how much was either previously or subsequently taken in at a French port. For this reason Governor Knowles declared that naval office lists could not be depended upon to give the correct produce of the island.[23] The naval officer at Kingston, profiting entirely by the fees of his office, found it a great temptation to wink at malpractices. It was no use discharging officials,

[23] Gov. Knowles to Board of Trade, Jan. 10, 1753, C.O. 137: 25, X 122.

said the governor, for the defect was in the institutions.[24] In view of the foregoing criticism of public accounts, it is probably not too much to say, that the British West Indies were a good deal less productive than was commonly supposed, and that the British planters' fear that their French rivals were undermining them in British markets was well founded. After repeated recommendations by Governor Knowles that custom-houses be established at the Jamaica outports, it was finally directed by order in Council of June 29, 1758, that, in addition to Kingston, Savanna la Mar, Montego Bay, and Port Antonio should be ports of entry with regular customs offices.[25]

24 Gov. Knowles to Board of Trade, Nov. 18, 1752, C.O. 137: 25, X 113. "It is well known," wrote Knowles, "that great Quantitys of French Sugars and Rum are introduced into this Island and Shipped home for Great Brittain annually. Some particular Persons My Lords may be benefited by this Indulgence of Ships Loading at the Out Ports but the Island in general receive great prejudice." To Board of Trade, Jan. 12, 1754, C.O. 137: 27, X 198. Again he wrote: "As it is a practice with Masters of Ships who load at the Out Ports of this Island, to clear at the Offices in Kingston before their Cargoe is on-board, or even before they know for a certain what it will consist of; therefore in course, what they give in to the Collector to be inserted in their Clearance as their Loading, is nothing but guess-work; so that what they Report at Home is often (most times) quite different both as to the quantity & quality." To Board of Trade, Dec. 31, 1754, C.O. 137: 28, Y 55. See also Gov. W. H. Lyttelton to Board of Trade, July 9, 1763, *ibid.*, 33, Cc 9. In view of this evidence, it has seemed wise not to print in this work tables, which the author has compiled from the Ledgers of Imports and Exports, purporting to give the amount and value of exports from each British West India colony to England from 1698 to 1765. The records are certainly unreliable and, for the latter half of the period at any rate, give too favorable an impression of the productiveness of the English sugar islands. This criticism applies also to the tables printed by Bryan Edwards.

25 Rose Fuller, agent for Jamaica, to Board of Trade, Nov. 20, 1757, approving the board's proposal to increase the ports of entry. C.O. 137: 30, Z 8. Robert Henley and Charles Yorke, attorney general and solicitor general, to Board of Trade, May 18, 1757, giving their opinions that the Crown had power to increase the number of ports of entry in Jamaica, *ibid.*, Z 1. Order in Council, June 29, 1758, C.O. 137: 31, Aa 27.

In the Windward Islands, after many attempts by both England and France to colonize them, Dominica, St. Lucia, St. Vincent, and Tobago were at this time recognized as neutral. Several hundred Frenchmen with slaves were settled, nevertheless, in these islands, and it was reported, in 1753, that North Americans were supplying them with lumber, horses, frames for houses, and all kinds of provisions. Sugar, cotton, cocoa, and coffee cultures were in progress here under the French, who sold great quantities of molasses and rum to the North Americans. The latter were thus aiding the extension of French supremacy in the Caribbean.[26]

Of the Northern Colonies, Pennsylvania, as well as Rhode Island and Connecticut,[27] was heavily concerned in intercourse with the French. The master of a Philadelphia vessel, clearing from Jamaica in ballast in 1752, was asked why he did not take the commodities of the island in return for his cargo. He answered that he was obliged to take specie because he could not afford to buy Jamaica produce for the Philadelphia market, where sugar, rum, and molasses, when he left, were cheaper than in Jamaica. He declared it was generally reported that William Allen & Co. of Philadelphia, in league with Abraham Taylor, the collector, who was one of the company, carried on a very considerable trade regularly with Léogane and St. Eustatius. This company always undersold traders who imported from the British West Indies, and there was a general complaint among the merchants

[26] Commodore Pye to Sec. John Cleveland, Barbadoes, Mar. 31, 1753, Adm. 1: 306. A thorough investigation of the admiralty papers would probably yield much material on American commerce and colonization.

[27] William Shobrook's memorial to Newcastle (ca. 1754), giving his observations on illicit trade in America, Brit. Mus. Add. MSS. 34729, f. 348. The Naval Office Lists for Boston for this period reveal some trade with the Dutch and Danish Islands which, in reality, was probably trade with the French. Cargoes recorded as coming from the British West Indies were in many cases very light. C.O. 5: 849.

of Philadelphia of this clandestine commerce. One powerful concern was in this way depriving other firms of the power of buying British sugar. Allen & Co., through the operations of their agent William Humphreys, were fast monopolizing the sugar business in Philadelphia. The price of molasses at Léogane was three dollars per cask of sixty gallons, or about three and a half pence per gallon; the same sold in Philadelphia for eighteen pence per gallon. The Northern consumer was not supplied "at any considerable" rate cheaper by the clandestine than by the fair trader. The illicit merchant was thus making exorbitant gains and usurping control of the market to the prejudice of the fair trader.[38]

The Irish market also was a source of worry to the British planters' association. Its agent, John Sharpe, in February, 1752, called the Board of Trade's attention to the fact that Portuguese sugar from Brazil was being imported directly into Ireland in British ships without their first stopping in any port of Great Britain. In both the Navigation Act of 1660[29] and Molasses Act[30] it was provided that no restriction should be put upon the direct importation into Ireland of Spanish or Portuguese sugar. After 1671, British plantation sugar, on the contrary, could enter Ireland, as a rule, only by way of England.[31] Thus foreign sugar, the planters felt, was put on a better footing than British in Ireland. They declared the British sugar trade was being damaged thereby, and urged the enactment of a law that no sugar or molasses could be imported into Ireland which had not first been landed in England. Several Portuguese mer-

28 "Copy of a Declaration concerning a Clandestine trade between Philadelphia and the French sugar Colonies," Sept. 18, 1752. In Gov. Knowles to Board of Trade, Nov. 18, 1752, C.O. 137, X 113.

29 12 Car. II, c. 18, sec. 1, 3, 5, 6, 8.

30 6 Geo. II, c. 13, sec. 13.

31 Beer, *The Old Colonial System*, Pt. I, Vol. I, 94 *et seq.*

chants were called before the Board. They affirmed that the price of Portuguese sugar was in general so high that it would not compete with English, and that importing it was a losing trade. Portuguese brown sugar was at times, however, taken in exchange for English woolens. For seven years before the Spanish War of 1739, Portuguese sugar generally undersold English, but since the war, the importation of Brazil sugar into Ireland had greatly decreased and there seemed to be no danger of its increasing. To the chagrin of the aggrieved planters, the board decided, owing to strained political relations at the time with Portugal, to postpone consideration of the planters' bill.[32] The monopolizing propensities of British planters were once more held in check by imperial authorities.

The period of the Seven Years' War reveals on a larger scale the same illegal habits in West India commerce which characterized the previous conflict. The extensiveness and universality of trade with the French, even in war time, indicate how interdependent British North America and the French islands were and how hopelessly opposed the Molasses Act was to the natural course of commerce.[33]

The American colonies began the war with the expressed resolution to prevent the exportation of provisions to the French.[34] An order in Council of June 30, 1756, approved instructions to the colonial governors

[32] Draft of a bill received from John Sharpe by Board of Trade, Feb. 14, 1752, C.O. 323: 13, O 103; B. T. Jour., Feb. 19, 20, 1752, C.O. 391: 59. The planting interest was led by Sharpe, Long, Maitland, and Douglas. The merchants who testified on the Portuguese trade were Burrell, Main, Mollorty, Gore, and Chase.

[33] The history of illicit trade in this period has been ably treated by Mr. G. L. Beer in his *British Colonial Policy, 1754-1765,* and it will be hardly possible to do more here than emphasize the bearing of that commerce on the British West Indies.

[34] Gov. Shirley to Sir Thomas Robinson, June 20, 1755, C.O. 5: 15.

directing them to do their utmost to hinder the French being supplied with provisions and warlike stores.[35] In September, an embargo was laid in Ireland on all vessels bound to neutral ports with provisions, and governors were instructed to enforce a like embargo in the West Indies and North America.[36] Since customs offices were not subject to the control of governors, the Commissioners of the Customs were expected to compel the former to carry out the embargo.[37] It was thought advisable that convoys be furnished to protect supplies in transit from North America to the British West Indies but, unfortunately, no provision was made for them.[38] The restrictions on commerce in provisions were reaffirmed, in 1757, by act of parliament.[39] Thus every legal precaution was taken to prevent the recurrence of the treasonable commerce that was so widespread in the previous war.

Reports, however, were soon made that flour and butter had been shipped from Antigua to Curaçoa where the French secured all kinds of English provisions.[40] At St. Eustatius, shortly afterward, five Irish ships

[35] C.O. 323: 13, O 136.

[36] Henry Fox to Board of Trade, Sept. 27, 1756; circular letter from Board of Trade to the governors, Oct. 9, 1756, C.O. 5: 7, ff. 132-133. Vessels were to leave colonial ports only under heavy bonds that their cargoes should be landed in British possessions. The declared aim was to stop supplies to the French via the Dutch and other neutral West Indies. Governors were warned to beware of vessels allowing "Collusive Captures" of their cargoes.

[37] Board of Trade to Henry Fox, Oct. 9, 1756, C.O. 5: 7, ff. 130, 135; Fox to Board of Trade, Oct. 12, 1756, C.O. 323: 13, O 147.

[38] Board of Trade to Henry Fox, Oct. 1, 1756, C.O. 5: 7, ff. 127-128; Fox to Board of Trade, Oct. 2, 1756, C.O. 323: 13, O 144.

[39] 30 Geo. II, c. 9. Board of Trade to Nicholas Harding, Mar. 4, 1757, C.O. 5: 7, f. 161.

[40] Gov. Charles Hardy of New York to Board of Trade, Mar. 11, 1757, enclosing depositions of eyewitnesses of the trade at Curaçoa. C.O. 5: 1067, Ll 97; copy in C.O. 5: 18.

arrived with provisions.[41] The war sent prices of West India produce and of northern provisions to exorbitant levels, which strongly tempted Northerners to renew trade to the French islands, where sugar was cheap and provisions high on account of the interruption of trade with France.[42] Notwithstanding special acts of the New York assembly to the contrary,[43] an extensive trade in provisions for the French was carried on by way of St. Eustatius and the Danish Islands, whereby, according to Burnaby, New York merchants acquired great riches.[44] Governor Hardy attempted to break up this illicit trade not only from his own colony but from Elizabeth and Perth Amboy, New Jersey.[45] New York masters were able to buy or forge certificates from customs officers in

[41] Hardy to Board of Trade, June 19, 1756, C.O. 5: 1067, Ll 43.

[42] New Haven newspaper quotations of prices at New York, in September, 1755, include the following:

Muscovado Sugar........	50s. per cwt.
Molasses................	2s. per gal. by the hhd.
West India Rum	3s. 3d. per gal.
Wheat..................	6s. per bu.
Flour	17s. per cwt.
Beef	50s. per barrel
Pork	£3 10s. per barrel
Salt....................	2s. per bu.
Indian Corn	2s. 3d. per bu.

C.O. 5: 1066, Kk 76.

[43] Acts of 1755 and 1756: C.O. 5: 1067, Ll 52, 1068, Mm 59; Gov. Hardy to Board of Trade, Dec. 29, 1756, C.O. 5: 1067, Ll 82; Lieut. Gov. DeLancey to Board of Trade, Mar. 15, 1758, C.O. 5: 1068, Mm 50. Both officials reported the publication of embargoes on neutral and enemy trade.

[44] Thomas Robinson to Bradock, Apr. 16, 1755, C.O. 5: 7, f. 7; Hardy to Board of Trade, Oct. 13, 1756, C.O. 5: 1067, Ll 55, in which the governor said it was impossible for New York to enforce her embargo when the neighboring colonies had allowed theirs to lapse. Burnaby, *Travels*, ed. R. R. Wilson, pp. 118, 128, 129, n., quoted by Beer, *Brit. Col. Pol., 1754-1765*, p. 130.

[45] Hardy to Board of Trade, June 14, 1757, C.O. 5: 1068, Mm 8. He also complained of much direct trade from Holland, especially via Connecticut and Philadelphia. To Board of Trade, July 15, 1757, *ibid.*, Mm 13.

the small British sugar islands to the effect that they had there deposited their northern cargoes, which really had been carried directly to St. Eustatius for the French. Bread and flour were plentiful in Martinique and Guadeloupe, while in St. Christopher and Nevis, according to Gilbert Fleming, the lieutenant general, there was not a sufficient supply for a fortnight's consumption.[46] By 1759, prices of breadstuffs and meats even in New York had risen greatly, a fact that was attributed to extraordinary exports to the foreign West Indies.[47]

Massachusetts, on several occasions, declared embargoes on the export of provisions to the French, and, at times, prohibited such export even out of the colony.[48] Whenever shipments were allowed, heavy bonds were required to secure their being landed in British dominions. Nevertheless, many cargoes of fish and other provisions left Salem and Marblehead for a very profitable market at St. Eustatius where, before sales were made, the bonds were falsely canceled. New Englanders regarded the abolition of trade with the French islands as a grievance of which they deserved "full redress from the Court of Great Britain."[49] Governor Hardy of New York said that officers of the navy hesitated to make seizures of suspicious vessels where the ships papers appeared correct for fear of prosecutions. In one instance, however, during his passage from New York to

46 Lieut. Gen. Gilbert Fleming to Hardy, St. Christopher, May 23, 1757, C.O. 5: 1068, Mm 12. Fleming entreated Hardy to release bread and flour for the Leeward Islands as they were in distress for want of them.

47 In the autumn of 1759, the following prices prevailed in New York: wheat 7s. per bu., flour 20 to 23s. per cwt., bread 16s. 6d. per cwt., beef 45 to 65s. per bbl., pork 70 to 80s. per bbl. C.O. 323: 14, P 19.

48 *Acts and Resolves*, 1755-1756, pp. 170, 171, 175, 181, 183, 201, in C.O. 5: 887, Hh 65, 75; Embargo Act of May 25, 1758, in C.O. 5: 18.

49 Extracts from the correspondence of Massachusetts merchants, in January and May 1757, relative to trade with St. Eustatius, found on vessels seized by Lieut. Gov. DeLancey, C.O. 5: 1068, Mm 11, Mm 15.

Halifax, Hardy ordered the seizure of a Salem schooner returning from St. Eustatius, and obtained its condemnation at Halifax upon evidence from the crew that she had carried provisions to the Dutch and had brought a certificate of landing her cargo at a port she had never entered in her voyage. Hardy urged that more seizures be made upon examinations of crews, no matter how protected the ships might be with proper papers.[50] Lieutenant Governor DeLancey of New York seized a number of New Englanders engaged in illicit trade and offered privateers increased rewards for all such seizures and convictions.[51]

No phase of illicit trade with the French West Indies could compare, however, in extent and boldness with that conducted between New England and Santo Domingo via Monte Cristi, a little port in the northwest corner of Spanish Santo Domingo. Evidence concerning this trade was apparently first revealed in May, 1757, when Martin Garland, an English prisoner who escaped from Santo Domingo, described it to Judge John Chambers of the New York supreme court of judicature. While seven months at Cap François (in French Santo Domingo), he declared that fourteen English vessels from Rhode Island and Boston arrived there with provisions, which they sold to the French for molasses, sugar, indigo, coffee, and cotton. It was the practice of Northerners, upon arriving, to put their men on shore at Monte Cristi, and there get a Spanish captain, crew, flag, and passport, and then proceed to Cap François. With a French cargo they returned to Monte Cristi, took on the English captain and sailors, and proceeded to Boston or Rhode Island. The Northerners' agent at Monte Cristi at that time was Gambauld, a Frenchman who had a wife and

50 Hardy to Board of Trade, July 15, 1757, C.O. 5: 1068, Mm 13.
51 DeLancey to Board of Trade, Jan. 5, 1758, *ibid.*, Mm 42.

family in New York. Provisions also arrived in Santo Domingo from Philadelphia and Jamaica. Occasionally, Spanish vessels brought the British supplies for the French.[52]

A typical instance of a Rhode Islander's methods of trade with the French in war time was reported by Governor Hardy, in June 1757, in connection with the capture of the sloop *Speedwell* of Newport, Thomas Cornell, master. As supercargo in this vessel was John Boutin, a French protestant who had lived with a wife and children at Rhode Island five years, but was unnaturalized. During these years, he had been employed in the trade from Rhode Island to Santo Domingo, and to St. Eustatius and other Dutch islands. On the previous 24th of March, he had arrived in the *Speedwell* at Monte Cristi with a cargo of 51 boxes of spermaceti candles, 6000 bricks, 11000 shingles, 1000 staves, 1100 hoops, besides small adventures in onions, cutlery, cheese, seven or eight cwt. of bread, and five barrels of beef. At Monte Cristi, Boutin sold all the cargo to Don Manuel Destrades, except the bricks and staves, which were delivered at Havana. Then, with a Spanish pilot and four other Spaniards, the vessel proceeded to Port Dauphin. Boutin going by land met the boat there. All of the English crew, except John Murphy, serving as captain, were left at Monte Cristi. At Port Dauphin, the *Speedwell* took in sugar, rum, and indigo. On her return to Monte Cristi, however, she was captured by Captain Spellen, a privateer of New York, to which port she was

[52] Deposition of Martin Garland before John Chambers, New York, May 31, 1757, C.O. 5: 1068, Mm 7. Garland's statements were corroborated by James Lilley who, having recently been at Monte Cristi, told of a Rhode Island sloop whose crew waited there while the vessel traded at Port Dauphin on the French side of the island. Sworn June 1, 1757, before Lieut. Gov. DeLancey, *ibid.* These depositions were enclosed in a letter from DeLancey to Board of Trade, June 3, 1757, *ibid.*

carried. Boutin, when seized, was in the employ of Ebenezer Richardson, Thomas Richardson, and Lemuel Wyatt of Newport. These employers had intended that Boutin, as their agent, should settle at Monte Cristi and collect great quantities of sugar and molasses, and they had agreed to supply him with cargoes. Boutin had expected to obtain the sugar and molasses from Port Dauphin and other French ports, upon which he was to have been paid a ten per cent commission. Boutin was instructed, in case of capture, to prosecute the privateer or naval officers who seized him. Governor Hardy believed that the *Speedwell* was but one of several vessels engaged in a similar trade.[53]

DeLancey became greatly alarmed at the enormous extent of this trade, and permitted Captain Sears, a privateer, to cruise off Block Island and the east end of Long Island in order to intercept some of the Rhode Islanders. But, missing them, Sears put into Rhode Island where he was not only threatened but imprisoned for a few days on a frivolous pretense.[54]

In the West Indies the British squadron at Jamaica succeeded in a great measure in putting a stop to commerce in provisions between the Dutch and the French. Early in 1759, efforts were also undertaken to break

[52] Hardy to Board of Trade, June 14, 1757, enclosing depositions of John Boutin and John Murphy taken before Lieut. Gov. DeLancey, June 9, 1757, and also the copy of a letter from the Richardsons and Wyatt to Boutin, dated Newport, Mar. 7, 1757, C.O. 5: 1068, Mm 8, 9, 10.

[54] "On his return he told me that their rage against him was in part [due] to his having chased a very rich vessel from Holland with tea & other goods not allowed to be imported thence into the Plantations. It is well known that much of the Dutch trade is carried on to Rhode Island & Connecticut, & thence through the Sound to this City both by Merchants there and here." To prevent this he suggested that a small naval vessel be commissioned to cruise through the Sound from Sandy Hook to Block Island. DeLancey to Board of Trade, New York, July 30, 1757, C.O. 5: 1068, Mm 14.

up the trade at Monte Cristi. On February 5, 1759, H. M. Sloop *Viper*, Captain Usher, under orders from Admiral Cotes, entered Monte Cristi and spoke with twenty-nine Northern sloops and schooners lying there engaged in the trade. A list of these vessels with particulars concerning their activities was prepared and sent to the Board of Trade with the opinion of Lieutenant Governor Moore of Jamaica that serious damage to the West Indian trade from French privateers could be terminated only by the suppression of the Monte Cristi trade.[55] In May, Captain Edwards in H. M. S. *Assistance* seized eight Northern vessels near Monte Cristi laden with French sugar and rum and carried them into Jamaica, where they were all condemned in the vice-admiralty court for trading with the enemy.[56] For the year 1759, Admiral Cotes estimated that more than two hundred Northern ships had loaded at Monte Cristi.[57]

Flags of truce were also quite as pernicious in carrying provisions to the enemy. A few of them had been seized, but the trials proved so expensive that captains were afraid to touch them.[58] At Philadelphia, especially, com-

[55] Lieut. Gov. Henry Moore to Board of Trade, Mar. 28, 1759, enclosing a list of 29 vessels spoke with by H. M. Sloop *Viper* in Monte Cristi harbor, Feb. 5, 1759, C.O. 137: 30, Z 43, 44. This list was submitted to the Commissioners of the Treasury, May 24, 1759, C.O. 391: 66.

[56] Gov. George Haldane of Jamaica to Board of Trade, June 9, 1759, inclosing affidavits of June 1 respecting these vessels. Capt. John Harper stated that, during the spring of 1759, he had seen over 100 Northern vessels at Monte Cristi trading for French sugar, molasses, and rum. Other estimates of the numbers of such vessels seen there were: Capt. Benjamin Tucker, over 120; Nathaniel Davis, 140; Thomas Mansfield, 150; Andrew Boin, 160; John Northam, 60 to 90; Nicholas Tracy, over 100; Constant Freeman, 125. Cooke, Tilly, and Woodworth were Northern agents at Monte Cristi; there were also some Spanish middlemen. C.O. 137: 30, Z 59. See also a letter of Admiral Cotes dated Jamaica, Aug. 25, 1759, C.O. 323: 14, P 20. The Board of Trade made a representation to the king on the Monte Cristi trade, Aug. 31, 1759, C.O. 391: 66.

[57] Cotes, Dec. 6, 1759, C.O. 323: 14, P 20.

[58] *Ibid.*

missions for flags of truce were purchased to a scandalous extent. Estimates of the amount of coin drained from Philadelphia by this and the Monte Cristi trade by the end of 1759 were placed at £100,000 sterling; not a vessel, it was said, took out less than £1500 to £2000. Not only money and foodstuffs but also British manufactures were carried to the French. French sugar was on the whole favorably received in the Philadelphia market, but provisions there were becoming scarce.[59] Similar effects of the trade were reported from New York.[60]

Commerce with the French West Indies had thus got far beyond the stage of being merely illegal, that is, a violation of the Molasses Act, which everyone in the North now seems to have ignored, it had become traitorous. The action now taken against it by the government was based, not upon any special policy of relieving the scarcity of supplies in the British sugar islands and securing to them the colonial and British markets, but upon the military necessity of gaining the mastery over the French West Indies as well as Canada. On August 23, 1760, William Pitt sent a circular letter to the colonial governors to report to him at once the state of any existing trade with the enemy, and to bring those engaged in it to the most exemplary and condign punishment, and to take every precaution to prevent its recurrence.[61] In-

59 Extract from a letter from Philadelphia, Dec. 1759, C.O. 323: 14, P 20. The cargo of the snow *Hercules*, a Northern flag of truce captured on entering Port Louis, Santo Domingo, contained 693 barrels of flour, 19 of pork, 10 of butter, 10 casks of bread, 17 of raisins, 1 of nails, 12 pieces of Osnabrig, 2 tierces of "Slops" (seamen's breeches), 1 bag of shoes, 6 of sundries, 33500 shingles, 2800 staves, 222 hoops, 1736 boards, 134 planks, and 123 joists. Cotes, Aug. 25, 1759, *ibid.*

60 Extract of a letter from New York, Dec. 16, 1759, *ibid.* George Colebrook to a Lord of the Treasury, Feb. 18, 1760, C.O. 323: 14, P 17. He stated that food was so high at New York that the royal forces could be provisioned cheaper from London; rates of exchange on specie at Philadelphia were also alarmingly low.

61 C.O. 5: 215.

quiries were instituted in most of the colonies and reports submitted to Pitt in which the existence of trade with the French West Indies was disclaimed altogether, minimized, or fully admitted and stoutly defended.

The first reply appears to have been from Governor Boone of New Jersey, who assured Pitt that, owing to the great activity of the navy and privateers, no trade of the kind was now carried on from his province.[62] Acting Governor Colden of New York also promptly reported that he had directed every official to exert himself in the discovery of persons engaged in this trade and bring them to punishment. But in his endeavors to get evidence he found that no one could be induced to inform against any particular person, though he knew that both New York and Philadelphia merchants had been too generally concerned in illegal trade. The method used in this commerce, he said, was to ship large quantities of provisions to New England, for which the merchants gave bonds and returned proper certificates of their being landed. French sugar was brought back to New Jersey and New England and thence imported into New York with proper cockets of having been legally imported. There was also much unrecorded trade from outlying ports in the province. He believed the only way to prevent this trade was to station cruisers properly along the coast. Colden deemed it worth while to state, however, that, as was averred, the trade was highly advantageous to Great Britain because of the great quantities of British manufactures exported, in value far exceeding the provisions, and by the large returns in sugar. "Some pretend they can demonstrate this, against the force of all contradiction."[63]

[62] Gov. Thomas Boone to Pitt, Oct. 23, 1760, C.O. 5: 19.

[63] Cadwallader Colden to Pitt, Oct. 27, 1760, C.O. 5: 19. Colden attempted to measure the amount of New York's commerce with the

The New York assembly to which Colden referred the matter of illicit trade, adopted four resolutions defining their policy toward trade with the foreign West Indies. (1) They favored sending provisions to the Spanish and Portuguese Islands in exchange for wines. (2) They thought it unjust to prohibit colonial trade to the neutral islands while it was known such trade was carried on from Great Britain. (3) Adopting the point of view of consumers rather than that of the British planters, the assembly reiterated, apparently with approval, the merchants' argument that the importation of foreign sugar into the colonies and Great Britain was for the highest advantage of British commerce and for the benefit of the mother country. (4) The trade to Monte Cristi had led to extraordinary exportations of British manufactures to the Northern Colonies, ultimately, therefore, benefiting England. Colden supplemented these resolutions in favor of trade with the foreign sugar islands by observing that the Northern Colonies could not pay for their consumption of British manufactures by their own produce exported only to the British colonies. The English sugar islands consumed but a small part of the provisions raised in North America. The result of the whole trade of North America, taking it in every direction, was barely sufficient to pay the balance due Great Britain. The consumption of British manufactures would increase in proportion to the ability of the colonies to buy them; but to curtail America's earning power would diminish that consumption. He concluded with this significant remark:

French islands by extracting from the customs accounts the quantities of provisions ostensibly exported during the war to New Haven, New London, and Rhode Island, and then the quantities of sugar ostensibly imported from those ports and from Perth Amboy under the denomination of prize sugar and British sugar from Guadeloupe. The presumption is strong that the bulk of New York's illicit trade was conducted in this manner. Colden to Pitt, Nov. 11, 1760, with enclosed accounts, C.O. 5: 19.

"It is difficult to prosecute with success against the bent of the people, while they are under the prejudice to think that the Sugar Islands have gained a preference inconsistent with the true interest of their mother country, & when a prosecution fails of success it is of prejudice to the service it was designed to promote."[64] Public sentiment in America was not in this respect far removed from the opinion of many in England that the sugar islands had gained a preference inconsistent with the true interest of Great Britain.[65] Until the end of the war, the customs records of New York continued to be falsified and ship masters' bonds were canceled repeatedly without proper warrant.[66]

Governor Bernard and a committee of the Massachu-

[64] Colden to Pitt, Dec. 27, 1760, enclosing the resolutions of the assembly, and a report of the council's committee (Dec. 24) containing 22 folio pages of depositions concerning illegal trade. C.O. 5: 20.

[65] The feeling in New York of public approval of trade with the French West Indies was shown in the treatment accorded two informers in 1759. George Spencer had informed Colden of illicit trade with Monte Cristi and had started eight suits in the vice-admiralty court against illegal traders. Whereupon, Spencer was ill-treated at a coffee house, arrested upon a bond obtained from one of his creditors, carried to a tavern where a mob of sailors and others forced him into a cart and drew him through a great part of the city while he was hooted and pelted with mud and filth. He was rescued from the mob by the governor, concealed for a time in a gentleman's back entry, and finally taken to jail. There he was held for debt while his admiralty suits were dismissed with heavy costs to himself. Spencer to a printer, May 20, 1760; same to Gen. Jeffrey Amherst, Nov. 29, 1760; testimonies by Francis Lewis and George Harrison against Spencer, Dec. 23, 1760, C.O. 5: 20. Another informer, Augustus Bradley, had a similar experience. Bradley to Amherst, Dec. 5, 1760, wherein he complains of the futility of prosecutions "as that set of People in the West Indies as well as New York, have hitherto by some Stratagem or other, prevented the Justices and Magistrates from ever taking Notice of any Information on behalf of the Crown." C.O. 5: 20.

[66] John Tabor Kempe, attorney general of New York, to Lieut. Gov. Colden, Nov. 14, 1763, on the difficulty of prosecuting masters engaged in illegal trade but whose papers were correct. C.O. 5: 1071, Pp 2, 3. See also Colden to Board of Trade, Dec. 7, 1763, C.O. 5: 1071, Pp 4; Colden to Board of Trade, Feb. 9, 1764, *ibid.*, Pp 19.

setts council, in answer to Pitt's circular, reported that the colony was quite free from illicit trade, and that, of only two flags of truce fitted out for Santo Domingo during the war, neither had engaged in trade.[67] Bernard later admitted, however, that the revenue officer made no distinction in his accounts between the sugar, molasses, and rum imported from the British West Indies and the same from the French islands, and said that a large amount of those commodities, even when entered from English islands, were of French origin. The governor was in doubt as to just what proportion of Massachusetts' imports from the West Indies came from the French, but he leaves the impression that, in war time, it was the major part.[68] The distillation of rum from molasses, wrote Bernard, was a very necessary part of their trade, and if it should be obstructed either by a severe execution of the present laws or by the enacting of new ones, the consequence would be felt by exporters of British manufactures to America. Then, mindful of how much molasses was of French origin, and accepting the current view that whatever increased the purchasing power of New England would increase the sale of British manufactures, he added: "Even illegal trade, where the balance is in favour of the British Subject, makes its final returns to England." With writs of assistance, however, and by frequent inspections from naval and customs officers, illicit importations were being effectively

[67] Gov. Francis Bernard to Pitt, Nov. 8, 1760, enclosing report of the council's committee (Oct. 31), C.O. 5: 19. But May 5, 1761, Bernard reported the case of a Massachusetts sloop engaged in the sugar trade at Monte Cristi. She and others also traded at New Orleans. C.O. 5: 20.

[68] Bernard to Board of Trade, May 17, 1762, C.O. 5: 891, Ll 37, enclosing an account by the provincial revenue officer of the impost and tonnage received at Boston on rum, molasses, and sugar, with the quantities of each and the rates of duties, between May 1758 and May 1762. See also Bernard to Board of Trade, Apr. 29, 1763, C.O. 5: 891, Ll 55, 56. For the table see Beer, *Brit. Col. Pol., 1754-1765*, p. 118, n.

checked. But the merchants made loud complaints "with great show of reason" against restraints to which Rhode Islanders as yet were strangers. Neither Bernard nor the mercantile class approved, therefore, of the way the revitalized colonial administration was manifesting its power.[69]

The merchants of Boston organized, in fact, a movement to nullify the use of writs of assistance and to destroy the court of vice-admiralty and with it the vitality of the laws of trade. Frequent actions were brought at common law against decisions determined in the vice-admiralty court or in the custom-house. In August, 1761, Bernard reported that five suits were pending in the common law courts, where juries, prejudiced in behalf of illegal trade, would probably acquit smugglers. In three actions Barons, a suspended collector at Boston, sued for damages, first, the surveyor general, Lechmere; second, Cradock who had been placed in Barons's position; and third, Paxton, the customs officer who had informed against Barons.[70] The danger, threatening the integrity of the whole customs service had the provincial courts awarded damages to a dismissed officer, was averted by the three actions not being brought to trial.[71]

Two other actions, brought in the movement to paralyze the vice-admiralty court, were Gray *vs.* Paxton and Erving *vs.* Cradock. In the former, Gray, the provincial treasurer, tried to recover the colony's share of a condemned ship, which share had been used on several occasions by the vice-admiralty court to pay the expenses of trials. The object was to discourage prosecutions, by compelling the governor and customs officers to bear the

69 Bernard to Board of Trade, Sept. 5, 1763, C.O. 5: 891, Ll 67, 68.

70 *Ibid.*, Aug. 6, 1761, C.O. 5: 891, Ll 21-24; Beer, *Brit. Col. Pol., 1754-1765*, p. 119.

71 *Ibid.*

expense, but even the provincial courts decided that the province should stand the charges. The second action was this: John Erving, a member of the council, whose vessel had been seized by collector Cradock for contraband trade, had been allowed by the vice-admiralty court, upon his own petition, to compound for his offense at £500. This sum, which Erving paid into the court, was equally divided between the king, the governor, and the collector. Thereupon Erving brought action against Cradock for damages, and in the inferior provincial court was awarded nearly £600. Appeal was taken to the superior court. The question was important, for if this judgment were upheld, it would be possible for any illicit trader to recover at common law with damages whatever fine he had paid into a vice-admiralty court, and vice-admiralty jurisdiction would be nullified. While the judges of the superior court maintained that decrees of the court of vice-admiralty had equal force with judgments at common law, the jury awarded Erving over £550. However, upon appeal to England, where the government was to bear the expense of the suit, Erving, in order to avoid answering the appeal, acknowledged that he had received full satisfaction. Thus the Boston merchants, in their determination not to be confined by the Molasses Act or by embargoes on trade with the British West Indies, were on the verge of destroying the whole revenue system and the vice-admiralty court without which the former could not subsist.[72]

[72] Bernard to Board of Trade, Aug. 6, 1761, C.O. 5: 891, Ll 21-25. These cases are well stated in Beer, *op. cit.*, pp. 119-121. Bernard's interpretation of this serious crisis through which the imperial administration was passing reveals most clearly the American attitude toward West India trade. "These Actions," he wrote, "have an immediate tendency to destroy the Court of Admiralty and with it the Custom house which cannot subsist without that Court. Indeed the Intention is made no secret of in the two cases above-mentioned, that were tried in the inferior Court, the chief Subject of the harangues of the Council for the plaintiff (and some

Contemporary with the movement against admiralty jurisdiction in Massachusetts was the well-known opposition to the use of writs of assistance. The unsuccessful effort led by Otis, in 1761, to obtain a pronouncement of illegality against them in the superior court was supported by the same commercial group who fought the vice-admiralty court.[73] In the hands of Paxton these writs proved a most unpopular but efficient weapon against illegal commerce.[74] So great a proportion of Boston's trade, however, was with the foreign West Indies, that the campaign for its suppression led to a notable decrease in shipping at Boston.[75] It was in this

of the Judges too) were on the expediency of discouraging a Court immediately subject to the King & independent of the Province & which determined property without a jury; and on a necessity of putting a stop to the practices of the Custom house officers, for that the people would no longer bear having their trade kept under restrictions, which their neighbours (meaning Rhode Island) were entirely free from. And one Gentleman who had had a considerable hand in promoting these disturbances, has been so candid as to own to me, that it was their intention to work them up to such a pitch as should make it necessary for the Ministry to interpose and procure them justice (as they call it) in repealing or qualifying the Molasses Act, & obliging the Neighbouring provinces to observe the same restraints which this is to be kept under. In regard to both of these points, if they were solicited in another manner, there would be much to be said on their behalf.'' The agitation, he said, though confined to Boston, was likely to extend to the other colonies.

73 See Beer, *op. cit.*, pp. 122-123.

74 Bernard to Pownall, May 9, 1761. Reporting the discovery of a ship trading to New Orleans, Bernard remarked: ''I must add that this discovery . . . is entirely owing to Mr. Paxton, whose diligence in discovering illicit trade, is rather increased, than abated since the persecution Mr. Barons raised against him.'' C.O. 5: 891, Ll 16.

75 Bernard wrote the Board of Trade, Sept. 5, 1763: ''The shipping of Boston has decreased of late: This is partly owing to the increase of the Trade of other towns in the Province and partly to the illicit trade which is carried on in Rhode Island Connecticut &c with greater Security than it can be here.'' C.O. 5: 891, Ll 67, 68. Even if hard times did not prevail over a wide area at this time, a local depression, in so important a center of the colonial press and public opinion as Boston was, must have had a powerful influence on American politics.

atmosphere of commercial depression that the germinal ideas of the American Revolution found root. Thus the enforcement of the Molasses Act at the close of the French War, after thirty years of obsolescence, prepared the way for the movement that ended in revolution.

While the records of the corporate colonies, Rhode Island and Connecticut, are more meager and formal in their allusions to West India trade, it is certain that, of all the Northern provinces, their industries were most dependent upon intercourse with the French sugar islands. And it was here also that the imperial machinery was least able to enforce the laws of trade.[76] Governor Fitch of Connecticut in a reply to Pitt's circular of 1760, exasperating in its brevity and apparent ignorance of what so many other correspondents had reported, declared that in spite of inquiries he had been unable to discover any illicit trade and did not believe that Connecticut was in any way concerned with it.[77] The letter is chiefly valuable in revealing the temperament of an elected governor, whose constituents were widely reputed to have been deeply involved in illegal trade with the French West Indies.

Governor Hopkins of Rhode Island, in a letter to Pitt, admirable for its frankness and courage to defend a trade elsewhere called "pernicious," admitted an extensive trade between his people and the French. Fifty Rhode Island privateers had brought in a large number of French prisoners. These had occasioned the fitting out of about thirty flags of truce principally to Santo Do-

[76] Bernard wrote to Pownall, May 9, 1761: "These practices will never be put an End to till Rhode Island is reduced to the subjection of the British Empire; of which at present it is no more a part, than the Bahama Islands were when they were inhabited by the Buccaneers." C.O. 5: 891, Ll 16.

[77] Gov. Thomas Fitch to Pitt, Nov. 26, 1760, and Apr. 25, 1761, C.O. 5: 20.

mingo. Such vessels had carried much lumber and British dry goods, and returned with some French sugar but for the most part with molasses. Also other vessels, under pretense of clearing to and from Jamaica, had engaged in a considerable commerce with the French islands. Of late, however, Rhode Islanders in both these branches of trade had suffered from a great number of seizures by the royal navy at Monte Cristi, Jamaica, New Providence, and elsewhere in the West Indies. It had always been thought lawful to trade to Monte Cristi, except in prohibited commodities, and vessels had been cleared for that port from all the Northern customhouses. Hopkins concluded with the familiar and sound defense of America's demand for access to an international market in the West Indies, if she was to find a sufficient vent for her surplus products and earn the wherewithal to buy British manufactures. Denied such a market, she would be forced to manufacture for herself.[78]

[78] "By the best Computation I have Seen, the Quantity of Flour made in these Colonies yearly is such, that after all the English Inhabitants, as well of the Continent as of the Islands are fully Supplied, with as much as they can consume with the year, there remains a surplusage of at least One Hundred Thousand Barrels. The Quantity of Beef and Pork remaining after the English are in like Manner Supplied is very large. The Fish, not fit for the European Market, and the Lumber produced in the Northern Colonies, so much exceed the Market found for them in the English West Indies, that a vast Surplusage remains that cannot be used there. How natural it is for the Proprietors of these Commodities to seek some Market for them, and what Risques they will run to find it I need not mention. From the Money and Goods produced by the Sale of the Surplusages, with many others of less Consequence, sold by one Means or other to the Spaniards, French, and Dutch in America, the Merchants of those Northern Colonies are principally enabled to make their Remittances to the Mother Country for the British Manufactures consumed in them. And this is most certain that the Quantity of British Manufactures consumed in these Colonies, are only limited by the Ability of the Inhabitants to pay for them.

"Supposing this Intercourse of the Colonies with the Spaniards, French, and Dutch, intirely Stopped, the persons concerned in producing

New Hampshire's West India trade was, according to Governor Wentworth, confined entirely to the British sugar islands.[79] His admission, however, that he might have received great sums of money had he issued commissions for flags of truce, indicates that here also there was a strong disposition to trade with the French, which probably manifested itself in a clandestine way.

Pennsylvania, next to Rhode Island, was the province most heavily involved in trade with the enemy by flags of truce. It was reported to Thomas Penn, in 1759, that Governor Denny had made over £1200 in granting flags of truce, and that the Delaware river swarmed with shallops unloading illegal cargoes.[80] When he began the practice, about May 1759, Denny sold commissions for 300 or 400 pistoles, but finally, just before his removal in the autumn of that year, he issued great numbers of blank flags of truce at £20 or less apiece, and these were sold from hand to hand at various prices. In consequence, a very large part of the Philadelphia merchants were engaged in commerce with the French islands. The most eminent lawyers defended the trade. Even the vice-admiralty judge decreed, in the only two cases brought before him, that French goods purchased by British subjects were not legal prize for the royal navy; the libeled ships and cargoes were delivered to their owners, and the captors ordered to pay the costs of suits. Under these circumstances, several ships laden with French sugar, which had been sent into Philadelphia by English

these Surplusages will of Course change the Manner of their Industry, and Improvement, and, compelled by Necessity, must set about making those Things they cannot live without, and now rendered unable to purchase from their Mother Country.'' In conformity with Pitt's orders the assembly resolved to put a total stop to all intercourse with the French, and to that effect he had published a proclamation. Gov. Stephen Hopkins to Pitt, Dec. 20, 1760, C.O. 5: 20.

[79] Gov. Wentworth, to Pitt, Dec. 9, 1760, C.O. 5: 20.

[80] Thomas Penn to Pitt, Sept. 12, 1759, C.O. 5: 19.

cruisers, were dismissed by the captors without prosecution. When, however, warships and privateers began making seizures in the West Indies and secured condemnations in the English islands, illicit trade from Pennsylvania was considerably checked. Here, as elsewhere, illegal trade was carried on by false clearances to and from Jamaica or other British West India ports. Much French produce was also entered under clearances from Providence, New York, Port Lewes, and Newcastle. The naval office lists are, therefore, utterly unreliable. Bewildered amidst a public sentiment almost unanimous in its approval of trade with the foreign West Indies, the conscientious governor knew not where to turn for aid in enforcing the laws of trade.[81]

In the Southern continental colonies there was much less commerce with the foreign West Indies. Governor Sharpe of Maryland was assured by the customs officers that they did not believe any clandestine trade had been carried on from that province.[82] Lieutenant-Governor Fauquier of Virginia resisted many tempting offers to induce him to authorize flags of truce; instead he sent French prisoners to England on tobacco ships. There was a small trade in corn to the neutral islands, disguised by false clearances, and a slight export of bullion in an open trade to Monte Cristi.[83] From South Carolina, according to Governor Bull, there was some trade with the French via Monte Cristi and also to Pensacola. Until 1761, it was impossible to secure condemnations of

81 Gov. James Hamilton to Pitt, Nov. 1, 1760, C.O. 5: 19. Flags of truce were sometimes heavily insured in London partly to cover the large amounts of coin carried to the French. John Tomlinson to James West, Feb. 15, 1760, C.O. 323: 14, P 18.

82 Gov. Horatio Sharpe to Pitt, Feb. 27, 1761, C.O. 5: 20.

83 Lieut. Gov. Fauquier to Pitt, Oct. 28, 1760, C.O. 5: 19. Burnaby stated that Fauquier was one of the few governors who refused to issue flags of truce. *Travels,* p. 73.

vessels with French produce coming from Spanish ports. Governor Bull suggested that, if the duties levied by the Molasses Act upon foreign sugar, molasses, and rum were rigorously exacted by customs officers and the clandestine landing of them discouraged by seizures, such high duties and seizures would to a great extent put an end to objectionable trade.[84] Georgia reported no trade with the enemy.[85]

In the West Indies the increasing number of seizures by the navy reduced illegal trade between the English and French islands to small dimensions. Bermuda was still useful to smugglers because of the many coves where there were no custom-houses.[86] Evidence was found upon a captured vessel belonging to Barbadoes that it and others owned there were concerned in illicit trade with the French. Several merchants of Barbadoes were prosecuted and one, at least, was fined and imprisoned.[87] Jamaica was the center of much naval activity, and the vice-admiralty court there, under Judge Thomas Bullock, was crowded with prize cases.[88] Curaçoa was still, in 1760, to some extent a depôt for the French, but, on the whole, the British had neutral trade under control.[89]

Ireland's trade with the French and Dutch sugar colonies was of long standing,[90] but was greater than ever

[84] Gov. Bull to Pitt, Feb. 18, 1761, C.O. 5: 20. Bull observed that some idea of the quantity of molasses smuggled into the Northern Colonies might be derived by comparing the great quantities of rum, of Northern manufacture, exported with the much smaller amount of molasses imported, as recorded in the customs accounts.

[85] Gov. Wright to Pitt, Jan. 26, 1761, C.O. 5: 20.

[86] Gov. Popple to Pitt, Feb. 17, 1761, C.O. 5: 20.

[87] Gov. Pinfold to Board of Trade, 1760, C.O. 28: 32, Ff 1.

[88] Vice Admiral Cotes to Bullock, 1760, C.O. 137: 32. Cotes complained of Bullock's delays in trying cases. See also address of council and assembly of Jamaica to the king, Feb. 27, 1761, C.O. 5: 20.

[89] A St. Christopher correspondent to Pitt (?), Feb. 27, 1760, C.O. 5: 20.

[90] Postlethwayt, *Dictionary of Commerce* (1757), "Sugar Colonies, British"; Cary, *Essay on the State of England in relation to its Trade,*

during the French War. In 1757, great quantities of Irish beef reached the French, and many Dutch vessels, laden with beef at Cork, were destined for the French Islands.[91] Several of these ships, in 1758, were seized and condemned in the vice-admiralty court at Antigua.[92] The Irish market was largely supplied with French sugar obtained at a considerably lower price than the English product.[93] When, in 1759, the royal navy broke up trade in provisions with the neutral islands, Ireland as well as North America developed an extensive trade with Monte Cristi, and, in its suppression in 1760-1761, the Irish together with the New Englanders suffered heavily.[94]

With the loss of Canada and the Mississippi, it was certain that the French West Indies would be de-

Bristol (1695), p. 11. Ireland was forced to find in foreign ports a market for its surplus provisions that was denied it in England.

91 Gov. Pinfold of Barbadoes, to Board of Trade, Jan. 7, 1758, C.O. 28: 31, Ee 16.

92 Gov. Thomas of the Leeward Islands to Board of Trade, May 18, 1758, enclosing an interesting correspondence between Dominick Farrell, a merchant in Waterford and his agent, Francis Peisly, at St. Eustatius, and also a letter from the governor of St. Eustatius to the intendant at Martinique, C.O. 152: 29, Cc 22, 23, 24. Farrell stated that "there is no less than 50 or 60,000 Barrells of beef gone & going from different parts of this Kingdom" to St. Eustatius. He insured at 4 per cent. Newport and Hobbs was another Irish firm heavily engaged in this trade. Dutch vessels from Amsterdam to St. Eustatius were in the habit of stopping in Ireland. Thomas and Adrian Hope and Abraham Terberch of Amsterdam were promoters of the trade. Nicholas Heylegar and John Aertsen were large factors in the Dutch island. Returns to Ireland were in French sugar. Farrell competed with other Irish firms for this trade. *Ibid.*, Cc 24.

93 Joseph Massie, *State of the British Sugar-Colony Trade* (1759) p. 16.

94 Waddell Cunningham, a New York merchant and part owner in the snow *Prince of Wales*, deposed, before the New York council, December 20, 1760, that his vessel and several others had been engaged in trade between Belfast and Monte Cristi, French sugar being entered at Belfast as Spanish. He referred also to a ship that had cleared from London with flour and wine for Monte Cristi. C.O. 5: 20. See also a letter from Monte Cristi to one McCarty, enclosed in Cleveland to Robert Wood, Aug. 10, 1761. *Cal. Home Off. Pap., 1760-1765*, pp. 60-61.

pendent more than ever upon British North America for temperate zone supplies. In the year following the peace of 1763 the governors of Martinique and Guadeloupe exhibited a most cordial liberality toward vessels arriving from the British Northern Colonies.[95] A French royal decree of April 18, 1763, permitted, under strict regulations, the import into the French islands of certain commodities, such as came from North America, in exchange for molasses and rum. In accordance with this decree, proclamations were published August 16 and 18 in Guadeloupe and Martinique.[96] The British sugar islands became greatly alarmed at the prospect of a renewal of an intercourse that would sustain the French plantations and deprive them of the necessities of life.[97] In spite of naval patrols and a reinvigorated customs administration, the North Americans were not slow in accepting these invitations to an alluring market. In addition to the old centers of illegal trade, Cayenne in French Guiana attracted considerable commerce from Philadelphia, New York, and other Northern ports.[98] Moreover, the French inhabitants of Dominica, acquired

[95] Halifax to Board of Trade, Dec. 13, 1763, C.O. 323: 17, R 22.

[96] Copies of the proclamations, C.O. 323: 17, R 23, 30. At Guadeloupe, the proclamation was to go into effect Nov. 1, 1763, *ibid.*, R 27.

[97] Gov. Dalrymple of Dominica to Egremont, Sept. 4, 1763, *ibid.*, R 26; Gov. Thomas of the Leeward Islands to Egremont, Sept. 9, 1763, *ibid.*, R 24. The latter said that such a commerce between North America and the French ''will so interfere with the British Sugar Colonies, as to disable Them utterly from supporting their Plantations, under the present high duties, & expensive Culture of their worn out Lands.''

[98] Gov. Pinfold of Barbadoes to Board of Trade, Mar. 23, 1765, C.O. 28: 32, Ff 79-82. Pinfold enclosed papers, captured on a Philadelphia vessel, exposing a contract of June 1, 1764, to furnish the French governor and intendant at Cayenne with all kinds of supplies. Several ships were to be employed. See also Pinfold to Halifax, Mar. 23, 1765, C.O. 28: 50; Rear Admiral Tyrrel's reports of trade with the French, *Cal. Home Off. Pap., 1760-1765*, pp. 573-574; Hillsborough to Shelburne, May 7, 1768, concerning New York contracts to supply Cayenne, C.O. 5: 43, f. 194.

by the British in 1763, were notorious smugglers, through whom sugar of French origin entered England.[99] Antigua, also, in January, 1764, expressed alarm at the quantities of French sugar, molasses, and rum entered there and reëxported to England and the Northern Colonies.[100] Through Jamaica, likewise, with the return of peace, a considerable amount of French sugar was smuggled into the British Market.[101]

At the close of the war, therefore, the French and British policies toward West India commerce were tending in opposite directions. France regarded the old agreement of 1686, for the restraint of international trade in the new world, as having been annulled by succeeding wars. All previous regulations for the French islands, excepting Santo Domingo, were revised in the interest of free trade.[102] Great Britain, on the other hand, in the face of this embarrassing liberality, so gratifying to her Northern subjects, determined through parliament and by the aid of an improved civil and naval service, to confine colonial trade to British dominions.

99 Newcastle Papers, Brit. Mus. Add. MS. 33030, f. 253.

100 Resolutions of Assembly, Jan. 26, 1764, C.O. 152: 30, Dd 36.

101 Gov. W. H. Lyttelton to Board of Trade, July 9, 1763, C.O. 137: 33, Cc 19; same to naval officers and customs collectors, Jan. 4, 1764, *ibid.*, Cc 51; Lieut. Gov. John Dalling to Dartmouth, Apr. 11, 1773, *ibid.*, 37, Gg 29; "A Cane Planter" to Dalling, Mar. 29, 1773, *ibid.*, Gg 30; Dalling to Dartmouth, May 2, 1773, *ibid.*, Gg 33; Jamaica Act against smuggling, Dec. 1774, *ibid.*, Gg 49.

102 For British reports of the revised French policy and objections to it see: Hertford, ambassador at Paris, to Duc de Praslin, Dec. 12, 1764, C.O. 323: 18, S 33; Hertford to Halifax, Dec. 20, 1764, *ibid.*, same to same, Jan. 3, 1765, *ibid.*, S 36; copy of French decree of Dec. 9, 1764, *ibid.*, S 37; Halifax to Board of Trade, Jan. 14, 1765, *ibid.*, S 32; attorney and solicitor general to Board of Trade, Feb. 12, 1765, giving their opinions that the Treaty of 1686 was no longer in force, *ibid.*, S 44; advocate general to Pownall, Feb. 15, 1765, stating the opposite opinion, *ibid.*, S 45.

CHAPTER XIV

THE PEACE OF PARIS

On the 9th of December, 1762, when that memorable debate took place in parliament on the peace which Bute and Bedford concluded with the crafty Choiseul, it is recorded that, outside, a turbulent crowd roared in concert its disapproval of the treaty.[1] Inside the House of Commons, however, where the treaty was carried by an overwhelming majority, we are told that many gentlemen from the West Indies had seats and that the number of their votes was "very formidable."[2] The contrast between the popular discontent over the peace of Paris and the opposition of the powerful West India interest to British expansion deserves some explanation. The English public was keenly disappointed at the almost insignificant additions of territory acquired in the West Indies, in view of the sacrifices borne and the conquests made, which included all the French sugar colonies except Santo Domingo, and Cuba and Manila. Manila, however, had been captured too late (October 6, 1762) to enable

[1] Julian S. Corbett, *England in the Seven Years War*, London, 1907, II, 363-364.

[2] A writer of 1760 stated that "Many Gentlemen of the *West Indies* have seats in the British House of Commons." *Remarks on the Letter Address'd to Two Great Men*, London, 1760, pp. 46, 47. Mauduit, the Massachusetts agent, wrote in 1764 that the West Indies had "a very formidable number of votes in the House of Commons." *Mass. Hist. Soc. Coll., Series I,* Vol. VI, 193. Franklin wrote to Collinson, Apr. 30, 1764: "For Interest with you we have but little. The West Indians vastly outweigh us of the Northern Colonies." Franklin, *Writings* (ed. Smyth), IV, 243. Cited by Beer, *Brit. Col. Pol., 1754-1765*, pp. 136 n, 158 n.

the British ministers, even had they been so disposed, to exchange it for Porto Rico. But there was some talk of exchanging Cuba for Porto Rico.[3] Nor did the imposing gains in Canada and India compensate for the failure to secure for the British Empire an area adequate to supply its demand for sugar. The failure of the peace in this respect is the more to be regretted, for it is conceivable that a great addition to the empire in the West Indies, say of Guadeloupe and Martinique, would have legalized much of the illicit commerce from North America, the control of which was a contributing cause of the American Revolution. In the clauses of the treaty relating to the West Indies, Bute and his colleagues let slip the opportunity of relieving England of one of its heaviest economic burdens, the exorbitant cost of sugar, and lost the possible chance of preserving an empire. Cuba was exchanged for Florida, Martinique and Guadeloupe were returned to France, and England renounced all claim to the strategically important island of St. Lucia, receiving instead the little islands of Dominica, St. Vincent, Grenada, the Grenadines, and Tobago. The total area acquired was about 700 square miles, or about 448,000 acres, the majority of which land was too mountainous for sugar culture.

The planting interest opposed the acquisition of Guadeloupe,[4] unless sugar from the old British islands was given a preference in British markets.[5] Such an extension of power in the West Indies as the public of England

[3] *Bedford Corr.*, III, 96, 119, 139; Almon, *Biographical, Literary, and Political Anecdotes*, London, 1797, II, 72, 73; Beer, *op. cit.*, p. 140 n.

[4] Referring to Pitt's plan of taking Martinique, Bedford wrote to Bute, July 9, 1761: "I suppose the sugar planters will no more desire (it) should be retained by us than they did in relation to Guadeloupe." *Bedford Corr.*, III, 25; Beer, *op. cit.*, p. 158 n.

[5] *Reflections on the True Interest of Great Britain.* C.O. 28: 50; Beer, *op. cit.*, p. 158 n.

needed and expected could not have been made without the support of British sugar planters. Their attitude in 1762 may be better understood by a review of their policy toward British expansion in the preceding half century.

As early as 1720, when England was contemplating the colonization of the neglected island of St. Lucia, William Gordon, in a memorial to the Board of Trade setting forth the objections to the idea, declared: "The settling of Santa Lucia would encrease the Sugar Trade of which there is already more produced than the Markets of Europe can well consume: And by Consequence very much Damage, Impoverish, and even Ruin the Already Settled Sugar Colonys." But Gordon observed that this objection might be overcome by restraining the grantees of St. Lucia from cultivating sugar; the island might be settled, and at much less expense, with cocoa works.[6] In the following year, the Board of Trade instructed Lord Belhaven, governor of Barbadoes, with reference to the proposed occupation of Tobago, that the vacant island should be settled only by consent of the council of Barbadoes and its "settlement so restrained as not to interfere with the Produce of His Majesty's other Charibbee Islands," and "that all Persons who shall settle there, be effectually restrain'd from planting of any sugars." Cocoa, anatto, and indigo were recommended for culture. Similar instructions were issued at this time concerning St. Lucia and St. Vincent.[7]

Not only did it seem desirable to prevent the settlement of new sugar colonies, but on one occasion, at least, the old island of Barbadoes put such obstacles as it could in the way of sugar production on the newer English planta-

[6] William Gordon, "Advantages and Disadvantages of settling St. Lucia," July 14, 1720, C.O. 28: 16, V 36. Cf. Robertson, p. 269, *ante*.

[7] Board of Trade to Carteret, Sept. 14, 1721, C.O. 29: 14, ff. 246-249; Board of Trade to Belhaven, Nov. 1721, C.O. 28: 44, no. 12; Board of Trade to Carteret, Feb. 1, 1721/2, C.O. 29: 14, ff. 291-292.

tions. Thus an act of 1736 refers to the impoverished soil of Barbadoes as compared with that of the Leeward Islands, states that the latter were underselling Barbadoes in sugar and rum, and that the export of clay for whitening sugar in the Leeward Islands was helping them at the expense of Barbadoes. Therefore, to discourage its export, a tax of five shillings per pound was levied on all clay shipped from Barbadoes.[8]

Toward the middle of the century, it was perfectly apparent that the West Indians were manipulating legislation for the sole purpose of elevating prices in England. That this was the object of the act of 1739 for direct exports to Europe has been already pointed out.[9] William Beckford's proposal of a bounty of one shilling per hundredweight on all sugar exported from Jamaica to foreign markets had the same object in view.[10] The power of the planting class in both the Board of Trade and the House of Commons was well recognized in the parliamentary debates of 1744.

There was, on the other hand, an increasing number of Englishmen with some capital who were desirous of establishing new sugar plantations. The inordinate prices of 1750 could hardly have acted otherwise than to invite new capital and labor into the sugar industry. But either the British sugar islands were completely cultivated, or, as in the case of Jamaica, the arable land was nearly all monopolized. Aspiring planters were tempted, therefore, to acquire plantations in foreign islands in the expectation, probably, that they might smuggle their sugar, as so many were doing, into the British market. Thereupon, the old planters were aroused to destroy this

[8] Hall, *Acts of Barbadoes*, London, 1764, no. 170.

[9] See p. 185, *ante*.

[10] Beckford to James Knight, June 18, 1743, Long Papers, Brit. Mus. Add. MSS. 12431, f. 125.

menace to their monopoly. At "a General Meeting of Sugar Planters and Merchants interested in and trading to the Sugar Colonies," a committee headed by John Sharpe was appointed to lay before the Board of Trade three proposals. These were: first, that an act of parliament should restrain English subjects from taking up lands or settling in any islands belonging to foreign powers; second, that colonial governors be instructed to endeavor to get laws passed to stop the import of French sugar; and third, that an act of parliament should prevent Portuguese sugar from being imported directly into Ireland.[11] The draft of a bill to effect these objects, which Sharpe submitted to the board, also stated that many old British sugar planters were deserting their settlements and, with their negroes and effects, were taking up lands in foreign sugar colonies "to the great Prejudice and Injury of the British Sugar Islands and the Trade thereof."[12] The pressing need of more land both for the planters on the exhausted soil of the older islands and for newcomers could not have been more clearly revealed. And yet, though the proposed bill was dropped, the Board of Trade ordered a circular letter to be written to the governors of Jamaica, Barbadoes, the Leeward Islands, Bahamas, and Bermuda directing them to use their utmost endeavors to prevent British subjects taking up lands in any foreign islands or colonies, and also to prevent any illicit importation of foreign sugars.[13]

The passage of such a measure as the planters had proposed undoubtedly would have met with bitter denunciation from the British public. For in the following year two petitions were brought before the House of

[11] B. T. Jour., Feb. 7, 1752, C.O. 391: 59.

[12] Draft of the bill, Feb. 19, 1752, C.O. 323: 13, O 104; B. T. Jour., Feb. 19, Mar. 5 and 10, 1752, C.O. 391: 59.

[13] B. T. Jour., May 27, 1752, C.O. 391: 59.

Commons, one from the refiners, grocers, and other dealers in sugar in London, Westminster, and Southwark, and the other from the dealers of Bristol, wherein it was alleged that "at this time the common people of England were laid under a most intolerable kind of tax, by the prodigious rise of the price of Muscovado sugars imported from Jamaica, which made all sugars manufactured in Great Britain double the price of what they were manufactured for by the French and other Nations."[14] The aim of these petitions was either to oblige the great proprietors of land in Jamaica to cultivate greater quantities of ground for sugar, or, for the petitioners to gain leave to import sugar from other countries when the price of Jamaica sugar rose above a certain level. It was alleged in the petitions "That the sugar planters in Jamaica found their interest much more in importing small than large quantities of sugar into Great Britain from that island."[15] In the evidence offered in the Commons in support of the petitions it was stated that the quality of British brown sugar at that period was lower than at any time in living memory, while its price, ranging from thirty-five shillings to forty-five shillings per hundredweight, was much higher than that paid for the same grade of sugar by continental consumers. In 1739, sugar sold as low as twenty-four shillings to twenty-five shillings per hundredweight. This rise accounted for the large quantities of French

[14] *Parl. Hist.*, XIII, 1293, Mar. 8, 1753; *Commons Journal*, XXVI, 659. Feb. 23, 1753, the House of Commons had received from the Board of Trade 17 papers dealing with the agrarian situation in Jamaica, *ibid.*, XXVI, 611.

[15] *Parl. Hist.* XIII, 1293. The consideration of these petitions in parliament may be traced in the *Commons Journal*, XXVI, 659, 664-665, 703, 709, 775, 811. A complete account of what occurred was soon published entitled: *An Account of the late Application to Parliament, from the Sugar Grocers,* etc., *of the Cities of London and Westminster, the Borough of Southwark, and the City of Bristol.* 48 pages, 8vo, London, Brotherton, 6d., 1753.

and other continental refined sugars being smuggled into England and Scotland, which sold at much lower prices than those for which London refiners could sell. The prices alleged were attested by refiners, bookkeepers, and sugar brokers who were brought to the bar and examined. At such high prices, the public were consuming, for the most part, brown sugar, "for the common People cannot afford to pay for refining a Commodity already so dear." Thus the refining industry in England was badly crippled; this affected eighty refineries in London, and at least twenty in Bristol, besides those in Chester, Liverpool, Lancaster, Whitehaven, Newcastle, Hull, Southampton, and several in Scotland—hardly fewer than one hundred and twenty in all. Reckoning that a family of six with nine servants were employed at each refining house, the number of people immediately afflicted in this way was eighteen hundred, besides other manufacturers depending upon them. It was this distress of the refiners that first prompted an application to parliament for relief. They "proved" that at least three-fourths of all the sugar imported from the colonies was used for refining. In 1751, 41,000 tons were imported. Incidentally, the petitioners complained of the existence of refineries in the sugar islands, of which there were seven houses in Jamaica containing sixteen pans, four negroes to each, five in Antigua, two in St. Christopher, and several in the other islands. Colonists, it was said, should be employed solely in the production of raw sugar for British refiners. The only prospect of relief from the deficient supply of British West India sugar, unless parliament acted, lay in the possibility that the French and Dutch would find means to supply the deficiency by smuggling sugar upon the English coast.[16] Correspondence was

[16] *An Account of the Late Application* etc. In demonstration of the deficient supply of sugar it was observed that the annual average importa-

produced proving that in the previous six months 140,000 lbs. of Dutch loaf sugar and 30,000 lbs of powdered sugar had been smuggled into Scotland; Sunderland had smuggled quantities of Dutch sugar, and French sugar from New York had been detected in the entries at London.[17]

Perhaps the most telling criticism of all against the planters lay in the "suspicions" that there was "doubtless" a "combination" among them in restraint of trade. Such suspicions were founded on the fact that, with an annual average importation from 1748 to 1751 of about 900,000 cwt. of sugar at 30s. per cwt., the crop had yielded £1,350,000, while an importation of only 820,000 cwt. at 40s. per cwt. in 1752 had yielded £1,640,000. Thus the planters profited more by a smaller importation, and saved on freight, insurance, and duties.[18] It is difficult to escape the conclusion that the refiners of Great Britain, through whose industry so much had been done to purify a food that had become a necessity, deserved from parlia-

tion into England during 1727-1739 was 865,908 cwt., whereas the importation in 1752 was only 820,000 cwt., while exportations from England to Europe during 1739-1751, as compared with the period 1727-1739, showed an annual average decrease of 14,126 cwt. *Ibid.*, 16. "With regard to the Home Consumption, the Quantity is greatly deficient; for by the Use of refined Sugar, and the Increase of refining Houses, the Consumption is greatly increased, for *Muscovado* Sugar being refined, is rendered more wholesome as well as palatable, and greatly diminished in Quantity, all the Molasses being separated from it, and the Filth contained in it when brought from the Plantations." *Ibid.*, 14. Brokers testified that not 300 hogsheads remained unsold, whereas formerly at that season they had known 10,000 hogsheads remaining unsold. As a result, several refiners had been idle for a considerable time. *Ibid.*, 17. German Protestants were largely employed in the English refineries. *Ibid.*, 22-23.

17 *Ibid.*, 22 n, 39-40. From Dec. 1, 1752, to Mar. 1, 1753, the average price for brown sugar in London was 44s. per cwt.; letters from Bordeaux, Rochelle, and Dunkirk showed that for the same period the prices in those places averaged 19s. per cwt., reckoned in English weight and money. Amsterdam prices were also much lower than the English. *Ibid.*, 26.

18 *Ibid.*, 23-24. The calculations were based on official statistics.

ment such relief as would secure them an adequate supply of raw material for manufacture.[19]

In their own defense the Jamaica planters prepared a strong representation of grievances and of the lamentable state of their island, which Alderman Beckford moved might be laid before the House of Commons. But this motion was rejected. Beckford, however, "who had a great concern in this affair," undertook to defend the planters. Papers on the state of Jamaica, together with a report from the Board of Trade, were referred to a committee of the House, which recommended two resolutions that were agreed to by the Commons. First, that the peopling of Jamaica with white settlers and the cultivation of its lands was the most proper measure to increase the trade and navigation of the empire; and second, that the efforts of Jamaica to secure these ends had proved ineffectual.[20] Thereupon, a motion was passed to frame a bill to accomplish the settlement and cultivation of Jamaica's idle lands.[21] A bill to effect these objects, by a redistribution of lands if need be, was reported, but after several debates it failed to pass into law.[22] While the planting interest seems to have been thoroughly alarmed by the agitation of this year, its

[19] The refiners prayed the House to assist them against the planters "by prescribing speedily, some limits to their immoderate Gain, arising from their present exclusive Privilege, and from want of cultivation in Jamaica." *Ibid.*, 25. The subject of uncultivated holdings in Jamaica was discussed at length in proof of the monopolistic policies of the great planters. The pretended efforts of Jamaica to secure more settlers were believed to be insincere. Barbadoes and the Leeward Islands had equally shared in the profit from the neglect of Jamaica and were thought to be in league with the latter in the restraint of trade. *Ibid.*, 29-34.

[20] *Parl. Hist.*, XIII, 1293. Mar. 8, 1753.

[21] *Commons Journal*, XXVI, 664-665, Mar. 9. John Pitt, Mr. Hume, Charles Townshend, Mr. Oswald, and Lord Dupplin were appointed to prepare the bill.

[22] *Ibid.*, XXVI, 811, May 9; *Parl. Hist.*, XIII, 1293.

political strength was proved by the fact that its monopoly of the British sugar market was left unimpaired.[23]

The popular impression, however, that Great Britain was supporting the burden of a sugar monopoly was not to be overcome. In 1754, there appeared in London a very able pamphlet entitled *A Short Account of the Interest and Conduct of the Jamaica Planters.*[24] The author asserted that the interests of the planters were opposed to the interests of Great Britain. The planters really objected to the development of Jamaica, whereby the price of sugar would be lowered, and their power in parliament blocked any attempt to relieve England of the burden of monopoly.[25] He then described in detail the futile efforts to settle Jamaica and the existence there of one million acres of arable land, all patented but neither occupied nor cultivated. The data on this situation had been made public by the parliamentary investigation of the preceding year. The writer's conclusion, in spite of his exaggeration of certain charges, appears to have been well founded. "The personal interest of a

[23] Sharpe and several West India merchants and planters attended the Board of Trade, April 3. Sharpe told the board that if foreign sugar were allowed to enter England, it would entirely ruin the sugar colonies, and put a stop to all further cultivation and improvement. But as it was unlikely that such a measure would pass, he did not care to trouble the lords with arguments on the subject. As for the complaint against colonial refining, he said the planters had no objection to its suppression provided this was effected in the Northern Colonies as well as in the sugar islands. B. T. Jour., C.O. 391: 60.

[24] *A Short Account of the Interest and Conduct of the Jamaica Planters. In an Address to the Merchants, Traders, and Livery-men of the City of London.* London, Printed for M. Cooper . . . 1754. Price 3d. 21 pages.

[25] *Ibid.*, 3. The absentee planters were spoken of as "those whose estates are in another quarter of the world, who, by their artful management, keep this country and *Ireland* under a constant yearly tax of several hundred thousand pounds, who accumulate great estates by means of this tax, so very burdensome to almost every individual in this country, and are thereby enabled to support contests [for seats in parliament] in some of the richest and most populous cities in this country."

rich planter at Jamaica," he said, "is contrary to the interest of every true Briton, whether in a national or personal light."[26]

With the opening of the French War, in 1756, the agitation temporarily subsided. It was naturally expected that the results of the war might bring relief from the sugar monopoly by the acquisition of one or more of the larger French islands.[27] When, in 1759 and 1760, the success of the British seemed assured, public feeling toward the planters again found expression through several pamphlets by the economist, Joseph Massie. In one of these he maintained that the profits of the sugar planters were twice as much per acre as landholders in England received, and that during the preceding thirty years the kingdom had suffered vast losses through the sugar colony trade.[28] In 1760, Massie published a broadside entitled *A Computation of the money that hath been exorbitantly raised upon the people of Great Britain by the sugar planters in one year, from January 1759 to January 1760; shewing how much money a family of each rank, degree or class hath lost by that rapacious monop-*

[26] *A Short Account of the Interest and Conduct of the Jamaica Planters*, 21.

[27] In a "Scheme for the taking of Santo Domingo," Aug. 25, 1756, Thomas Cole stated that island produced nearly as much sugar as all the English islands put together. Besides, there was an enormous commerce between Santo Domingo and British North America, which the conquest of that island would legalize. "Therefore if we can take this island we should not only add a large yearly sum to the revenue of the customs, but command all the Markets of Europe for sugar, indigo &c." Newcastle Papers, Brit. Mus. Add. MSS. 33029, ff. 349-350.

[28] *A State of the British Sugar-Colony Trade*, etc., . . . *Most humbly submitted to the Consideration of the Honourable House of Commons.* By J. Massie. London, 4to, 90 pages, 1759. An anonymous pamphlet appeared the same year entitled *An Inquiry into the Causes of the present High Price of Muscavado Sugars; with some Remedies humbly proposed for a grievance so considerable. In a letter to a Member of Parliament.* London, Cooper, 8vo, 4d., 1759.

oly having continued so long, after I laid it open, in my State of the British Sugar-Colony Trade, which was publish'd last winter.[29] This proclaimed that "their Monopolizing of Sugar has continued about Twenty Years, in which Time . . . the said Planters have exorbitantly raised upon the People of *Great Britain* no less than EIGHT MILLIONS of POUNDS Sterling, over and above very good Profits." The loss to Great Britain, in the year 1759, through such excessive profits, Massie put at £840,000, which would more than pay and clothe for one year an army of 40,000 foot soldiers. Massie supported his charges with other broadsides during the following year.[30]

Thus the conviction grew that Great Britain was suffering a great and unnecessary annual loss through the limited area of sugar culture and the conditions of monopoly surrounding the sugar trade. A most reasonable remedy, the apparently inescapable answer to public demands, lay in a substantial expansion of British dominion in the West Indies. It would seem that the arguments presented to Pitt in 1760 for the acquisition of Guadeloupe were unanswerable. From the standpoint of the British public it is difficult to understand how Canada could in any sense have been regarded as of equal value with the sugar islands. Guadeloupe, it was pointed out, could produce at least 100,000 hogsheads of sugar annually, or nearly as much as all the British islands put

[29] London, 1760, 1 folio, Brit. Mus. 1890, e. 4 (19).

[30] Joseph Massie, *To the Printer of the Gazetteer* [*Computations on the profits of merchants and tradesmen, etc.*] (. . . *A state of the exports from the British Sugar-Colonies. . . . A re-stated account of the . . . charges . . . on British Muscavado Sugar,* etc.), London, 1760. Brit. Mus. 8244, ee. 42; Joseph Massie, *General Propositions relating to the Colonies* (*an exemplification of these general propositions by the colonies . . . of North America, and by the sugar colonies on the West Indies Islands*), 4to, London, 1761.

together.[31] The reply, that Jamaica could, if properly cultivated, afford more sugar than England wanted, was unconvincing, for England had waited thirty years in vain for the theory to be put in practice. The planters' opposition to Guadeloupe was in the interest of the British sugar monopoly.[32] The unwarranted assumption existed that *both* Canada and Guadeloupe could not be taken, that a *choice* between them must be made. That furs were as important as sugar could hardly be maintained; "the consumption of sugar is daily increasing both in America and Europe, and become one of the necessaries of life." Furs, moreover, would never give rise to as much shipping as sugar. The greater strategical value of Canada was, however, more difficult to disprove, "Acquiring Canada, dazzles the eyes, and blinds the understandings of the giddy and unthinking . . . : yet it is easy to discover that such a peace might soon ruin Britain," for the withdrawal of the French would promote the independence of America.[33] On the other hand, it was perceived that an increase in the number of sugar islands would solve the most vital commercial need of North America as well as of England; the Northern Colonies would get access to a far more adequate tropical market and England would obtain sugar enough for herself and, possibly, for export. It would give the British empire in America a proper balance between temperate zone and tropical colonies; heretofore the empire had been overweighted on the temperate zone side.[34] But the old British planters dreaded Guadeloupe as a rival.

[31] *A Letter from a Gentleman in Guadeloupe to his Friend in London,* Aug. 1760. In Almon's *Anecdotes of Pitt,* London, 1792, III, 215.

[32] *Ibid.,* 209-211. [33] *Ibid.,* 211-212.

[34] "It is our sugar islands that raise the value of North America, and pours in such wealth upon the *mother-country.* The more we have of those islands, America becomes from that cause the more important and valuable, and England the richer. In America we have more than enough; in the

"Thus Guadeloupe, one of the greatest acquisitions Britain ever made, acquires many powerful enemies from private views, and has nothing to plead but her public utility an advantage often found too feeble an opponent to the private interest of a few.'"[35]

In *A Letter to a Great M . . . R* [Pitt] *on the Prospect of a Peace,* published in April 1761,[36] the importance of Canada to England was again refuted, and the reasonableness and necessity of retaining the French sugar islands defended. The argument for their retention was based largely upon the benefits that would result to the trade between the Northern Colonies and the West Indies. The prevalence of illicit trade with the French and the draining of coin from Jamaica and the other English islands was fundamentally due to the deficiency of the British sugar supply. At least six-sevenths of the annual crop, which was estimated at not over 70,000 hogsheads, was consumed in Great Britain; the quantity left for North America and Ireland was altogether insuffi-

sugar islands a great deal too little: the nearer they can be proportioned to one another the better for both, and the more trade and wealth for England. The fur trade does not employ the hundredth part of the shipping and seamen, that the sugar trade does.'' *Ibid.,* p. 211. ''Guadeloupe can furnish as much sugar, cotton, rum, and coffee, as all the islands we have put together, and consume a vast quantity of British and American produce, from which trade the shipping and naval strength of Britain must greatly increase: without any allowance for the cinnamon trade. . . . The consumption of sugar is daily increasing both in Europe and America, and we cannot at this day serve ourselves with that article; . . . In a word, it is most obvious to every impartial eye, that the increase of the sugar islands is particularly the interest of Britain: she is there too weak.'' Such colonies would never revolt and their proprietors would bring their wealth to England. *Ibid.,* 213-214.

35 *Ibid.,* 214.

36 *A Letter to a Great M . . . R* [Pitt] *on the Prospect of a Peace; Wherein the Demolition of the Fortifications of Louisbourg is shown to be absurd; The Importance of Canada fully refuted, . . . By an unprejudiced Observer. . . .* London, Printed for G. Kearsly, at the Golden Lion, in Ludgate Street, 1761, 148 pages. Brit. Mus. E. 2053, 114.

cient.[37] "The cause of complaint arises from the disproportion of our sugar colonies to those of our northern colonies, the latter entirely subsisting upon the former, which is their only market, and can alone enable them to pay for the manufactures of Great Britain."[38] All apprehensions of the rise of manufacturing in the Northern Colonies arose from the same disproportion between them and the sugar islands. The true policy for England, therefore, lay in the acquisition of as many as possible of the larger French islands; to take the neutral islands—Dominica, St. Lucia, St. Vincent, and Tobago—would help, but would not give Great Britain control of the European sugar market.[39]

That the acquisition of Guadeloupe alone would have done much, if not entirely enough, to rectify the deficiency of sugar in the British empire is indicated by the value of her exports to England in 1761, which was £603,269, two-thirds of which represented sugar. Canada's exports to England in the same year amounted to only £14,015.[40] The supposition that Canada would create a market for English manufactures was opposed by the practical certainty that an expansion of British territory in the West Indies would so encourage North American trade as to more than compensate for the return of Canada. It was futile to attempt to measure the importance of West India islands by the amount of manufactures they consumed, which was less than tem-

37 *A Letter to a Great M . . . R* [Pitt] *on the Prospect of a Peace,* 79-82.

38 *Ibid.,* 82.

39 *Ibid.,* 86. The proposal to retain all the conquered French islands was supported also in the *Reflections on the True Interest of Great Britain. By a planter in Barbados,* C.O. 28: 50, summarized in Beer, *Brit. Col. Pol., 1754-1765,* pp. 150-151.

40 [William Burke], *An Examination of the Commercial Principles of the late Negotiation,* London, 1762; Beer, *op. cit.,* p. 149.

perate zone colonies required. Their importance in this connection lay largely in the sugar and molasses, which, taken by Northerners in payment for supplies, made possible the purchase by the latter of British manufactures.

The arguments in favor of the retention of Canada and the return to France of the sugar islands appeared in a number of pamphlets which have been effectively used by Mr. Beer in his study of the peace of Paris. Reasonable as many of the advocates for Canada were, they failed to recognize what must have seemed to the majority of Englishmen a very tangible need, a radical reduction of the price of sugar and the recovery for England of some share in the European sugar market. Thus a pamphlet, attributed to the Earl of Bath, urged the retention of Canada as the only basis for a durable peace, and regarded the West Indies as insignificant compared with such security as Canada would bring.[41] In a reply, William Burke suggested that security was illusory if it facilitated, as he believed it would, the independence of America. A greater degree of security in both America and the West Indies would result from leaving to the French a Canada bounded by the St. Lawrence and the lakes, and retaining Guadeloupe. Such an arrangement would also answer infinitely better the economic wants of both England and the Northern Colonies.[42] Benjamin Franklin's answer to Burke, emphasizing America's need of Canada in its inevitable expansion—and discounting the apprehensions of the rise of manufactures and independence, paid slight attention to the condition on which temperate zone settlement would for a long time depend,

[41] *A Letter Addressed to Two Great Men*, London, 1760; Beer, *op. cit.*, p. 143.

[42] William Burke, *Remarks on the Letter address'd to Two Great Men*, London, 1760; Beer, *op. cit.*, pp. 143-146.

namely, access to adequate markets.[43] In reply to Franklin, another pamphlet, attributed to William Burke, declared that the political advantages of having Canada were illusory, whereas the release of the sugar islands involved the sacrifice of a substantial trade, which the French had regarded as the most valuable branch of their commerce.[44] While the importance of the West Indies as producers of exports to England was recognized, Burke endeavored to correct the impression that they did not create markets for manufactures, by pointing out that, as a market for slaves and for produce from the Northern Colonies, it was really the West Indies that enabled Africa and North America to purchase British goods.[45] In his comparison of the colonies north of Maryland with the West Indies, he showed the former, by their tendencies to compete with Great Britain, to be of doubtful value, while he thought the latter group were ideal colonies.[46]

These views of the superior value of sugar colonies to the empire at that time were supported also in memorials from Barbadoes[47] and St. Christopher.[48] Another me-

[43] *The Interest of Great Britain Considered,* London, 1760; Beer, *op. cit.,* pp. 146-147.

[44] *An Examination of the Commercial Principles of the late Negotiation,* London, 1762, pp. 3, 4, 16-18; Beer, *op. cit.,* pp. 147-148.

[45] *Ibid.,* p. 29; Beer, *op. cit.,* pp. 148-149.

[46] *Ibid.,* pp. 64, 65, 66; Beer, *op. cit.,* pp. 149-150.

[47] *Reflections on the True Interest of Great Britain. By a planter in Barbados.* C.O. 28: 50; Beer, *op. cit.,* pp. 150-151.

[48] Feb. 27, 1760, C.O. 5: 20. "St. Christopher is by far the best [British] Sugar Island, both for the Quality of the Sugar, and the Quantity it produces in Proportion to Extent; but it is very small indeed, and not one Inch uncultivated, and must wear out. . . . Before the War, the Demand from Great Britain, and Ireland, and North America was greater than His Majesty's West India Colonies could furnish, and should there be an Increase of Territory in North America, the Demand will certainly increase in Proportion." The author also observed that the French islands produced a greater variety of products than the English islands. He advocated British expansion in the West Indies.

morial urged that, although Guadeloupe should not be retained in preference to Canada, the neutral islands ought unquestionably to be retained: (1) because of the validity of England's claim; and (2) because "The English want more sugar land to plant not only to supply foreign markets, but also to encrease the quantity for home consumption, and thereby reduce the price of a commodity now become of general and necessary use."[49]

It is clear from the foregoing evidence that there was a widespread and well founded belief in England that the empire was in pressing need of more sugar colonies. Whatever else peace might bring, the public was justified in expecting a substantial expansion of British power in the West Indies. Had Pitt remained in control of the negotiations, it is incredible that he would have disregarded this expectation of the nation. That Bute and Bedford sacrificed one of England's most vital interests, by returning Guadeloupe, Martinique, and St. Lucia to France, seems explicable only by their subserviency to the powerful planter class of Great Britain.[50] Consistent with their policy of the previous half century, the aim of which was to perpetuate monopoly prices for sugar in England, the British planters as a class opposed the acquisition of any considerable area of sugar land. The retention of Canada, on the other hand, offered the pleasing prospect of an enlarged colonial market for sugar.

In the debate on the treaty in parliament, December 9, 1762, the Earl of Hardwick denounced its terms in the House of Lords,[51] and Pitt in the Commons made a scathing criticism of the Bute ministry. "They seem to have lost sight of the great fundamental principle," he said,

49 An unsigned memorandum, April 13, 1761. Newcastle Papers, Brit. Mus. Add. MSS. 33030, f. 2.

50 See p. 335, note 4, *ante*.

51 *Parl. Hist.*, XV, 1251-1258.

"that France is chiefly, if not solely, to be dreaded by us in the light of a maritime and commercial power. And therefore, by restoring to her all the valuable West India islands, and by our concessions in the Newfoundland fishery, we had given her the means of recovering her prodigious losses, and becoming once more formidable to us at sea." Since the peace of Aix-La-Chapelle, France had gained a decided superiority over England in West India commerce. Britain stood in need of tropical supplies which only the retention of both Guadeloupe and Martinique, or at least one of them, could satisfy. The advantages to be gained in the West Indies were not conjectural but certain and immediate. Pitt insisted, furthermore, upon the benefits that would accrue to the North American and African trades through an enlargement of the British West Indies, all of which trades would center in Great Britain. But, if the French islands were restored, he gave warning that a revival of illicit trade between America and the French islands would result, and the British would lose a great part of the benefit of their colonial commerce. Though England had retained both Guadeloupe and Martinique, her conquests were such that there was still left an abundance with which to display her moderation.[52]

It was said in defense of the treaty that France would never be brought to any considerable cession of the West Indies, and that her power and increase there could never become formidable, because the existence of her settlements would depend more than ever upon trade with British North America.[53] But the fulfilment of this ministerial hope rested upon the ability of England to restrain by the Sugar Act of 1764 America's trade with the French—a policy of coercion that led straight to

52 *Parl. Hist.*, XV, 1261-1266.

53 *Ibid.*, 1272.

revolution. In losing something of the sugar trade, it was claimed that England lost little else than a luxury.[54] But this view was contrary, as we have seen, to the facts as stated by many contemporary writers. At last, the treaty was approved by the Lords and, in the Commons, by the overwhelming vote of 319 to 65.[55]

The independent public was clear and strong in its opposition to the terms of peace.[56] Yorkshire, Surrey, London, and other places refused bribes to support the treaty by addresses to the Crown.[57] New York merchants complained in March 1764 that, since the peace, British planters were less able than ever to supply the demand for sugar.[58] English consumers found no relief from the exorbitant prices of sugar. Their feeling was well expressed in the following note from "A Middling Tradesman" on the sugar monopoly, which appeared in a London newspaper in 1766: "It is not a misery which affects a few counties of Britain only; it is a cloud which invelopes the whole British dominions. It touches both rich and poor, but is felt only by the poor and the people of a middling station, with whom three-pence or four-pence in the shilling makes a great difference at the end of the year. So great a difference in the expense of housekeeping might prevent some bankruptcies and ease all. Why then will not the ministry oblige the nation by planning a proper law for the patriot purpose of repressing the malpractices of engrossers, monopolizers, and forestallers, in the sugar-trade? A proper regulation in this interest would certainly encrease the revenue."[59] Several treatises on the sugar trade and industry, which

[54] *Parl. Hist.* 1272. [55] *Ibid.* [56] *Ibid.*, 1275.

[57] *Ibid.*, 1274. The text of the peace, articles viii, ix, and xix of which refer to the West Indies, is given on pp. 1291-1304.

[58] Memorial of New York merchants to the House of Commons, Mar. 9, 1764, C.O. 5: 1071, f. 170.

[59] *Lloyd's Evening Post*, XVIII, 354, Apr. 12, 1766.

appeared in the years 1763 to 1765, demonstrated the need for new sugar colonies by their accounts of the enormous expense of sugar culture on the partially exhausted soil of the long settled English islands.[60]

Associated with the feeling of disappointment at the insignificant area acquired was a desire to make the most of the ceded islands. The suggestion that England develop the sugar industry in "East Louisiana" was not seriously considered.[61] Even the proposals to settle the new colonies, small as they were, alarmed the rich planters. But "Why," said a writer of 1764, "should we fear that sugar and rum will become drugs? Although more are produced, more will be consumed. Our conquests alone in North America will not fail to encrease the demand for them."[62] But the ceded islands, in fact,

60 John Campbell, *Candid and impartial considerations on the nature of the sugar trade; the comparative importance of the British and French Islands in the West Indies: with the value and consequence of S. Lucia and Grenada truly stated. . . .* 8vo, 228 pages, London, 1763; *Thoughts on trade in general, our West Indian in particular,* etc. [Signed *Ignotus*], 1763; *Some observations, which may contribute to afford a just idea of the nature, importance, and settlement, of our new West-India colonies,* 8vo, 51 pages, London, 1764, containing tables of plantation expenses; H. L. Duhamel du Monceau, *Art de raffiner le sucre,* folio, 78 pages, Paris, 1764, a treatise that was soon known to British planters; Samuel Martin, *An Essay upon plantership, . . .* fourth ed., 8vo, 62 pages, Antigua and London, 1765, a valuable treatise that ran through several later editions. Perhaps the most curious treatise of the period was Dr. James Grainger's *The Sugar-Cane: a poem, in four books; with notes,* 4to, 2561 lines, London, 1764. This poem was written at Basseterre, St. Christopher, in 1763, and describes the culture and manufacture of sugar, and the damage caused by rats, ants, and other insects. A 12mo edition appeared in London in 1766, and another copy is contained in Grainger's *Poetical Works,* Edinburgh, 1794. Dr. Samuel Johnson reviewed the poem favorably in the *Critical Review,* p. 270, and Dr. Percy in the *London Chronicle.* See Boswell's *Johnson,* ed. J. B. Hill, 1887, I, 481, and *Dict. Nat. Biog.*

61 *Propositions for Improving the Manufactures, Agriculture and Commerce of Great Britain,* 8vo, 132 pages, London, 1763.

62 *Considerations which may tend to promote the Settlement of our New West-India Colonies,* London, 1764, p. 33.

offered few opportunities to newcomers or to experienced planters seeking relief from soil exhaustion and exorbitant land values in the older islands. Grenada, in 1763, had eighty-two French sugar plantations, many of which were purchased by Englishmen.[63] The Grenadines were unsettled and of little importance.[64] Dominica had been settled by the French contrary to treaties, and contained 1400 whites who had 5000 to 6000 slaves. Cotton, coffee, and cocoa were raised there, but the island was too mountainous for sugar, and the Jesuit lands were the only ones for sale. St. Vincent was inhabited by about 2000 wild negroes, descendants of a shipload wrecked on the coast. These were, in 1763, under the influence of French missionaries and settlers, and were tilling the soil. St. Vincent was adapted for sugar, though none had yet been grown there. Tobago was unsettled. All these islands were, in 1764, united into one government and were known as the Windward Islands.[65] No person was permitted to purchase over three hundred acres in Dominica or more than five hundred acres in any one of the other islands. Clearances at the rate of five acres a year were required until half the arable land should be cleared, under penalty of £5 per annum for every acre not cleared. For every hundred acres cleared the proprietor was obliged to keep thereon one white man, or two white women, under penalties of £40 for each man and £20 for each woman deficient. An annual quit-rent of sixpence for every acre was to be paid to the Crown. Government land that was cleared was expected to bring £5 per acre,

63 *Considerations which may tend to promote the Settlement of our New West-India Colonies*, 6. Grenada exported, from Mar. 25, 1762, to Jan. 17, 1763, 4273 hhds., 190 tierces, 55 barrels of sugar, or 4,126,363 lbs., 22 hhds. of rum, and 510 hhds. of molasses. C.O. 323: 16, Q 65.

64 *Ibid.*, p. 7.

65 *Ibid.*, 8-11. In 1872, Dominica was classed as a Leeward Island.

and uncleared £1 per acre.[66] Good lands in St. Christopher, one of the best of the older islands, often sold at £100 sterling per acre, whereas in the new colonies an acre might, with proper management, be cleared for less than £20, including every expense and allowing even for the mortality of slaves.[67]

Ten years of experience in the ceded islands proved them, however, of little value to England. They already contained a considerable French population that remained under British rule. The native Caribs, in St. Vincent especially, proved recalcitrant and resisted the survey and sale of their lands. British troops, in 1772, were sent to St. Vincent, and, in 1773, the Caribs agreed by treaty to acknowledge the supremacy of England and accept reservations in the north end of the island.[68]

But British capital was not attracted to the new colonies. On April 29, 1772, Pulteney championed a bill in the House of Commons to encourage foreigners to lend money upon the security of freehold and leasehold estates in the West Indies, in order to secure "the more speedy and extensive cultivation" of the sugar islands. The bill was warmly opposed by several persons interested in the West Indies. Pulteney answered by saying: "If the opposers of this Bill mean by public purposes the purpose of keeping up the price of sugar, and by these means to enrich a few old planters at the expense of this nation, I must own that their plan is the more public spirited of the two. Mine is calculated for rendering that commodity more plentiful, and therefore cheaper, by facilitating the cultivation of the newly acquired plantations."

66 *Considerations which may tend to promote the Settlement of our New West-India Colonies*, 19-20.

67 *Ibid.*, 41.

68 C. P. Lucas, *Historical Geography of The British Colonies*, II; C. Atchley, *The West Indies*, pp. 212-213.

Money in the West Indies could not be had, he said, for less than eight per cent; from foreigners it might be borrowed for five per cent if it were made clear by law that the foreigners could recover their capital. Pulteney owned an estate in the Windward Islands and said the planters there wanted such a bill giving foreigners power to secure the sale but not possession of estates. Burke and Cornwall supported the measures. But on a division it was defeated by 36 to 29.[69] Pulteney laid the matter before the House again on December 11.[70] On January 22, 1773, Oliver Nugent, a merchant of Dominica, testified to the need of capital in the ceded islands, the lack of which was seriously retarding the production of sugar. In Holland and Genoa, he added, money might be had for five per cent. Pulteney's bill, extending to a foreign mortgagee the right of application to chancery for a forced sale to recover a loan, was finally passed into law.[71]

The new islands were advertised as affording relief to old British planters whose profits were declining under the costly conditions of planting in the older islands. That there were many such "of the middling and poorer sort" who would have welcomed access to virgin soil is evidenced by the considerable number who were acquiring estates in the Danish islands[72] and in the Dutch colony of Surinam.[73] But unfortunately, the sugar lands obtained by the peace of Paris were altogether too limited to offer such planters adequate relief under the British flag. The aspirations and interests of this class had little

69 *Parl. Hist.*, XVII, 482-484. 70 *Ibid.*, 642-644.

71 *Ibid.*, 686-690. The act is 13 Geo. III, c. 14.

72 *Considerations that may tend to promote the Settlement of our New West-India Colonies*, p. 32. Many Englishmen who were driven from the Leeward Islands by poverty to Santa Cruz "are now become from the lowest degree of indigence, possessors of good estates."

73 Gov. Bernard of Massachusetts wrote the Board of Trade, Sept. 5, 1763, that in Surinam "Many English have considerable Estates." C.O. 5: 891, Ll 67, 68.

or no weight with the rich proprietors who guided West India policies. For had the former prevailed, the diplomacy applied to the West Indies in the eighteenth century would not have differed so radically as it did from the policy of expansion supported by the American slavocracy in the following century. Spain, France, and Mexico yielded to the United States, through negotiation or force, vast areas for the spread of cotton culture. The sugar industry of Great Britain in the eighteenth century possessed all the incentives and potentialities for a similar expansion. Anti-slavery sentiment, moreover, which developed a check on American expansion, exercised, in 1763, no such influence in British politics. The dwarfing of this great British industry was unnatural. The reason for it lay in the exorbitant profits which a position of monopoly conferred on the influential planters who controlled West India affairs.[74]

[74] The attitude of great planters toward new settlements is well stated, in 1764, in the following passage: "It has been remarked by many persons, that some of the planters of our old islands have painted in very strong colours the difficulties and discouragements attending the establishment of our new ones. . . . But when difficulties are exaggerated, from party prejudices, or selfish and interested views, the motives are ungenerous, and deserve to be exposed. It is easy to see that the settlement of our new colonies will impair the value of our old. Such lands in them as are not of good quality and well situated, will not be worth cultivation. They will be displanted, and converted into provision, and pasture grounds. Many inhabitants of the middling and poorer sort, will withdraw themselves, and be induced to remove where fertile lands may be easily procured, which are new, and watered with frequent showers. This will much weaken our leeward islands, already weak indeed, and render the wages of managers, overseers, and servants, much higher. It is not wonderful, therefore, if some of our wealthy planters who have a large landed interest in the old colonies, should not be studious to promote the new. Private advantage will bias some, and it is to be feared, jealousy and envy may actuate others. Men accustomed to affluence and authority are not usually pleased to see their inferiors rise above dependence, and possess, by lucky circumstances, superior fortunes to their own. Nor are these the only foundations of discontent and objection. Some affect to condemn the enterprize they mean to embark in, and depreciate those very lands they wish to procure; whilst

West India planters apprehended that tropical expansion would mean not only increased production and a lowering of the price of sugar but social upheaval and decay in the older communities such as Barbadoes and the Leeward Islands. Families long established there had, in the absence of industrial rivalry, grown conservative and naturally shrank from innovations that would expose them to the competition of virgin soils. Even absentees cherished a certain sentiment and pride in their great and venerable estates in the West Indies. Their detachment from the practical operations of planting, however, deterred them from participating in a forward movement. Under the protective system their revenues were large and assured and they had risen to social prominence and political power in the home land. Why should they be disturbed? Jealousy and envy, perhaps, actuated some to withhold from their inferiors the opportunity of rising in wealth and position. New colonies, they thought, would cause the abandonment of worn lands or their conversion into pasture and provision grounds, the draining of population, the increase of wages for managers, overseers, and servants, and general social decay. That such fears were well founded seems probable in the light of similar changes which occurred in the eastern part of the American cotton belt previous to the Civil War.[75] The great planters discouraged the settlement, therefore, of even the small islands acquired by the treaty, exaggerating the hardships and anxieties of a planter's life. A

others again, who hoped to obtain undue preferences, and to monopolize large tracts, with a view of forstalling the market, or of selling again at advanced prices what they never meant to cultivate, now find themselves disappointed, and their designs effectually prevented, by the present impartial plan of government. But juster sentiments, it is hoped, will, at length, prevail over such selfish considerations.'' *Considerations which may tend to promote the Settlement of our New West-India Colonies*, pp. 35-36.

75 See F. L. Olmstead, *Seaboard Slave States*, 1856, pp. 576-577.

few proprietors were prepared, it is true, to develop the new lands on a large scale or to speculate in them at least, but the government's policy was now contrived to promote small holdings and prevent the monopolization of land.

The peace of Paris marked a momentous crisis in the history of the empire. Through it the planting interest came triumphant. Its position of monopoly was practically undisturbed; Great Britain and America were still exposed to exploitation by an interest whose aims were well understood. The opportunity of adjusting a balance in the American empire had been thrown away, and the means of reducing the exorbitant cost of sugar ignored. Colden had officially declared, in 1760, that it would be difficult to confine America to the British West India markets while the people believed that "The Sugar Islands have gained a preference inconsistent with the true interest of their mother country." But his warning went unheeded. The treaty of Paris, the Sugar Act of 1764, and the administrative reforms of Grenville revealed a firm determination to restrict America to the same old markets which time and again had been proved inadequate for either England or America. It is not surprising that murmurs about "the inconvenience of being British subjects" grew louder in the Northern Colonies. The West India planting interest had laid substantial foundations in the realm of economic life for that great discontent which culminated in the American Revolution.

BIBLIOGRAPHICAL NOTE

Bibliographies

Andrews, C. M., and Davenport, F. G. Guide to the Manuscript Materials for the History of the United States to 1783, in the British Museum, in Minor London Archives, and in the Libraries of Oxford and Cambridge. Washington: Carnegie Institution, 1908.

Andrews, C. M. Guide to the Materials for American History, to 1783, in the Public Record Office of Great Britain; Vol. I, The State Papers; Vol. II, Departmental and Miscellaneous Papers. Washington: Carnegie Institution, 1912, 1914.

Cundall, Frank. Bibliographia Jamaicensis, etc. Kingston, Jamaica, 1902.

de Dampierre, Jacques. Essai sur les Sources de l'Histoire des Antilles Française (1492-1664). Paris, 1904.

Roth, H. L. A Guide to the Literature of Sugar. London, 1890.

Thorpe, Henry. A Bibliography of Sugar in the *Sugar Cane.* January 8, *et seq.* 1884.

Watt, Robert. Bibliotheca Britannica. Edinburgh, 1824. 4 vols. Volume IV under the title "Sugar" gives an extensive list of the earlier writings on sugar.

Meyer, H. H. B. Select List of References on Sugar Chiefly in its Economic Aspects. Library of Congress, Washington, 1910.

Manuscripts in the Public Record Office

As these manuscripts are fully described in Andrews's Guide, the following is simply a list of the classes of manuscripts used to which abbreviated references are usually made in the footnotes.

Reference	
C.O. 1:	Colonial Office Papers, Class I.
C.O. 5:	America and West Indies, Original Documents.
C.O. 5:855-891,	New England, Original Correspondence, 1689-1766.
C.O. 5:1303,	Rhode Island, *ibid.*, 1702-1782.
C.O. 5:1050-1071,	New York, *ibid.*, 1710-1764.
C.O. 5:1288-1296,	Proprieties, *ibid.*, 1696-1773.
C.O. 5:840-850,	Massachusetts Shipping Returns.
C.O. 7:	Antigua, Correspondence with the Secretary of State.
C.O. 8:	Antigua, Acts of the Assembly.
C.O. 28:	Barbadoes, Original Correspondence with the Board of Trade.
C.O. 29:	Barbadoes, Entry Books.
C.O. 30:	Barbadoes, Acts of the Assembly.
C.O. 31:	Barbadoes, Proceedings of Council and Assembly.
C.O. 33:15-17,	Barbadoes, Naval Office Lists, 1711-1764.
C.O. 137:	Jamaica, Original Correspondence with the Board of Trade.
C.O. 139:	Jamaica, Acts of the Assembly.
C.O. 140:	Jamaica, Proceedings of Council and Assembly.
C.O. 142:14-17,	Jamaica, Naval Office Lists, 1702-1769.
C.O. 142:31,	Jamaica, List of Landholders, 1754.
C.O. 152:	Leeward Islands, Original Correspondence with the Board of Trade.
C.O. 154:	Leeward Islands, Acts of the Assembly.
C.O. 155:	Leeward Islands, Proceedings of Council and Assembly.
C.O. 157:1,	Leeward Islands, Naval Office Lists, 1704-1715.
C.O. 176:	Montserrat, Acts of the Assembly.
C.O. 185:	Nevis, Acts of the Assembly.
C.O. 187:1-2,	Nevis, Naval Office Lists, 1720-1729.
C.O. 240:	St. Christopher, Acts of the Assembly.

C.O. 318: West Indies, Trade and Miscellaneous Papers.

C.O. 323: Plantations General, Original Papers.

C.O. 324: Plantations General, Entry Books and Miscellaneous Papers.

C.O. 388: Board of Trade, Commercial Series I, Papers and Consular Reports on British Foreign Trade.

C.O. 389: Board of Trade, Commercial Series II, Entry Books.

C.O. 390: Board of Trade, Commercial Series, Custom House Accounts.

C.O. 391: Board of Trade Journal.

Treas. 70: Books and Papers of the Royal African Company. Vol. 1263 contains estimates of the numbers of slaves exported from Africa. The series includes a few volumes of correspondence from the company's agents in the West Indies.

B. T. 6: 1-19, Modern Board of Trade, Papers and Accounts relating to Africa, chiefly from the reorganized African company (1750-). Vols. 9-12 contain valuable testimonies on West India slavery given before a committee of the Privy Council in 1788.

Treas. 38: 253, 256, 258-268, 270, 362-364, Revenue Accounts, Miscellaneous, England. These bundles contain various accounts of West India trade and production, also accounts of the four and a half per cent and enumeration dues.

Treas. 64: 273-275, *Ibid.*

Custom House Accounts. Ledgers of Imports and Exports. See Andrews's Guide, II, p. 116. The charts I-VII, IX-XII are based on tables compiled from these accounts.

Adm. 1: Admiralty, Secretary's Department, In-Letters. Admirals' despatches from West India stations. Vols. 305-307, 3814-3819 contain many references to illicit West India trade.

Manuscripts in the British Museum

Among the very large number of manuscripts relating to the West Indies, many of which are cited in the footnotes, especial mention should be made of the Edward Long Papers (Add. MSS. 12402-12440), including manuscript histories of Jamaica by James Knight, one of the colony's agents, and by Dr. Henry Barham. This extensive collection relates particularly to Jamaica in the eighteenth century. For comprehensive lists of manuscripts relating to the West Indies in various British archives other than the Public Record Office, see Andrews and Davenport's Guide.

The most authoritative printed materials are in the Calendar of State Papers, Colonial, America and West Indies (1574-1708), the Acts of the Privy Council (Colonial Series), edited by W. L. Grant and James Munro, Vols. I-IV (1613-1766), and the Reports of the Royal Historical Manuscripts Commission. Nearly thirty collections of West India laws have been printed. They generally contain only acts in force at the dates of publication and hence are often of no historical value. Fairly complete collections of acts, including those obsolete or disallowed, are to be found only in manuscript in the Public Record Office. Printed proceedings of West India assemblies are rare. The Journals of the Assembly of Jamaica, 1663-1826, 15 vols., Jamaica, 1811-1829, is the only notable collection of its kind. The only collections approaching completeness are the manuscript minutes of councils and assemblies in the Public Record Office. Printed books, personal narratives, pamphlets, agricultural treatises, botanical, medical, and other works relating to the West Indies are legion in number and accessible for the most part only in the British Museum, the Bodleian Library, and the Library of Congress. Reference has been made in the footnotes to such printed materials of this kind as have proved useful. Practically no histories treating the development of the West Indies through any considerable period of time have yet been written. Edward Long, History of Jamaica (1774) and Bryan Edwards, History of the West Indies (1793) are primarily literary descriptions of the islands in the periods when the authors lived; as such they are of permanent value but as histories they fail to answer modern inquiries. Among the very few scientific studies that have appeared mention should be made of Nicholas Darnell Davis, Cavaliers and Roundheads of Barbados, 1650-1652. With some Account of the Early History of Barbados, Georgetown, Demerara, 1887; Vere Langford Oliver, History of the Island of Antigua . . . from the first Settlement in 1635 to the Present Time, 3 vols., folio, London, 1894-1899, an edition limited to one hundred and fifty copies and largely genealogical in content; and Frank Cundall, Studies in Jamaica History, Kingston, Jamaica, 1900. C. Atchley's volume on the British West Indies in C. P. Lucas, Historical Geography of the British Colonies, Oxford, 1905, is a reliable manual based on printed material and has short bibliographies for each island. The British West Indies have been far more neglected by modern historians than the French Antilles. The field, however, is rich in interest and opportunities for students of economic and social history.[1]

[1] A partial list of printed and manuscript materials for the history of the British West Indies and the sugar trade may be found in a dissertation by the author on "The Development of the British West Indies, 1700-1750," pp. 232-410, in manuscript in the Yale University library.

APPENDICES

APPENDIX I

THE POPULATION OF THE BRITISH WEST INDIES

While the slave population of the British West Indies steadily increased from about the year 1650, the number of whites shows remarkable fluctuations without much tendency to increase after the opening of the eighteenth century. In fact, a decrease in the white population of the islands is not infrequently noted during the later period. Many reasons for this situation may be mentioned. First, climatic conditions impelled a great many to migrate to the cooler climes of North America. Others of an adventurous disposition, including many criminals and outcasts from England, encouraged by the turbulence of the age and at times by the government itself, joined the buccaneers and pirates till well into the eighteenth century. But the most powerful factor in preventing the settlement of poor whites lay in their increasing inability to compete in tropical industry with the capitalists and their slaves who entered the field in the last quarter of the seventeenth century. Further, it must be borne in mind that an untold number of the offspring of Englishmen were mulattoes. These and their children were lost to the white race. Finally, if we may cautiously draw conclusions from annual lists of baptisms and burials among the white people, it would seem that in general births did not keep pace with deaths.

Barbadoes increased in population with great rapidity during the reign of Charles I and during the Commonwealth and Protectorate. Many who arrived in this period were either republicans or royalist refugees. In 1636, after ten years of actual settlement, the island contained about 6000 English.[1] In 1645, there were 18,300 effective men, of whom 11,200 were proprietors; there were also 6400 negroes.[2] At that time allotments of five, ten, twenty, and thirty acres of land were made to settlers, while to indentured servants plots of three, four, or five acres were granted upon the completion of their services.[3] In 1650, the population was supposed to be 30,000.[4] As early as 1652, Barbadoes was overcrowded and its resources overtaxed, for in that year Colonel Thomas Modyford wrote of it: "This Island of Barbadoes cannot last in an height of trade three years longer especially for suger, the wood being almost already spent, and therefore in prudence a place must be presently thought upon, where this great people shall find maintenance and employment."[5] He urged the settlement of Guiana as an outlet for Barbadoes. Ligon's estimate of 50,000 souls for Barbadoes, besides negroes to a number double the Christians, is surely an exaggeration.[6] English merchants, in 1656, supposing there were at least 25,000 Christians in the island, reckoned on a market for 12,000 dozens of shoes in Barbadoes.[7]

Migration from Barbadoes had assumed respectable

1 Capt. Sir Thomas Warner, Governor of St. Christopher, to [Sec. Windebank], Sept. 10, 1636. *C. S. P. Col., 1574-1660,* p. 240, no. 21.

2 Sloane MSS. 3662, f. 62 *et seq.*

3 N. Darnell Davis, *Cavaliers and Roundheads of Barbados, 1650-1652,* Georgetown, British Guiana, 1887, p. 71.

4 *A briefe Description of the Island of Barbados,* Trinity College, Dublin MSS. G. 4, 15.

5 Modyford to John Bradshaw, Barbadoes, Feb. 16, 1652. C.O. 1: 11, no. 41; *C. S. P. Col. 1574-1660,* p. 374, no. 41.

6 Richard Ligon, *A True and Exact History of the Island of Barbados,* 1st. ed. London, 1657, p. 43.

7 *C. S. P. Col. 1574-1660,* p. 446, no. 9.

dimensions by 1670. A writer about that time observed that a great many people had moved away from the island during the past year. "The falling of the Profits of the Suger Trade," he explained, "hath proportionally diminished the number of the Inhabitants of the Suger Collonys Barbadoes being since 1665 fall'n from 14000 Fighting Men to lesse than 8000 and they continue still going away, 2000 people having gone away within a year past, whereof the soberest & the Richest to the parts about New England some to the French plantations, & the looser sort out of hopes of plunder to Jamaica."[8] A census of Barbadoes prepared by the governor, Sir Richard Dutton, in 1683, giving the resources of each parish, records a total for the whole island of 4156 families and householders, 17,187 free persons, 2381 unfree persons and servants, 46,602 slaves, 6761 men able to bear arms, 90,517½ acres possessed and useful, and 358 sugar works.[9]

In the eighteenth century, census returns for Barbadoes have been found for ten different dates. For the purpose of comparison it is worth while, perhaps, to print summaries of them. The census for 1715 occupies nearly a whole manuscript volume in which a small booklet is devoted to each parish, giving the names of the white inhabitants, men, women, boys, and girls, with their ages.[10] While the totals are not given for that year, two characteristics are apparent: (1) the small number of persons over fifty years of age—most of them being from twenty to forty years, and (2) the small number of chil-

[8] *The true State of the Manufacture of Sugars within our Plantations,* etc. Brit. Mus. Egerton MSS. 2395, f. 638.

[9] Sir Richard Dutton, The State of Barbados; Official Copy prepared for the Council of Trade preparatory to the Government coming under their Lordships' control, Phillipps MSS. 8797, printed in Schomburgk, *History of Barbados,* London, 1848, p. 82.

[10] Gov. Lowther to Board of Trade, July 20, 1716, C.O. 28: 16, T 136.

dren to each white family, it being considerably less than in the British Northern Colonies. Governor Robinson, in 1747, stated that church wardens had not returned census lists, that he could not compel them, and that those sent in were imperfect. However, by the parish registers for the past eight years he found the number of inhabitants rather increased.[11] In the second half of the eighteenth century the white population tended to fall off, a tendency that became more marked in the early nineteenth century. Slaves, on the other hand, steadily increased and, owing to the poll tax they were subject to, were probably more numerous than is admitted by church wardens' lists.

Barbadoes contained, in 1710-1712, 1309 plantations, 409 windmills, 76 cattle mills, 24 pot kilns, and 2471 horses.[12] Such census data as is available for Barbadoes in the eighteenth century is condensed in the following table:

Year	Whites			Able to bear arms	Total	Free Mulattoes and Negroes	Slaves
	Men	Women	Children				
1710[13]							52337
1712[14]	3537	3529	5462	3438	12528		41970
1724[15]				4812	18295		55206
1734[16]	4917	5822	7374	4708	16113		46362
1748[17]	3840	5209	6203	3716	15252	107	47025

11 Gov. Thomas Robinson to Board of Trade, Feb. 20, 1746/7, C.O. 28: 27, Bb 57.

12 G. Lillington to Board of Trade, Aug. 1, 1710, C.O. 28: 13, R 75; also in C.O. 28: 43, f. 76; Lowther to Board of Trade, Aug. 16, 1712, C.O. 28: 14, T 15.

13 Lillington to Board of Trade, Aug. 1, 1710, C.O. 28: 13, R 75.

14 Lowther to Board of Trade, Aug. 16, 1712, C.O. 28: 14, T 15.

15 Gov. Worsley to Board of Trade, Oct. 18, 1724, C.O. 28: 18, W 51.

16 Gov. Lord Howe to Board of Trade, Apr. 25, 1734, C.O. 28: 24, Aa 16.

17 Gov. Henry Grenville to Board of Trade, June 22, 1748, C.O. 28: 29, Cc 9. Grenville thought these figures too small and states that the number of whites could not be less than 25,000 and the blacks 68,000. He gave the militia as consisting of 4898 whites and 2741 blacks. In 1747, Robinson estimated the whites at 20,000 and the blacks at 70,000. C.O. 28: 27, Bb 57.

Year	Whites: Men	Whites: Women	Whites: Children	Able to bear arms	Total	Free Mulattoes and Negroes	Slaves
1753[18]							69870
1757[19]	4608	5549	6615	4176	16772		63645
1762[20]					18419		70000
1768[21]					16139	448	66377
1773[22]					18532		68548
1783[23]	4466	5903	5798	4361	16167	838	57434
1786[24]					16167		62115
1787[25]					16127		64405
1805[25]					15000		60000
1812[25]					13749		69132
1829[25]					14959		82902

The settlement of Jamaica in the second half of the seventeenth century while not so rapid as that of Barbadoes was more permanent in character.[26] Being the

18 Schomburgk, p. 86.

19 C.O. 28: 31, Ee 8.

20 Gov. Charles Pinfold to Board of Trade, June 1, 1762, C.O. 28: 32, Ff 25. Pinfold added that "by the Parish Registers comparing the Births and Burials, the Inhabitants are within ten Years decreas'd 442—in which are not computed those who have left the Island—the small Pox raged with great Violence in 1759—the Decrease in that year is near 300."

21 C.O. 28: 51, A 22.

22 Schomburgk, p. 86.

23 Certified by Francis Workman, Dep. Sec'y, Sept. 17, 1783, C.O. 28, Bundle 42.

24 Schomburgk, p. 86.

25 *Ibid.*

26 From estimates gathered mainly by Long the progress of Jamaica in population may be seen from the following table:

Year	Whites	Negroes	Total population
1658	4500 (a)	1400 (a)	5900
1664	6000 (a)		
1665			17280 (a)
1670	8000 (a)		
1673	8564 (a)	9504 (a)	18068
1677	9000 (a)		
1694	7000 (a)		
1703	3500 (b) (men)	45000 (b)	
1722	7100 (a)	80000 (a)	87100
1730	7658 (b*)	74525 (b*)	83048
1734	7644 (c)	86546 (c)	94190

largest island in the British group—it is very nearly the size of Connecticut—there is no evidence of overcrowding or of migration of those once settled. The varieties of soil, the broad savannas, forests, streams, and mountains offered a wider range of opportunity than Barbadoes or the other smaller islands. In spite of natural resources favorable to a large society with diversified industries Jamaica was, by the opening of the eighteenth century, practically monopolized by a slavocracy who, in the interest of high prices, appear to have restricted the cultivation of sugar. To many British expansionists Jamaica must have been regarded, by 1750, as a sore disappointment. But with the exception of Edward Long and a few enlightened ones the planters were well contented, save for an occasional suspicion that they might be disturbed in their snug position.

Year	Whites	Negroes	Total population
1739	10080 (a)	99239 (a)	109319
1746	10000 (a)	112428 (a)	122428
1754	12000 (d)	130000 (d)	142000
1762	15000 (e)	146464 (e)	161464
1768	17949 (f)	166914 (g)	184863
1778	18420 (h)	205261 (h)	223681
1787	25000 (h)	210894 (h)	235894

(a) Long, Jamaica, I. 376; Brit. Mus. Add. MSS. 12404, p. 299.

(b) Lieut. Gov. Handasyd to Board of Trade, May 30, 1703, C.O. 137: 5, E 41.

(b*) Gov. R. Hunter to Board of Trade, Dec. 24, 1730, C.O. 137: 19, S 124.

(c) *Hand book of Jamaica*, Kingston, 1899, pp. 34 *et seq.*

(d) Gov. Charles Knowles to Board of Trade, Dec. 31, 1754. C.O. 137: 28, Y 55.

(e) Gov. W. H. Lyttelton to Board of Trade, July 9, 1763, C.O. 137: 33, Cc 19. The total population includes 865 free negroes, Indians, and mulattoes.

(f) Long, Jamaica, I. 376.

(g) Census of Jamaica, Sir William Musgrave Papers, Brit. Mus. Add. MSS. 8133, ff. 95-96.

(h) Long Papers, Brit. Mus. Add. MSS. 18273, f. 93.

Though the numbers of whites slowly increased in the eighteenth century down to 1791, there were occasional setbacks which, from a military point of view, were alarming. Thus in 1703, the lieutenant governor wrote: "Our number of Slaves Augment dayly but to my great grief the Number of white men dayly decrease. At my first coming to the Government, I did Compute by the Rolls that the Island had 4500 men, but Cannot find now above half."[27] In 1721, the fighting strength of Jamaica was estimated at 3000.[28] Several of the governors reported the difficulty of obtaining correct, if any, census returns from the parishes. However, in 1730, Governor Robert Hunter transmitted figures for each parish, making the total population of Jamaica as follows: 2171 masters and mistresses, 3009 white men servants, 984 white women servants, 1484 white children, 136 free colored or mulatto men, 321 free colored or mulatto women, 408 free colored or mulatto children, 74,525 slaves, and 55,341 cattle.[29]

From these returns it appears that of the white population of Jamaica at that period over a half were white indentured servants,[30] three-quarters of whom were men.

[27] Lieut. Gov. Handasyd to Board of Trade, Oct. 5, 1703, C.O. 137: 6, G 19.

[28] Gov. Nicholas Lawes to Board of Trade, Apr. 20, 1721, C.O. 137: 13, P 126.

[29] Hunter to Board of Trade, Dec. 24, 1730, C.O. 137: 19, S 124. Hunter himself estimated the slaves at 100,000 and the cattle at 200,000 "which I take to be the truest having had it from such Gentlemen as have been Conversant in these affairs for many Years, but as the above List has been but Just returned to me I think it my Duty to give You an Exact Copy. . . . In this is only to be understood Neat Cattle and those that are daily Penned who can be counted and I believe there are near as many wild in the Woods as are mentioned above."

[30] In 1739 and again in 1768 Long estimated that indentured servants composed about one-third of the white population. Brit. Mus. Add. MSS. 12,404, p. 299.

Also the number of white children, which would be extraordinarily small for a frontier community in the temperate zone, corresponded to the great preponderance of males over females among the white adults. While the mulattoes must have been quite numerous by 1730 only a very small proportion had apparently been freed. The number of free negroes was almost insignificant. The average of thirty-four slaves to a master is much lower than the number which travelers of the time commonly observed. Finally, the number of cattle, which was an underestimate, indicates the extent to which stock-raising was carried in Jamaica.

In 1740, Jamaica's colonial agent, James Knight, compiled a table of the number of negroes and cattle in each parish based on the returns of a tax of 12d. per head on slaves and 3d. on cattle, the total being 99,239 negroes and 84,313 cattle.[31] In 1741, Governor Trelawny complained of the difficulty of getting accurate information but estimated the whites to be 10,000, militia 4000, and blacks 100,000.[32] Previous to 1739, the increase of settlements was considerably retarded by the Maroon Wars which were very destructive of life and property. Following this year came a period of general prosperity unattended by negro rebellions, and the white population of Jamaica slowly but steadily increased from 10,000 in 1740 to about 30,000 in 1791. This date marks a climax in the increase of white people in Jamaica. The Napoleonic Wars were about to open and the anti-slavery movement of Clarkson and Wilberforce was well under way. These were the palmy days of Jamaican history. An account of the population and industrial strength of

31 Papers of James Knight, Brit. Mus. Add. MSS. 12434, f. 14.

32 Trelawny to Board of Trade, Nov. 21, 1741, C.O. 137: 23, W 47.

each parish in Jamaica calculated from the poll tax roll authorized in December 1768 gives the following totals for Jamaica: 135,773 cattle, 67,852 persons employed in the cultivation of coffee, cotton, ginger, pimento, &c., 99,062 persons employed in the producing of sugar, making a total of 166,914 laborers; the whole number of sugar plantations was 648, of which 369 were equipped with cattle mills, 235 with water mills, and 44 with wind mills; the total quantity of sugar made in 1768 was 68,160 hogsheads; the number of negroes required to produce 100 hogsheads of sugar varied from 102 in St. James' parish to 250 in the parish of Port Royal, the average being 166.[33] It thus appears that, in 1768, about 60 per cent of the labor of Jamaica was employed in the making of sugar, about 40 per cent in producing coffee, cotton, ginger, pimento, etc., and live stock. It will also be noticed that more than half of the mills were operated by cattle or mules, instead of by wind or water. Even where wind or water power were available many planters preferred cattle mills as being more reliable. This, however, created an extraordinary demand for live stock and in some cases unnecessarily increased the cost of production.

For the number of mulattoes in the eighteenth century there is but little data. In 1730, it was estimated that there were 865 free negroes, Indians, and mulattoes.[34] In 1763, Lyttelton reported 4000 free negroes and mulattoes,[35] and in 1787 the same were estimated at

[33] The account is signed by John Knill and is among the Papers of Sir William Musgrave: Revenue of Customs, Brit. Mus. Add. MSS. 8133, ff. 95-96.

[34] Gov. Hunter to Board of Trade, Dec. 24, 1730, C.O. 137: 19, S 124.

[35] Gov. W. H. Lyttelton to Board of Trade, July 9, 1763, C.O. 137: 33, Cc 19.

10,000.[36] Apparently very few negroes or mulattoes were emancipated in the first half of the eighteenth century while in the last half there was an extraordinary increase in the number who obtained their freedom.

An unofficial survey of Jamaica's population and resources, made about 1786, at the period of its greatest prosperity under the old régime, credited the island with 20 parishes; 18 churches and chapels; 36 towns and villages; 1061 sugar estates; 2018 other settlements; 255,710 negro slaves; 224,500 head of cattle; about 25,000 whites; and nearly 10,000 free negroes and mulattoes.[37]

While the total population in the settlements composing the Leeward Islands was small, the few analyses that were made of those miniature societies in the eighteenth century may prove interesting. A census made in 1720 is presented in the following table:

YEAR 1720[38]

	Free persons					Servants—Free and Unfree				Persons fit to bear arms in militia	Negroes
Island	Men	Women	Boys	Girls	Total	Men	Women	Boys	Girls		
Antigua	739	819	744	652	2954	471	140	45	42	1109	19186
St. Christopher	645	694	626	575	2540	163	54	28	15	755	7321
Nevis	331	426	206	312	1275	33	18	13	4	378	5689
Montserrat	486	492	295	320	1593	64	10	9	12	444	3772
Anguilla	133	164	112	139	548	..	..	..	..	121	879
Spanish Town	92	86	90	103	371	..	..	..	..	88	364
Tortola	39	48	61	55	203	..	..	..	..	53	266
Total	2467	2729	2134	2156	9484	731	222	95	73	2948	37477

36 Bryan Edwards, *West Indies*, ed. London, 1801, p. 922. Edwards estimated the Maroons, in 1787, at 1400.

37 Long Papers, Brit. Mus. Add. MSS. 12414, ff. 22-23, and 18273, f. 93.

38 Gov. Hamilton to Board of Trade, Aug. 22, 1720, C.O. 152: 13, Q 46.

By 1724 these islands had increased in population slightly, as shown in the next table.

YEAR 1724[39]

	Whites	Blacks	White Militia
Antigua	5200	19800	1400
St. Christopher	4000	11500	1200
Nevis	1100	6000	300
Montserrat	1000	4400	350
Anguilla	360	900	85
Spanish Town	340	650	78
Tortola	420	780	100
Total	12420	44030	3513

The considerable increase of population at St. Christopher was due to the occupancy of the former French half of the island by settlers mostly from Nevis and Montserrat, and this explains the decrease of population in the latter islands.

By 1745 the four major islands of the Leeward group contained the following number of inhabitants:

YEAR 1745[40]

	Whites	Slaves	Militia
Antigua	3538	27892	1417
St. Christopher	2377	19174	845
Nevis	857	6511	309
Montserrat	1117	5945	411

Detailed reports on the population of each parish in Antigua were made in 1729, 1753, and 1756, the results of which for the whole island are summarized in the following table:

39 Gov. Hart to Board of Trade, July 12, 1724, *ibid.*, 14, R 101.
40 Gov. Mathew to Board of Trade, Apr. 15, 1746, C.O. 152: 25, Y 154.

NUMBER OF ALL FREE PERSONS AND WHITE SERVANTS IN ANTIGUA

Year	Men	Women	Boys	Girls	Slaves	Militia
1729[41]	1337	1096	563	561		
1753[42]	1433	1123	439	466		
1756[43]	1392	1117	458	468	31428	3200

In Antigua between 1720 and 1756 it appears, therefore, that while the adult white population slightly increased the propagation of white children had actually fallen off. If the figures can be trusted it may be concluded that the white society in Antigua was dying out and that the maintenance or slight increase in numbers was due to immigration. The slaves during the period nearly doubled in numbers. Also the militia more than doubled its ranks.

St. Christopher, until the peace of Utrecht, was partly under the control of France. In 1713, the French half of the island with 2000 whites and 12,000 blacks was ceded to England. Several of the best families and most of the slaves remained.[44] Census returns for the parishes of St. Christopher were made occasionally in the eighteenth century and are summarized for the whole island in the following table:

POPULATION OF ST. CHRISTOPHER

Year	White Men	Men able to bear arms	Women	Children	Boys	Girls	Slaves
1717[45]	799						7973
1720[46]	645	755	694	1244	654	590	7321
1724[47]		1200					11500

[41] Gov. Mathew to Board of Trade, etc., 18, T 61. [42] *Ibid.*, 27, Aa 84.

[43] *Ibid.*, 28, Bb 75. In 1778, there were about 37,000 slaves in Antigua. B. T. 6: 10, p. 496.

[44] Campbell, *Candid and Impartial Considerations on . . . the Sugar Trade*, London, 1763, p. 43.

[45] Gov. Mathew to Board of Trade, Aug. 31, 1734, C.O. 152: 20, V 46.

[46] Gov. Hamilton to Board of Trade, Aug. 22, 1720, C.O. 152: 13, Q 46.

[47] Gov. Hart to Board of Trade, July 12, 1724, C.O. 152: 14, R 101.

Year	White Men	Men able to bear arms	Women	Children	Boys	Girls	Slaves
1729[48]	1117		994	1586			14663
1734[45]	1115		1118	1648			17335
1756[49]	992	888	1253	538	265	273	21891

Thus in St. Christopher, which was perhaps the most fertile of the British sugar islands, we observe a very decided increase in white and slave population between 1717 and 1734. This was due to the settlement of the lands in the former French half of the island. But from 1734 to 1756 the number of whites slightly decreased and there was a marked falling off in the number of white children, while the number of slaves increased about twenty-five per cent. It is apparent that the white race was not perpetuating itself in the island of St. Christopher.

In the island of Nevis similar tendencies may be observed from the returns:

LIST OF ALL THE INHABITANTS OF NEVIS

Year	White Men	White Women	White Children	White Boys	White Girls	Total Whites	Slaves
1729[50]	373	390	533			1296	5646
1756[51]	394	369	355	178	177	1118	8380

Here also the white race revealed a serious weakness in the matter of self-perpetuation. Between 1729 and 1756, the number of white adults in Nevis declined nearly fifteen per cent, and the number of white children dimin-

[48] C.O. 152: 19, T 57.

[49] Gov. George Thomas to Board of Trade, Feb. 20, 1756, C.O. 152: 28, Bb 77.

[50] C.O. 152: 19, T 51, 68.

[51] Gov. Thomas to Board of Trade, Feb. 20, 1756, *ibid.*, 28, Bb 76. In 1733, Nevis had 6330 slaves and 372 militia. Gov. Mathew to Board of Trade, Aug. 31, 1734, *ibid.*, 20, V 46. In 1750, there were 7701 slaves and in 1751, 7772. Treasurer's Account, *ibid.*, 28, Bb 54, 55.

ished over fifty per cent. The slaves, on the other hand, increased in the same period about fifty per cent.

Montserrat, where Irish settlers predominated in the population, returned figures which indicate a somewhat hardier type of white people. Reports on the population may be summarized as follows:

POPULATION OF MONTSERRAT

Year	White Men	White Women	White Boys	White Girls	White Children	Total Whites	Slaves
1719[52]						386	4192
1720[53]	486	492	295	320	615	1593	3772
1724[54]						1000	4400
1729[55]	294	284			475	1053	5855
1745[56]						1117	5945
1756[57]	480	412	268	270	538	1430	8853

The number of white children decreased between 1720 and 1756, and it would seem that here also the whites were not maintaining their numbers by propagation. The gradual increase of adults after 1724 was due to immigration. The slaves more than doubled their numbers in the period.

The small islands included in the Virgin group were, at the opening of the eighteenth century, largely used as islands of refuge for pirates and former buccaneers. But little definite information has been found concerning them. However, in the first half of the eighteenth century, the Virgin Islands were settled by agriculturists most of whom were small planters, poor debtors, and a few criminals and pirates, who had fled here from the

[52] Gov. Hamilton to Board of Trade, May 10, 1719, C.O. 152: 13, Q 40.
[53] Same to same, Aug. 22, 1720, *ibid.*, Q 46.
[54] Gov. Hart to same, July 12, 1724, *ibid.*, 14, R 101.
[55] Gov. Mathew to same, Aug. 31, 1734, *ibid.*, 20, V 46.
[56] Same to same, Apr. 15, 1746, *ibid.*, 25, Y 154.
[57] Gov. Thomas to same, Feb. 20, 1756, *ibid.*, 28, Bb 81.

larger and richer islands. Their population increased rapidly in this period and by the middle of the century they contained decent and respectable communities. Some idea of their growth may be obtained from the following tables:

YEAR 1717[58]

Island	Men (Planters)	Women	Children	Slaves
Anguilla	96	154	234	824
Spanish Town	53	60	204	308
Tortola	37	34	38	176
Crabbe Island	46			62

POPULATION OF TORTOLA AND OTHER VIRGIN ISLANDS [59]
YEAR 1756

Island	Men	Women	Children	Slaves	Small Arms	Cannon	Small Arms Wanted	Cannon Wanted
Tortola	181	138	146	3864	157	58	122	40
Spanish Town	127	95	174	1204	62	8	100	12
Jost Van Dyke	54	49	87	472	22		50	4
Peter's Island	8	8	11	170	5		6	2
Camains & Scrubb Islands	16	9	21	140	9		9	
Guanna	5	7	10	160	3		5	
Beef Island	4	2	6	76	5		4	
Thatch Island	6	6	14	35	4		4	
Virgin Islands	401	314	469	6121	267	66	300	60

It seems conclusive that, comparing the number of children with the number of adults, the white race was not perpetuating itself in the Virgin Islands. The families appear extraordinarily small. The growth in number of white adults must have been due primarily to newcomers. Nevertheless, the increase in the number of

[58] Enclosures in Gov. Hamilton to same, Mar. 7, 1717, *ibid.*, 12, P 68, 70, 71, 72, 73.

[59] Gov. Thomas to Board of Trade, Feb. 20, 1756, C.O. 152: 28, Bb 83, 86.

slaves per capita from eight or nine in 1717 to fifteen or more in 1756 indicates a very marked improvement in the economic condition of the islands. From a military standpoint their means of defense seems to have been pitifully weak. Their slight commercial importance evidently did not appear to justify the imperial government in making adequate appropriations for their protection. Virgin Islanders might be considered a sort of frontier community in the British West Indies.

Comparisons of Births and Burials

The conclusion reached above that the white people in the British West Indies did not maintain their numbers by reproduction is confirmed by a comparison of the numbers of births and deaths in certain of the islands over long periods. Records of such were kept in the parish registers, and each year the governors were supposed to report to the Board of Trade the numbers that had been born or buried. Reports for Barbadoes and the Leeward Islands are available for a part of the eighteenth century; those for Jamaica were very imperfectly kept[60] and scarcely any have been found. In a few cases where the burial lists are detailed it is clear that the infant mortality was very high, also untimely deaths from smallpox and tropical fevers were very numerous. Where figures are available it appears that, on the whole, the number of deaths among the white people was greater than the number of births. Where "christenings" are given it is often stated that the "births" are of equal number. The following table for Barbadoes has been compiled from lists extending over a considerable period. Years where burials predominate are starred.

[60] Gov. Nicholas Lawes to same, June 21, 1718, C.O. 137: 13, P 13.

NUMBER OF BAPTISMS AND BURIALS IN BARBADOES[61]

Year	Baptisms	Burials	Year	Baptisms	Burials
1683	407	1062*	1761	524	493
1710[62]	201	140	1762	561	621*
1711-12[63]	449	352	1763	550	538
1715[64]	523	555*	1764	543	638*
1722-26[65]	1543	1434	1765	480	599*
1738-39[66]	578	344	1766	509	646*
1747-48[66]	366	483*	1767	506	492
1748-49[66]	514	344	1768	573	568
1749-50[66]	569	399	1769	591	659*
1750-51[66]	567	564	1770	600	618*
1751-52[66]	396	442*	1771	514	615*
1752[67]	556	680*	1772	578	598*
1753	543	638*	1773	603	463
1754	575	507	1774	693	586
1755	588	584	1775	585	497
1756	559	726*	1776	740	873*
1757	583	611*	1777	546	658*
1758	660	589	1780	683	1092*
1759	682	960*	1803	874	1087*
1760	538	437			
			42 years	22650	24192

For Antigua vital statistics are fragmentary. In three parishes, during the years 1721 to 1729 inclusive, the total number of baptisms was 1007 and the total number of burials 1461. In eight parishes, during the period 1733 to 1745, where records of baptisms and burials are

[61] Schomburgk, *Barbados*, p. 83 for the year 1683; C.O. 28: 11, P 85; C.O. 28: 13, R 75; 14, T 11; 15, T 151; 18, W 79; 25, Aa; 49, f 11; 29, Cc 10; 29, Cc 39; 29, Cc 92, 135; 30, Dd 10, 51, 62, 75, 100; 31, Ee 9, 27, 36; 32, Ff 2, 15, 26, 34, 56, 85, 100; 50; 52, B 4; 33, Gg 51, 60, 75; 34, Hh 5, 20, 29, 35, 43, 47; 35, Ii 10; 42.

[62] Jan. 1 to Aug. 1.

[63] June 11, 1711, to June 11, 1712.

[64] Apr. 24, 1715, to Apr. 24, 1716.

[65] Jan. 19 to Jan. 19.

[66] Mar. 25 to Mar. 25.

[67] Jan. 1 to Jan. 1.

complete for the same intervals, 597 baptisms are registered as against 822 burials.[68]

The records for St. Christopher, which are also very incomplete, give 1908 baptisms and 1924 burials in certain parishes for a part of the period 1721 to 1745.[69] For Nevis the records are practically complete and give for the period 1721 to 1745, inclusive, 241 baptisms and 382 burials.[70]

If we may assume that the numbers of baptisms and births are the same, then it is safe to conclude that, in the islands where data is available in the first half of the eighteenth century, the burials were for the most part considerably more numerous than the births.

The conclusion that white people were making but slow if any progress in acclimating themselves in the British West Indies is illustrated and in part explained by contemporary observers. One writer, in 1700, thought that Englishmen who came to Barbadoes ate too much meat and fish; he believed a pound a day was excessive and that four pounds a week would be conducive to better health.[71] In 1712, Jamaica's dearth of inhabitants was aggravated considerably, wrote Governor Hamilton, "by the late mortality that has been here; But at present the Island is again become more healthy."[72] Epidemics among cattle did not dissuade people from eating diseased

[68] These statistics were compiled from C.O. 152: 18, T 80; 16, S 64; 22, W; 23, X 24; 21, V 96; 25, Y 144.

[69] C.O. 152: 17, P 59; 16, S 79, 111; 21, V 97; 23, X 49, 75; 25, Y 146.

[70] *Ibid.*, 16, S 28, 95; 18, T 72; 21, V 102; 25, Y 144. A clergyman, Robert Robertson, writing from Nevis, December 31, 1729, said: "It does not lie in the Minister's Way to take an account of the Births; but as all the Children born here (except those of the Jews) are constantly Baptiz'd, an Account of the Christenings is much the Same as of the Births."

[71] *Considerations Concerning The Causes Of Pestilential Feavors In Barbados,* communicated to the Board of Trade by Mr. Pollexfen; received and read May 14, 1700, C.O. 28: 4.

[72] Gov. Hamilton to same, Jan. 19, 1711/2, C.O. 137: 10, N 6.

meat which killed great numbers of both whites and negroes.[73] Droughts also, in the smaller islands especially, were frequent and bred unsanitary conditions that must have promoted the spread of germ diseases. The death rate among children was especially high. For Governor Hart wrote from Antigua in 1723: "The Fever affects equally the Whites and Blacks, And is so violent that they become Delirious and Convuls'd: Tho' not many in proportion to the Sick dye above 15 years of Age."[74] St. Christopher, in 1723, suffered an epidemic of smallpox and "other Contagious Distempers." Acts were passed for putting the island in quarantine.[75]

Detachments of the royal army suffered terribly from tropical fevers. This was largely due to their wearing heavy clothing and being garrisoned along the coast where mosquitoes swarmed. The hardships of British soldiers in Jamaica are thus portrayed by Colonel Robert Hayes in a letter from Port Royal February 14, 1731. Arriving from Gibraltar February 7, he "brought the Regiment here in very Good health but [they] begin now to be very sickly. No Oven sure ever was so Hot, I find it affects my Eyes very much & still have the Gravel very much, and my Legs swell, We have yet only fourteen Companys Landed of Both Regiments, the Remaining six go to Port Antonio . . . We are an unexpected Guest. The Affairs of the Blacks I look upon to be quite a Bam, for I can find No Body that has either seen or felt them in a Wrathfull Manner; we shall very soon be dispersed about the Island . . . after that I no no Business I have here except to Sacrifice my health and Impoverish my Fortune."[76] March 11 he wrote: "Both men and Officers fall

[73] Gov. Lowther to same, May 20, 1715, C.O. 28: 14, T 81.

[74] Gov. Hart to same, Jan. 20, 1722/3, C.O. 152: 14, R 57.

[75] Same to same, June 8, 1723, *ibid.*, 14, R 69.

[76] Col. Robert Hayes to Maj. Soule, Port Royal, Feb. 14, 1730/1, C.O. 137: 19, S 118.

sick very fast, indeed I don't Expect to have half a Regim^{t} in 3 Months.''[77] Hayes himself died about a week afterward. Colonel Townsend who went to Port Antonio stated it to be ''a very unhealthy Place, so soon as the Rains begin to fall, which will be in the Month of May;'' and that ''many of both Regiments are at present ill of Feavers and Fluxes; I expect to have a bad account of them with in these 3 Months.''[78] ''This is the most Expensive disagreeable place under the Sun;'' wrote Colonel Cornwallis, ''I have as Yet kept my health very well but the sort of Distempers here are so sudden, that one has little Notice, Capt. Bellendine Tipp'd off by Very Short Notice, Our people begin now to be sickly thô 'tis at present the Healthiest Time Reckoned so you may Guess what will Happen in the Rainy Seasons.''[79] Five days later he added: ''Since my last we have buried the Maj^{r} of Our Regiment & have some others in great Danger, & I fear every Acc^{t} from this place will be worse & worse.''[80] He thought the distress was due to poor provisions and lack of shelter. Another letter reported an ''Abundance of Men are dead and more likely to dye, & Capt Molloy died at Spanish Town, & we have a report that several other of the Officers are dead in the Country.''[81] ''I would this Moment give my Commission to be in England,'' lamented Cornwallis, ''to represent what ill Usage so many of His Majesty's Subjects Met with, for the Sake of I fear a few People, for I can't be an Hour in the Day without hearing of some of the Regiment being either sick or dead.''[82] Another writer, reporting the deaths of so many officers, added that ''the

[77] Col. Robert Hayes to Maj. Soule, Mar. 11, 1730/1, *ibid.*

[78] Col. Townsend to Col. Cope, Mar. 2, 1730/1, *ibid.*

[79] Col. Cornwallis to Lord Cornwallis, Port Royal, Mar. 5, 1730/1, *ibid.*

[80] Same to same, Port Royal, Mar. 10, 1730/1, *ibid.*

[81] —— to a Merchant in London, Port Royal, Mar. 19, 1730/1, *ibid.*

[82] Col. Cornwallis to Lord Cornwallis, Mar. 20, 1730/1, *ibid.*

climate is too Inveterate an Enemy for them, in Short the Surviving Gentlemen are almost Lost.''[83] Under such conditions it is no surprise that it took until 1739 to suppress the rebellious Maroons of the interior.

The dangers to life in the towns appear to have been worse than in the country. In 1737 John Gregory wrote from Jamaica: ''I am sorry to acquaint Your Lordships there has been an Exceeding great Mortality among the Inhabitants. I have been Credibly informed there has been buried in the Town of Kingston within these four Months five Hundred White people some indeed of them Sailors but the greatest part dwellers [on] Shoar thô the whole Number of Inhabitants of that place are not Computed above Two thousand.''[84] Leslie, writing in 1740, said it was then estimated that the entire population of Jamaica died once in seven years. ''No doubt,'' he continued, ''the multitude that dies would soon leave the place a Desart, did not daily Recruits come over from Great Britain. Scarce a Ship arrives, but has Passengers who design to settle, and Servants for Sale. This is a constant Supply, and a necessary one; for notwithstanding their vast Numbers, the Island but slowly settles, and I don't know if 50 new Plantations have been made these Dozen Years past. But I persuade my self the Death of Thousands can be ascribed to nothing else than Intemperance. Sometimes at Sea the Living is but hard; and when they once get ashore, and find Plenty of Rum-punch, &c. they swallow it down with excessive

[83] A letter communicated to Lord Torrington, Port Royal, Mar. 20, 1730/1, *ibid.* Later in the century, English troops in the tropics took better care of themselves. Sir William Young reviewed the regiment in St. Vincent and observed it to be ''well disciplined, and nothing hurt by residence of near seven years in the West Indies.'' *A Tour through . . . Barbadoes*, etc., *1791-1792*, London, 1801.

[84] Lieut. Gov. John Gregory to Board of Trade, Nov. 25, 1737, C.O. 137: 22, V 57.

Pleasure, get drunk, expose themselves to the noxious Dews, are seized with Fevers, and die. If New-Comers were more careful to live moderate, and abstain from the Use of spirituous Liquors, they might live as happily and free from Disease as anywhere else.'' Few escape fevers when they first arrive ''and this is the Reason why so many die so soon after they come to Jamaica.'' Peritonitis (''Belly-ach'') was very common. There was a great field, especially among the negroes, for young doctors from England, some of whom became wealthy.[85]

St. Christopher and Montserrat were reputed to be somewhat healthier than the other English Islands. Indeed the resident governor thought ''The whole of these Islands is Certainly as healthy as any part of Europe. But we here think Montserrat and this Island healthyer than Nevis or Antigua. From an Observation of my Own, of some Years, I have found, that of the Militia of St. Christophers (and they are the most Irregular and most Mortal of the two Sexes) There dy'd but in the Proportion of one in five and Thirty in fifteen Months.''[86] However, among children, as we have earlier observed, there must have been about as great mortality as elsewhere in the West Indies.

Perhaps the most potent single thing in undermining the ability of white people to resist the inroads of tropical disease was the universal and excessive drinking of rum. ''Rum Punch is not improperly called Kill Devil,'' said Leslie, 'for Thousands lose their lives by its Means.''[87]

[85] Charles Leslie, *A New and Exact Account of Jamaica*, Edinburgh, 1740, pp. 50-52.

[86] Gov. Mathew to Board of Trade, Aug. 31, 1734, C.O. 152: 20, V 46.

[87] Charles Leslie, *Jamaica*, p. 32.

APPENDIX II

AN ACCOUNT OF THE NUMBER OF NEGROES IMPORTED AND EXPORTED AT JAMAICA EACH YEAR, 1702-1775[1]

Year	Number of Slave Ships	Number of Negroes Imported	Number of Negroes Exported
1702	5	843	327
1703	14	2740	481
1704	16	4120	221
1705	16	3503	1669
1706	14	3804	1086
1707	15	3358	807
1708	23	6627	1379
1709	10	3234	1275
1710	15	3662	1191
1711	26	6724	1532
1712	15	4128	1903
1713	19	4378	2712
1714	24	5769	3507
1715	10	3372	1039
1716	24	6361	2872
1717	29	1775	3153
1718	25	6253	2247
1719	27	5120	3161
1720	23	5064	2815
1721	17	3715	1637
1722	41	3469	3263
1723	20	6624	4674
1724	25	6852	3569
1725	41	10297	3368
1726	50	11708	4112
1727	17	3876	1555
1728	30	5350	935
1729	10	10499	4820
1730	43	10104	5222
1731	45	10079	5708
1732	57	13552	5288
1733	37	7413	5176

[1] Appended to a memorial from Stephen Fuller, agent for Jamaica, to the Board of Trade, London, Jan. 30, 1778, C.O. 137: 38, Hh 3, 4.

Year	Number of Slave Ships	Number of Negroes Imported	Number of Negroes Exported
1734	20	4570	1666
1735	20	4851	2260
1736	15	3943	1647
1737	35	8995	2070
1738	32	7695	2070
1739	29	6787	598
1740	27	5362	495
1741	19	4255	562
1742	22	5067	792
1743	38	8926	1363
1744	38	8755	1331
1745	18	3843	1344
1746	16	4703	1502
1747	33	10898	2376
1748	9	10430	2426
1749	25	6858	2123
1750	16	3587	721
1751	21	4840	713
1752	27	6117	1036
1753	39	7661	902
1754	47	9551	1592
1755	64	12723	598
1756	46	11166	1902
1757	32	7935	943
1758	11	3405	411
1759	18	5212	681
1760	23	7573	2368
1761	29	6480	642
1762	24	6279	232
1763	33	10079	1582
1764	41	10213	2639
1765	41	8951	2006
1766	43	10208	672
1767	19	3248	375
1768	27	5950	485
1769	19	3575	420
1770	25	6824	836
1771	17	4183	671
1772	22	5278	923
1773	49	9676	800
1774	79	18448	2511
1775	39	9292	1629
	2090	497736	137114

APPENDIX III

AN ACCOUNT OF THE QUANTITY OF SUGAR, RUM, AND MOLASSES EXPORTED FROM JAMAICA TO THE BRITISH NORTHERN COLONIES[1]

Lady Day to	Sugar				
Lady Day	Hhds.	Tierces	Bbls.	Casks	Lbs.
1726-27	292	..	188	64	....
1727-28	200	45	75	57	....
1728-29	259	29	221	75	6000
1729-30	214	27	101	59	....
	885	101	585	255	6000

Lady Day to	Rum					
Lady Day	Puncheons	Hhds.	Tierces	Bbls.	Casks	Gals.
1726-27	88	5	119	41	174	5000
1727-28	28	4	5	8	13	5000
1728-29	22	71	48	13	262	....
1729-30	99	..	..	4	275	500
	237	80	172	66	724	10500

Lady Day to	Molasses			
Lady Day	Casks	Hds.	Tierces	Gals.
1726-27	253	116	1401	5000
1727-28	338	258	415	15000
1728-29	1218	898	699	18000
1729-30	1645	2011	1190	400
	3454	3283	3705	38400

[1] Jno. Manley Cust. Ho. 18 Feb. 1731.

AN ACCOUNT OF THE QUANTITY OF SUGAR, RUM, AND MOLASSES EXPORTED FROM BARBADOES, ANTIGUA, NEVIS, ST. CHRISTOPHER, AND MONTSERRAT TO THE BRITISH NORTHERN COLONIES[1]

Lady Day to Lady Day	Sugar				
	Barbadoes Lbs.	Antigua Lbs.	Nevis Lbs.	St. Christopher Lbs.	Montserrat Lbs.
1726-27	645674	182076	19850	170627	19200
1727-28	650973	180673	79350	222409	52850
1728-29	896935	487853	131300	333144	28550
1729-30	549147	204432	81029	61890	19000
	2742729	1055034	311529	788070	119600

Lady Day to Lady Day	Rum				
	Barbadoes Gals.	Antigua Gals.	Nevis Gals	St. Christopher Gals.	Montserrat Gals.
1726-27	686214	63625		1400	3519
1727-28	607150	227953		6353	14228
1728-29	838643	384230	100	5700	8817
1729-30	589070	268058		4144	12280
	2721077	943866	100	17597	38844

Lady Day to Lady Day	Molasses				
	Barbadoes Gals.	Antigua Gals.	Nevis Gals.	St. Christopher Gals.	Montserrat Gals.
1726-27	45055	11199	5445	112186	6320
1727-28	43185	28266	13910	136993	17100
1728-29	57635	54187	22671	140480	15665
1729-30	71440	17374	30952	95034	20040
	217315	111026	72978	484693	59125

Custom Ho: London 22 Feb. 1731.
R. Parsons Comp. Gen.

[1] C.O. 323: 9, M 35.

APPENDIX IV

AN ACCOUNT OF THE SPECIES AND QUANTITIES OF GOODS EXPORTED FROM JAMAICA TO NORTH AMERICA FOR WHICH THE DUTIES IMPOSED BY 25 CAR: 2$^{d.}$ HAVE BEEN PAID; FOR SEVEN YEARS TO CHRISTMAS 1751, DISTINGUISHING EACH YEAR[1]

Year: Xmas to Xmas	Sugar: White at 5^{s} per cwt.			Sugar: Muscovado at 1^{s}/6 per cwt.			Other Goods: Ginger at 1^{s} per cwt.			Dying Wood at 6^{d} per cwt.			Indico at 2^{d} per lb.	Cocoa at 1^{d} per lb.	Cotton at ½^{d} per lb.
	cwt.	qrs.	lbs.	cwt.	qrs.	lbs.	cwt.	qrs.	lbs.	cwt.	qrs.	lbs.	lbs.	lbs.	lbs.
1744-45	...	.	..	10891	.	6	85	1	8	...	.	..	...	.....	2,669
1745-46	...	.	..	10453	2	10	57	2	14	...	.	..	...	.....	2,873
1746-47	...	.	..	8715	1	14	77	3	14	...	.	..	...	.....	.....
1747-48	...	.	..	9627	3	13	22	3	8	15	.	20	...	6,061	7,354
1748-49	...	.	..	5724	1	6	80	1	26	...	.	..	...	257	2,384
1749-50	156	.	17	20073	2	5	108	3	16	220	.	..	127	1,084	3,625
1750-51	297	1	5	22738	2	13	...	.	..	...	.	..	298	.....	15,164
	453	1	22	88224	1	11	433	.	2	235	.	20	425	7,402	34,069

[1] Treas. 64: 274. This and the following table give some indication of the volume of trade from the British West Indies to the Northern British colonies in the middle of the eighteenth century. From 1749 to 1751 it will be observed that the volume of commerce enormously increased, owing probably to the government's prosecution of illegal trade at that time. Because of the duties on intercolonial trade in "enumerated" commodities, the actual amount of such commerce was no doubt nearly always much greater than what was legally recorded.

APPENDIX V

AN ACCOUNT OF THE NET PRODUCE OF THE DUTIES OF 1s. 6d. PER CWT. IMPOSED ON MUSCOVADO SUGARS AND 5s ON WHITE SUGARS EXPORTED FROM ANY OF HIS MAJESTY'S SUGAR COLONIES IN AMERICA TO NORTH AMERICA FOR SEVEN YEARS TO CHRISTMAS 1751 DISTINGUISHING EACH YEAR[1]

Year: Xmas to Xmas	Jamaica	Barbados	Antigua	St. Christopher	Nevis	Montserrat	Total
	£	£	£	£	£	£	£
1744-45	388	507	210	136	48	15	1306
1745-46	373	529	190	70	6	13	1183
1746-47	311	416	50	75	8	7	869
1747-48	343	674	202	136	15	25	1398
1748-49	204	506	121	65	8	7	914
1749-50	735	955	182	130	19	17	2039
1750-51	847	918	248	132	15	16	2178
	3205	4508	1206	746	122	103	9891

Custom ho. London
22 March 1753

W. Parsons per Comp Gen.

[1] Treas. 64: 274. The omission of shillings and pence accounts for the apparent errors in the totals of the above table.

APPENDIX VI

NEW YORK'S SHIPPING TO THE WEST INDIES COMPARED WITH ITS SHIPPING TO OTHER PARTS, 1705-1716[1]

	Number of vessels from the port of New York to			
Year	Europe	West Indies	Neighboring plantations	Africa
1705	9	68	54	
1706	13	70	33	1
1707	18	55	39	
1708	13	62	40	1
1709	15	57	65	1
1710	18	59	82	1
1711	25	63	57	
1712	21	95	50	
1713	12	75	55	
1714	22	91	56	
1715	27	102	66	2
1716	31	124	63	
	224	921	660	7

[1] C.O. 5: 1051, Bb 98.

APPENDIX VII

THE AMOUNT OF SHIPPING FROM BOSTON, SALEM, AND NEW YORK TO THE WEST INDIES AND OTHER PARTS, 1714-1718

	24 June, 1714-24 June, 1717						24 June, 1715-24 June, 1718[1]		
	Cleared from Boston			Cleared from Salem			Cleared from New York		
	Ships, Sloops and other Vessels	Tons	Men	Ships, Sloops, &c.	Tons	Men	Ships, Sloops, &c.	Tons	Men
For British Islands in W. I.	495	27831	3830	59	2296	328	...	.....	....
For Foreign Plantations	58	2597	393	2	75	10	85	2595	603
For West Indies	23	1335	175	10	304	46	...	.....	....
For Newfoundland	45	1835	274	...	.....	....	9	395	67
For Europe	23	1865	210	7	9122	1152	9	615	122
For Madeira, Azores, &c.	34	1690	236	9	421	59	24	1395	282
	678	37153	5118	197	12218	1595			
For Great Britain	143	11956	1412	4	208	29	63	4382	638
For British Plantations on the Continent	390	11589	1883	31	1005	158	205	4234	897
For British Plantations on the Islands	....	.....	....	...	.....	....	250	8776	1904
For Bay of Campeachy for Logwood	25	1675	221	...	.....	....	...	.....	....
To Ports unknown	11	415	63	...	.....	....	...	.....	....
Total in 3 years	1247	62788	8697	232	13431	1782	645	22392	4513
i.e. Communibus Annis	415	20929	2899	77	4477	594	215	7464	1504

1199 of the aforesaid 1247 ships and vessels containing 58152 tons were plantation built.

[1] C.O. 324: 10, pp. 386-387.

APPENDIX VIII

QUANTITIES OF BRITISH WEST INDIA SUGAR SHIPPED TO GREAT BRITAIN, BRITISH NORTH AMERICA, AND FOREIGN MARKETS, 1728-1746[1]

ANNUAL AVERAGES

		Hhds. of 12 cwt.
1742	Imported into Great Britain	60,950
	Shipped to N. America and Foreign Markets	5,000
1728-1735	Imported into Great Britain	74,500
	Shipped to N. America	5,000
	Exported from Great Britain	10,580
	British home consumption	63,920
1735-1742	Imported into Great Britain	66,289
	Shipped to N. America	5,000
	Exported from Great Britain	5,964
	British home consumption	60,325
1742-1746	Imported into Great Britain	60,523
	Shipped to N. America	5,000

1 Long Papers, Brit. Mus. Add. MSS. 12404, f. 416.

APPENDIX IX

JAMAICA'S IMPORTS FROM BRITISH NORTH AMERICA FROM CHRISTMAS 1757 TO CHRISTMAS 1762[1]

Barrels of beef and pork	31,355	
Barrels of shad and mackerel	10,209	
Hhds. of salt fish	8,436	
Barrels of flour	128,740	
Tierces of bread	69,288	
Tierces of rice	18,674	
Staves and heading	7,241	M.
Shingles	8,632	M.
Hoops	737,000	
Boards	10,718	M feet
Also Indian corn, horses, stock, spermaceti and other candles, soap, tar, pitch, turpentine, lamp oil, etc. All of which imports were sold by Jamaica merchants per annum for about	£200,000	currency

---

Imports from Ireland in the same period comprised

Barrels of beef	77,415	
Barrels of pork	20,563	
Barrels of herring	25,824	
Firkins of butter	7,136	
Also tongues, tripe, lyng fish, potatoes, etc. All of which imports were sold by Jamaica merchants per annum for about	£140,000	currency

[1] Gov. Lyttelton to Board of Trade, July 9, 1763. C.O. 137: 33, Cc 19.

APPENDIX X

DOCUMENTS CONCERNING THE CASE OF THE SLOOP *ENTERPRIZE*, 1749[1]

1. Representation to His Majesty from the Governor, Council, and Assembly of Jamaica:

Most Gracious Sovereign—

We your Majesty's most Loyal and Dutifull Subjects the Governour Councill and Assembly of this Your Majesty's Island of Jamaica with all Deference and Humility approach your Royal Throne having had from a late Incident an Opportunity of discovering that since the Conclusion of the late War there hath been Revived and carried on with great application a most pernicious Intercourse and Commerce between the French of Hispaniola and your Majesty's Colony of Rhode Island; by means whereof very great Quantitys of Sugar, Rum and Molasses of the French produce hath been and still continues to be Introduced into Your Majesty's Northern Colonys in America, in fraud of the Acts of Trade and Navigation and more particularly of the Act passed by Your Majesty for the better securing and Encouraging the Trade of Your Majesty's Sugar Colonies in America as a Commerce of such a nature if continued, will not only be of most fatal Consequence to this Your Majesty's Island of Jamaica, but to the other of Your Majesty's Sugar Colonies and the Trade of Great Britain in General.

We Your Majesty's Loyal Subjects the Governour Councill and Assembly of this Your Majesty's Island of Jamaica Judged it to be our most indespensable Duty to your Majesty, to the rest of the Sugar Colonys, to our Mother Country; humbly to

[1] The following documents are in C.O. 5: 6, ff. 5-25.

represent to Your Majesty such Fact[s] and Circumstances, as upon Examination appeared unto us in relation to the premisses; together with what we apprehend, must be the Result therefrom (unless prevented by Your Majesty's great Wisdom) and most humbly to petition your most Sacred Majesty for your Royal Interposition in our Behalf.

That your Majesty's Commander in Chief of your Squadron here having sent in a Sloop called the Enterprize Commanded by on[e] Richard Mumford laden with French Sugars and Molasses from the Island of Hispaniola and bound for Your Majesty's Colony of Rhode Island, with the several Sea Papers and a Letter from a Merchant in Hispaniola to his Factor at Rhode Island relating to the Vessell and her Cargoe which were delivered to the Commandore by the said Mumford of which Your Majesty's Governor Council and Assembly here having Notice and the Papers and Letters being Communicated to them sent for and Solemnly Examined the said Mumford touching the Same and Information being further given to the Assembly by Edward Manning Esqr: one of the Members thereof, with respect to the Method of Carrying on the Trade and Intercourse between Hispaniola & Your Majesty's Northern Colonys the same thing reduced into Writing the Truth thereof was at the Instance of the said Edward Manning attested upon his Oath, a Copy of which Information, the Examination of Mumford and of the several Papers and Letter laid before the Governour, Councill and Assembly is annexed unto this our humble Representation and Petition, and with all Humble Duty we Crave leave to referr to them for Your Majesty's greater Satisfaction in the premisses.

That the Governour Councill and Assembly apprehend it appears from the Examination of Mumford the Information of Mr. Manning and the several Papers and Letter, whereof Copies are annexed (notwithstanding the said Act of the sixth of your Majesty Continued in force to this time); for the better securing and Encouraging the Trade of your Majesty's Sugar Colonys in America that since the Conclusion of the late War great Quantitys of French Sugars, Rum and Molasses, have been and Continue to be Exported from Hispaniola to Your Majesty's Colony of Rhode Island in a most fraudulent and Clandestine manner, and there is great Reason to believe, that very Considerable Clandestine Imports of French Sugars, Rum and Molasses, are made into other of Your Majesty's Northern Colonys.

That it appears to the Governour Councill and Assembly of this your Majesty's Island that such French produce is frequently sent not in return for the Native Commoditys of Your Majesty's Northern Colonys, but as to an Ordinary Market on a Meer Factorage Account laden too in Vessells Prohibited by the Laws of Trade and Navigation Viz[t]: French property most fraudulently concealed under the Legal Authentication of English Registers That returns for the Sale of Such French Sugars & other produce are of late Generally made by their Factors in Your Majesty's Northern Colonys in Specie to the great Impoverishment of those Colonys; and rarely in Commodities of their Growth they not being Importable into the French Sugar Islands save by particular Permission in the Instances of such few Commoditys as they cannot Supply themselves with from their own Northern Settlements.

That the Governour, Councill, and assembly can't but Represent to Your Majesty, what an Emolument such a Trade into the French Sugar Colonys, who not only find an advantageous Market for their Sugars in Your Majesty's Dominions but more particularly for their Rum and Molasses, which would be almost useless to them without a Circumstance, as prejudicial to Your Majesty's Sugar Colonys as it is of Benefit to theirs which would otherwise be incapable of vending a Considerable & now very Profitable Part of their Produce of very little value to them before the Increase of their Pernicious Commerce by the Continuance of it our Markets will be Glutted with Foreign Supplys and the Price of our own Produce must sink in proportion as the Quantity at Market is encreased without any additional Demand, which together with the heavy Duties laid on Sugars can't but end in the Ruin of the Planters & the whole Sugar Trade thereby fall into the Hands of the French, whose Extensive Commerce already renders them too Formidable in the American Seas.

That the Governour Council and Assembly are most Sensible

of the great Regard shewn by your Majesty and your Parliament to the Circumstances of the Sugar Colonys when the before mentioned Act of the Sixth year of your Majesty's happy Reign was passed for their Encouragement and that the same was Intended to prevent such Importations as were then usual from the Foreign Sugar Colonys by Rendering it more Eligible the Duties imposed by that Act being paid for your Majesty's Northern Provinces, to Supply their Markets with such Commoditys from your own Sugar Islands in America.

. . . [The revival of illicit trade since the war may be] attributed in some Measure to the Encouragement and Attention the French Governors give to the same Establishing such Rules as may render that Trade Regular, and of the greatest Benefit to their Sugar Islands. . . . [They conclude that the trade is carried on] in a very frudulent and clandestine manner that few or no Entrys are ever made of any French Produce Pursuant to the beforementioned Act tho very great Quantitys are continually Imported or the Dutys upon such Imports ever paid that in apprehension of the Governour Council and Assembly it can arise from nothing less than a general Connivance in the whole Government of Rhode Island, and the greatest neglect of Duty in the Officers of the Revenue in the rest of Your Majesty's Northern Colonys that the Delinquents are unnoticed and thence the almost daily Breaches of that Law, pass without any Legal annimadversion and by every fraud the force of it is eluded and Rendered of no Effect; and beg leave Humbly to Represent, that it would be impossible such a Commerce could be carried on, if the said Act was Strictly comply'd with, and made as it was intended to be the General Rule of it.

. . . [They suggest that to prove their conclusions an inspection be made of the custom house records of Rhode Island. The British sugar colonies] depend upon the having the British Markets entirely secured from such Clandestine Imports. . . . As every Regulation by way of Impost, has hitherto been without Effect; that nothing can ever give a Sufficient Check, to the Carrying on, or prevent the Pernicious consequences, & fatal Effects, attending the progress of this most Mischievous Commerce, but an Absolute Prohibition thereof, inhibiting all foreign

Imports of Sugar, Rum, Molasses & Paneles into any of your Majesty's Colonies in America. [They desire that the matter be laid before parliament. Passed the council November 21, 1749, and the assembly November 16, 1749.]

2. Lawrence Cholet to Jonathan Nichols:[2]

Ste Dominique le 7 Nov.r 1749.

Monsieur & Ami

La presente va par la Bateau l'Enterprize Capitaine Richd Momford, dans lequel j'ai chargé 58 Barriques creoles, 8 Barriques de Bordx & 36 Quarts Sucre bruit, le tous pour mon compte particulier, & dont ici inclus le Connoisement, il m'a falu mettre tout ce que le possede au monde, & meme faire mes Billets pour le Payment du Sucre que j'ai chargé dedans, afin de pouvoir l'expedier.

Vous savez que je vous ai toujours dit que cela me faisoit un torte considerable de trouver mon Oncle parti, mais il est encore plus grand que je n'ai pu vous le dire, puisque je n'ai aucun Interest dans toutes les affaires que j'ai fait avec vous, & que j'ai perdu dans toutes les autre affaires où je m'etois interessé, je n'ai donc d'autre Esperance que dans ce Voyage pour retablir un peu ma Fortune & je compte sur votre Amitié pour m'aider de tout votre Poir.

Il s'agit mon cher ami defaire vendre sur le champs tous les Sucres & le Bateau quel Prix qu'il vaile & m'envoyer le montant en Louis d'Or ou Piastres gourdes; les Louis d'Or vaudroit mieux par la premier Batiment que vous enverrez ici, afin de me mettre en etat de payer mes Billets, & d'entreprendre quelque autre chose, pour aller moi meme à Rhode Island, au mois de Mars prochain, je vous prie dont, mon cher Nichols de vous souvenir de toutes les Assurances d'Amitié que vous m'avex donné de m'aider dans toutes les Occasions; celle-ci est d'une consequence infinie pour moi: Car je serois obligé de me sauver du Paris, si je ne recevois pas le Montant de ce Bateau & Cargaison, dans le courant de Janvier prochain, pour lequel j'ai fait mes Billets.

[2] Nichols was deputy governor of Rhode Island. The frequent absence of accents in the French text is as in the original.

Vous devez—persuadé de toute mon Attention p[our] L'Éxpedition de vos Batimens ici, quois que je n'y sois pas interesé, je le ferai pour vous & non pas pour ceux d'ici qui sont trop ingrat à mon regard; quand j'irai à Rhode Island j'apporterai des permissions avec moi qui le Bregantin le George, pour les Bois qui nous avons commandé qu'il ne seroit point envoyé avec les Permissions que vous avez, parce que je travaille à faire augmenter le nombre des D[tes] permissions pour vous procurer un Commerce considerable, & soyez assuré que j'en viendrai a bout.

Ce Bateau & Cargaison me revient ici avec la Sucre de hore à trente mille quelque cens Livres, & c'est mon cher ami bien plus que je n'ai vaillant, jugez de ma Situation jusqu' à ce que vous m'envoyez le Montant.

L'Equipage qui le meme est toujours aux gages du Navire & vous les payerez en consequence du Rolle d'Equipage du dit Navire que je vous remets ci-joint, il n'y a que Montford avec qui je n'ai fait aucun marché, qui est sur le Compte du Bateau, je m'en raporte à vous pour le payer a que vous croirez juste, il a mis abord le privilege qu'il a voulu, ainsi il n'y a que les pages que vous aurez la bonté de regler & de lui payer; je crois que l'Usage de vous autres Capitaines est sufficient.

Cy joint le Compte de chargie pour l'Equipage que vous aurez la bonté de retenir sur leurs Gages du Navire & le porter au Credit du Bateau, parceque c'est moi qui l'ai payé.

	£
Il y a pour Harten Marquantoche, qui est mort	13. 2. 1
Jos. Brouk	16. 2. 5
Jacques Wilson	13. 2. 1
Guill[me] Coupre	13. 2. 1
	55. 9. 2

pds Ce qui fait—18.10—vieux Teneur	
en outre un Billet de Jemi Roche	21. 3. 6
un Billet de Cap[ne] Stevenson	178.
D[o] de Cap[ne] Monford	276.
Le toute ensemble se mont	493.13. 6 vieux Teneur

que vous retientrez aussi sur leurs Gages, ou sur les Sucres qu'ils ont abord, & que vous aurez la bonté de passer au Credit du Bateau L'Enterprize; vous voyez toutes les avances que j'ai été obligé defaire pour l'Expedition de ce Bateau, en fin Dieu soit loué, il n'y a que la prompte Remise que vous me ferez de ce Fonds qui peut me tirer d'affair aussi y compte—je en consequence de toutes les offres & promesse que vous m'avez fait.

Il y a dans le Bateau 2 B^ls^ Sucre bruit marqué I.S. qui sont pour le Compte de Legrettier, que je vous prie de vendre & en garder le montant á ses ordres.

J'ai mis abord pour vous 2B^ls^ Sitrons & un Bar^l^ d'Orange & Annannas pour Mad^me^ Nichols, je vous prie defaire part des Sitrons à Mrs Channing Chalender & Wanton le Collecteur, & Mr Aure ou restoit Mr Laville.

Je vous prie de dire à Madame Wanton, que je lui enverrai de Plats de Fayance par la premiere occasion, j'enverrai aussi la panotille à Madame Nichols à vos Enfans & à Mad^me^ Layton; car je me suis pas a portée ici de rien envoyer.

Ci inclus est aussi la Nottes de Privileges qu'il y a abord. Je crois que le Brigantin le George sera parti avant que je n'arrive à Leogane; nous attendons tous les jours quelqu' autre de vos Batimens.

Par ma derniere Lettre je vous ai marqué de vendre le Brigantin le Prince Frederic, avec tout ce qui en depend, je vous le reiterez, parce que je veux retirer le Montant de l'interet que j'y ai dedans qui est un 1/4 et un 1/12. Je vous ai aussi fait faire à Boston avec le Cheval de Madame Richards, je vous prie de me envoyer par la premier occasion.

Enfin mon cher Nichols, il ne me reste qu' à vous soushaiter une bone Santé & à Mad^me^ Nichols que j'assure de mon Respect & embrasse toute votre cher famile. J'ai l'honneur d'etre bien Sincerement Monsieur & cher ami

Votre tres humble &
tres obeis^t^ Serviteur

L^s^ Cholet.

3 and 4. Copy of the Clearance from Rhode Island to Jamaica for the Sloop Enterprize, August 19, 1748.

Custom House Rhode Island — In Persuance of an Act made Anno Decimo Georgii 2de Regis Entituled An Act for the better Encouragement of the Trade to his Majesty's Sugar Colonies in America.

No/3 An Accot of the Names, Ages, and Descriptions of the Men belonging to the Sloop Enterprize whereof Richard Mumford is Master Bound for Jamaica.

Num.	Men's Names	Stations	Ages	Stature	Complexion
1	Hector McIntosh	Mate	30	Tall	Dark
2	Francis Bradford	Sailor	23	Short	Light
3	James Roach	Ditto	24	Ditto	Ditto
4	John Whitson	Ditto	27	Ditto	Dark
5	James Price	Ditto	21	Short	Ditto
6	Joseph Brooks	Ditto	22	Ditto	Ditto
7	William Cooper	Ditto	25	Tall	Ditto
8	James Whalen	Ditto	24	Short	Ditto

These are to Certify that the above written is a true Copy of a List of Men belonging to the aforesaid vessel attested by the Master thereof and delivered into this Office Given under my Hand and Seal of Office this 19th day of August Annoque Dom: one Thousand Seven hundred and fforty Nine.

Vera Copia

Richard Mumford — J. Gould Pro Coll.

Newport By the Honourable William Greene Esqr Governor of the Colony of Rhode Island &c.,

No 4. Permission is hereby Granted to Richard Mumford Commander of the Sloop Enterprize Burthen fifty Tons or thereabouts Navigated with Six Men, Plantation Built and Duly Qualified to Sail for Jamaica he hath Enter'd and cleared at the Naval Office and all other His Majesty's Offices in this Place according to Law and for whom this shall be a sufficient Lett pass Given under my Hand and Seal at Arms this 19th day of August 1749. In the twenty third Year of His Majesty's Reign.

Signed

W: Greene

Passed at the Naval Office
Charles Bardin
pro Naval Officer

Vera Copia
Richard Mumford.

5. (N° 5) Received On Board the Sloop Enterprize on the Account & Risque of Mr. Lawrence Cholet Fifty Eight Hogsheads Eight Tierces and Thirty six Barrels Sugar which I promise to deliver to Jonathan Nichols Esqr: Merchant in Rhode Island or his order the Danger of the Seas Excepted Signed three Receipts of this Tenor and Date one of which being Accomplished the other two to stand void
Hispaniola October 26th, 1749 signed

Richd Mumford

Vera Copia
Richd Mumford.

6. Jonathan Nichols of Newport took oath that she [the *Enterprize*] was built at Taunton Massachusetts in 1749 and is owned by him solely, which oath was made before William Greene Governor, Joseph Wanton, Esquire Collector and principle Officer of his Majesty's Customs in said colony. Signed by W. Greene & J. Gould for Coll.

7. Deposition of E. Manning before the Assembly of Jamaica. Veneris 10 die Nov^{ris} 1749.

Mr. Manning in his place informed the House that in the Month of April last he was in His Majesty's Ship the Cornwall with Admiral Knowles at Leogan and there saw two English Vessels as he was informed and believes belonged to North America, That during the time his Majesty's said Ship the Cornwall lay at Leogan a large Brigantine or Snow arrived from some Parts of the Northern Colonies in order to trade there as he was likewise informed and believes, and that being on Shore at Leogan he there met with one of the Captains of the said

Vessels, at one of the French Merchants Houses Bargaining for Rum and Molasses and that the said French Merchant asked him the price of Molasses at Jamaica who acquainted him that it was about twelve pence a Gallon to which the said Merchant replied, That the Northern Men must pay pretty near the same there or they would not get their Loading for there were there at that time six or Eight Vessels trading for Molasses and Rum in the Cul de Sac and other parts adjacent and some of them had lain there for six or Eight Weeks. That thereupon he asked the said French Merchant if the English Vessels had free liberty of trading there from the Northern Colonies who informed him that the new General lately arrived there had Instructions from his Court to admit of the Trade with any Country or Colony without Exception, provided it was for the Interest & Encouragement to the Settlers of that Island and that it was for the Interest of the Colony to suffer any Vessels to Trade there who should take off Molasses and Rum those being Commodities, that before they never had any vent for and for the purchase of which they obliged them to bring Cash or such Produce of the Northern Colonies that they could not be supplied with from their own Settlements, And that thereupon he acquainted the said French Merchant that this was a very pernicious Trade to the Island of Jamaica and that he would give information of the same on his return there, to which the French Merchant replied, he hoped that he would not, as it would be of infinite prejudice to him he having the Consignation of most of those Vessels.

Which having taken down in Writing and read to the House Mr. Manning desired he might attest upon Oath the Truth of that Declaration and he Withdrawing was sworn accordinly.

signed

Sworn before me E. Manning

Jno. Pallmer.

8. Examination of Richard Mumford before the Assembly of Jamaica, November 9, 1749.

That on or about the first day of November instant he this Examinant in the Sloop Enterprize was lying at Anchor in

Tiberoon Bay when Commodore Townshend came in there, Who sent an Officer and some People on board to take charge of her, that at that time he had Seven White men on board named, John Whitson, James Price, James Roach, Joseph Brooks, Francis Bradford, William Cooper, and James Whally, that he this Examinant came from Rhode Island to Leogan and from Leogan to English Bay where he took in Sugar and Molasses of the Product of Hispaniola and from thence to Cape Tiberoon where this Examinant in his said Vessell lay at Anchor when the said Commodore Townshend came in, and this Examinant saith that when he left Rhode Island he had no cargo or Money on board his said Vessell belonging to his owner to purchase a Cargo but that the Sugars which was the Cargoe, he this Examinant took on board from the Merchant at Leogan was by him remitted (as this Examinant believes) for a debt due from him to this Examinant's Owner at Rhode Island and that this Examinant went on purpose from Rhode Island aforesaid to Leogan to receive the said Sugars and that the same were addressed to Jonathan Nichols Merchant this Examin[ts] Owner at Rhode Island, and that he this Examinant went with no other View or Intention, and that such Debt so due to the said Nichols was from one Cholet at Leogan aforesaid, and that this Examinant farther declares that he had on board at the time the said Commodore sent an Office[r] and People on board at Tiberoon Bay as aforesaid Fifty eight French Hogsheads eight Tierces and thirty six Barrels of Sugar and Seven Casks of Molasses, and that no part of the said Sugars were on this Examinants own account, but that the same was on the Account of the said Jonathan Nichols, and that the said Seven Casks of Molasses were on this Examinant's own Account, and that he carried with him from Rhode Island aforesaid about twenty pistoles in order to lay out the same for an adventure for himself and that this Examinant gave at and after the Rate of fifty Livres for each Cask of Molasses and that they each gauged about one hundred Gallons, and this Examinant farther declares that during the time he was lying at Leogan he there saw a Snow and a Brigantine belonging to Rhode Island but dont know upon what Account they went there particularly but apprehends and believes they went there to take in Molasses, that he this Examinant cannot declare how many Vessels there are trading between Rhode Island and Leogan this Examinant never having been at Leogan before or using that Trade and this Examinant declares that he doth not know neither doth he believe that the said Snow and Brigantine touched at any other Port than Leogan in their said Voyage being as he apprehends designed for that Port, that the Current Price of Sugars at Hispaniola (as he this Examinant believes) were about twenty Livres a Hundred and that the Duty upon the Importation of French Sugars at Rhode Island is as he believes five Shillings Sterling for every Hundred Weight and this Examinant believes that if such Duty were paid at Rhode Island on the said Sugars it would be worth the while of Merchants at Rhode Island to carry on such Trade otherwise they would not send Vessell[s] to trade in the manner they do, and this Examinant declares that he was born at Rhode Island, and is a Liver there and that in all his this Examinant's Time he never knew or heard of Merchants or other Persons paying any Duty on French Sugars there and he believes that there are five or Six Vessels in the Year or more trading between Rhode Island and Leogan that the Price of French Sugars at Rhode Island at the time this Examinant left the said Place was at the Rate of Sixteen or Eighteen Pounds that Currency per hundred Weight and this Examinant further declares that the said Jonathan Nichols is concerned as he believes in Seven or eight Vessels trading to different Parts but whether to Leogan particularly cannot tell but this Examinant believes that part of one of the said Vessels lying at Leogan aforesaid belonged to the said Jonathan Nichols and that the said Nichols is no Port Officer but a private Merchant at Rhode Island and that he this Examinant was not to return back to Leogan again as he knows of, and this Examinant further declares that he Cleared out his said Vessell from Rhode Island to Jamaica which is the Usual Way for all Vessels clearing out there when they intend to trade with the French but that at the time he so Cleared out he had no design or Intention to go to Jamaica but did intend (as he afterward perform'd) to go to Leogan and this Examinant believes that notwithstanding the Port Officers at Rhode Island had cleared out the said Vessell from thence to Jamaica and had a Cargo of French Sugars and Molasses on Board yet they would admit such Vessell at her Return to Enter upon Payment of the Duty and this Examinant declares that he had no permit to go to Leogan neither did he ever hear of any such for his Vessell, but that the said Jonathan Nichols this Examinant's Owner told him that he this Examinant in his said Vessell would be permitted and allowed to go with Safety to Leogan and hath also heard that Permissions have been Granted by the French Governours to several Vessels belonging to Rhode Island to go and trade with them and this Examinant declares that the Paper Writing produced and shewn to this Examinant purporting as this Examinant believes to be a Letter from the said Cholet to the said Nichols was delivered by the said Cholet to be by this Examinant delivered to the said Jonathan Nichols and that the several other Paper Writings now likewise shewn to this Examinant [(]True Copies of which are hereunto annext) Marked Number 1 No. 2 No. 3 & No. 4 are all original Papers belonging to the said Vessell and under the respective hands and seals of the Several officers thereto sett Subscribed and affixed & that the same are true Copies of the severall Originals this Examinant having carefully Compared the same with the respective originals and that the Paper Writing markt No. 5 hereto annext purporting to be a Receipt or Bill of Lading for the Cargoe of the said Vessell the original of which is of this Exam^t's own hand Writing and that the said Paper No. 5 is a true Copy of the Original this Exam^t having Compared the same with the said Original,

signed

Richd Mumford.

APPENDIX XI

THE BOARD OF TRADE'S INVESTIGATION OF COMMERCE BETWEEN BRITISH NORTH AMERICA AND THE FRENCH WEST INDIES, 1750

An investigation of illegal trade between the English Northern Colonies and the French West Indies was undertaken by the Board of Trade, in 1750, as the result of a memorial from the British sugar planters. The memorial contains fifty-nine signatures, three of them being those of colonial agents. It begins by complaining that the act of 1733 was too weak to suppress the trade between the Northern Colonies and the French West Indies. The more important parts of the document are as follows:

The Memorial of the Sugar Planters Merchants and others . . . trading to and interested in his Majesty's Sugar Colonies in America.[1]

. . . That the French sensible to the Advantages which would result to themselves from such an intercourse have already settled a Correspondence in Rhode Island established Factors there and bought and provided Vessels destined to carry on this Traffick between Hispaniola and Rhode Island . . .

That this is a Traffick not taken up Casually or by Chance but the Result of a well weighed and concerted Plan formed or at least approved by the Court of France (as may reasonably

[1] The original is in C.O. 323: 12, O 59; there is a copy in C.O. 5: 38.

be collected from the French Kings Instructions to his Governors in the Sugar Islands and their conduct in Consequence thereof) and intended to be steadily and regularly pursued and if it should succeed will prove the surest and most effectual means of distressing the British Sugar Colonies and raising their own that they could devise . . .

That there are still more interesting and important Reasons for not suffering French Rum or Molasses to be imported into the Northern Colonies or any of the British Dominions than even Sugar itself—It is well known to every one concerned in the Sugar Trade that the Profits of the Planter depend upon the Vent which he finds for his Rum and Molasses for if Sugar only and no Rum or Molasses could be produced from the Sugar Cane it would hardly pay the expense of Culture and making consequently as the Vent of Rum and molasses is stopt or increased the Sugar Colonies (whether French or English) must thrive or decline and as Rum is not allowed to be imported into Old France or any of its Colonies (because it interferes with Brandy which is the Product of the Mother Country) this evidently shows how much it is in the Power of Great Britain to check the Progress of the French Sugar Colonies and advance that of their own; for if the bringing French Rum or Molasses into any of the British Dominions were once effectually stopt the great Profit arising from the Rum and Molasses made in the French Sugar Colonies would be lost to them as they could find no Vent or at least no considerable Vent for it in any other part of the World This Point vigorously and steadily and invariably adhered to would in its natural tendency damp the Growth of Sugar in the French Islands and increase and exalt it in our own and might very profitably in the end prove a means of enabling the English to beat them out of all the foreign Markets in Europe and confine them to their own Consumption. . . .

That this Trade is of such a Nature that if it be not soon eradicated it will grow obstinate take such deep Root and spread so wide that the Cure by length of time will become very difficult if not impracticable.

On October 18, 1750, John Sharpe, the solicitor for the sugar colonies, appeared before the Board of Trade and read to the commissioners the above memorial. Its consideration, after numerous postponements, was taken up by the board on November 13 and continued on December 6, 7, 10, and January 10. The evidence presented on these occasions concerning the traffic of the continental colonies with the French sugar islands is recorded in the Journal of the Board of Trade, and is as follows:[2]

Tuesday, Novr 13 . . . Mr. Sharpe having observed to their Lordships, That as this affair had been already decisively determined in Parliament by the Act passed in 1733, for the Encouragement of the Sugar Colonies, which Act from the Spirit and Expression of it was plainly intended for a Prohibition, he should not enter into any Arguments to prove the Mischiefs of such a Trade, but should show that the Act had proved ineffectual, and propose an absolute Prohibition.

That he should show in what manner and in what Degree the Northern Colonies more especially Rhode Island had carried on this Trade. In times of War by Flaggs or Truce, sometimes with only one two or three Prisoners and sometimes purchasing prisoners for that purpose, and since the War in an open regular course. That it was a known fact that the Northern Colonies consumed a great quantity of French and other foreign Rum, Sugar Molosses &c And as well known that they never paid any Duties for it. That Vessels have cleared from Rhode Island for Jamaica, have gone to some other British Settlements have sold their Lumber for Specie refusing Rum and Molosses and have then gone to some foreign Settlement and bought a cargo of Rum and Molosses with that Specie and sometimes Linnens, Silks, East India Goods and other prohibited Merchandize.

. . . . .

[2] The selections that follow are from the Board of Trade Journal, C.O. 391: 57; there is a copy of the same entitled "Copy of All the Proceedings had and the State of the *viva voce* Evidence taken before the Commissioners of Trade and Plantations in the Year 1750 relating to the Trade carried on by the British Northern Colonies with the French Sugar Colonies" in C.O. 5: 38.

Thursday, Dec[r] 6th Whereupon Mr. Sharpe produced the following Depositions and Examinations which were read viz[t]

Depositions of Mr. Manning relating to Vessels trading from the Northern Colonies to the French Sugar Islands taken before the Assembly of Jamaica Nov[r] 10th 1749. Examination of Richard Mumford on the same Subject taken before the said Assembly 9th Nov[r] 1749. A French Letter and the Translation of it from Monsieur Le Cholet to Mr. Nichols dated at St. Domingo 9th Sept[r] 1749.

And then proceed to call several Gentlemen to give their Lordships an Account of what they know concerning the illicit Trade complained of.

Admiral Knowles acquainted their Lordships, That, in the Year 1737, being on the Leeward Island Station, in going into Bassatirre Road, he met a sloop which seemed lightly laden; that on sending on board her he found she was bound to Boston and had clearances to that port as if laden with Rum and Sugars, but that her real lading was empty English Casks which she was to fill at Eustatia with Rum and Sugar. That he afterwards went to Eustatia and saw 16 or 17 North American Sloops loading and unloading there, and that he wrote an Account thereof to Mr. Horace Walpole. That during his command at the Leeward Islands several of the vessels of the Northern Colonies not finding Cargoes ready for them at Eustatia had taken Dutch passes and gone to Martinique. That Cap[n] Tyrrel, one of the Cruizers of the Admiral's Squadron, had fired 52 shots at one such Vessel before she would bring to, as appeared by a letter from the said Captain, which the Admiral produced and read. That every Captain of his Squadron knows that these North American vessels supplied the French with Provisions otherwise he should certainly have taken Martinique. That there were at one time 42 sail of North American Vessels at Hispaniola with fictitious Flags of Truce and three at Port Louis when taken by the Admiral one of which Vessels belonged to Providence that having represented this to the Duke of Newcastle and received no Orders in Consequence he ordered his Squadron to take such vessels whenever met with That Captain Holmes accordingly took a Snow with a considerable cargoe and two Vessels were taken by Captain Hughes the cargoes of which were condemned at Boston in January 1746/7 That the Vessels carrying on this Trade generally refuse Rum and Sugars in pay for the Lumber &c they bring to the British Islands and will sell only for Specie for which Consideration they will even take less than the real Value of their Lumber &c. in order to buy Cargoes at Eustatia, for which purpose Specie is necessary as no freight of Lumber or Stave they can carry will be sufficient to buy a Cargoe of Rum and Molosses that the Colonies are hereby drained of all their specie. The Admiral being then asked by Mr. Sharpe whether he apprehended the French could be supplied with the same kind of lumber &c but from the Northern Colonies, answered, he believed not; for that when St. Piere a town in Martinique was burnt down, the Governor of that Island had by Proclamation invited the North American Vessels to trade thither that Provisions had been so scarce in the French Islands that the Governors of them have applied for leave to buy 100 Barrels of Flower in Jamaica. That he did not apprehend that Hispaniola could afford a sufficient quantity of Lumber and that Species of Woods growing in their Islands were different from those of the Northern Colonies; neither could they be supplied from Canada, the Navigation of the river St. Lawrence being so dangerous that Six Sail of Ships never made a safe passage up and down That the Danger and length of the Voyage is also a reason why the French could not be supplied with Lumber from the Mississippi, That he believed but could not positively assert that the Value of the Lumber and Provisions carried from the Northern Colonies, was sufficient to pay for the Rum and Molosses they buy of the French, but as these Commodities would be useless to the French if not disposed of in this manner, he did not believe the French would trade with the Northern Colonies if obliged to pay them in Specie; and therefore an Act forbidding the Importation of Rum and Molosses into the Northern Colonies would if faithfully executed reduce the French to extreme necessity. The Admiral being then asked, by Mr. Sharpe whether the said North American Vessels do not take off the French Sails Ozenbrigs &c. could not assert that of his own knowledge, but said he heard they took off great quantities of silks, dry Goods, &c.

Mr. Philip Pinner said there was no Wood in Jamaica of which Hogsheads and Puncheons could be made, and that the wood of Hispaniola was of the same Species and Qualities as that of Jamaica.

Captain Hughes said that during the time he was under the Command of Admiral Knowles in the West Indies he had been three times at Leogan, and had each time seen there a number of North American Vessels. That since the Peace he had been asked by Merchants whether he would take them off if they fell in his way.

Captain Tyrrel, who was late of the said Admiral's Squadron, and has a Plantation at Antigua, being asked whether Rum and Sugar had not been refused and Money insisted on by the North American Traders; answered that he had himself been obliged to draw Bills on England in order to procure Specie when they refused Rum and Sugars in Payment for Lumber. That a particular Instance of this happened to him in 1747, when the Master of a New England Vessel who had refused his Rum and Sugar was seen afterwards loading at Eustatia. That it is their Common Practice to insist on Specie because a Cargoe of Lumber carried to Eustatia, would not purchase a Cargoe of Rum and Molosses there. That by means of this Trade, our own Sugar Islands find no Market for their Rum &c that of the Northern Colonies being forestalled. That if this trade could be effectually stopped, the French in time of War could not be able to supply their Sugar Islands with any thing. That the Prosperity of the French Islands and the Ruin of our own must be the certain Consequences of these Practices if continued, that the King's Fleet had felt the effect of them as well as our own Islands This Circumstance Admiral Knowles confirmed and said that the Northern Colonies used to buy French Prisoners at a great Price of one another for a Pretence to go to the French Islands that he had at length been obliged to threaten the French Governors that he would send to England all French Prisoners if they delivered any English to the Northern Flags of Truce.

Mr. Crockat desired leave to ask if the French could not be

supplied with Lumber from the Mobile River which is to the Windward of Mississipi that he apprehended that Mobile would in time be capable of supplying all the French Islands with Lumber and therefore the Encouragement that must arise to that Settlement from making it difficult to the French to get North American Lumber must in the end be prejudicial to our own Sugar Colonies To this Admiral Knowles replied that the River Mississipi and Mobile were not more than 8 or 10 Leagues or under that the Timber growing on each is not the same as the Northern Colonies and that the difficult Navigation of each together with the Length and Expense of the Voyage would in his opinion render it impossible for them to be supplied from either of those Places.

Mr. Daniel Moore said he has been three or four months together at St. Eustatia and has seen great Numbers of Vessels laden with Horses and Staves which they sold for Molosses That great Quantities of Cordage and Canvas were also bought at St. Eustatia. That he has known People of New England to have sold their Cargoes at Barbadoes for ready Money to have bought Rum Hogsheads which they have filled with water cleared out as laden with Rum and then go to the French Islands and buy Rum That he has also been at Martinique when he has heard the French say that they should have no use for their Rum and Molosses if they could not sell it to the British Settlements. That the Dutch as well as French supply Rhode Island with Rum and Molosses of which he had heard several Instances. That he agreed with the other Gentlemen that these Practices would end in the Destruction of our Sugar Colonies and that if a stop was not put to them he would sell his own Plantations Being asked by what Means this Trade might be effectually prevented he said that one or two Men of War stationed at Rhode Island would be sufficient in those parts and that the Custom House Officers of New York and Pensilvania if they did their Duty might without the Assistance of any Man of War effectually prevent any Trade of this sort in those Provinces.

Mr. Maynard said that he was in Trade at Barbadoes from 1715 to 1747 That it is the constant custom of that island to make all agreements for Money That the People of Rhode

Island bring Horses thither which they sell for money That they then buy Molosses Casks and go to Martinique of which they make no Secret That the Rhode Islanders do generally but always refuse to take the Rum of Barbadoes in Payment for the Cargoes they bring thither.

Mr. Maitland acquainted the Board that he had been several Years at the Leeward Islands and sometimes at Eustatia That he has seen theré 30 Sail of North American Vessels at one Time That he has never seen Rum and Molosses loaded in such Vessels. That he has been told 300 Vessels called there in one Year That great Part of the Inhabitants of Eustatia are English That those English have set up Still Houses in which they distill French Rum and then buying a commission from the Governor of Martinique (which one voyage costs 500 pieces of Eight) they trade with said Rum which passes for English. That several People at the Leeward Islands have set up Cooperages at Eustatia and have considerable business. That the North American Traders dispose of the Cargoes they bring to the Leeward Islands for money if possible, whereby the Islands are so distressed for Specie that Merchants are often obliged to order their Correspondents here to send them out Spanish and Portugal Money. That Eustatia though small and mountainous, exports and imports more than any one of the Leeward Islands. That if this Trade were effectually stopt, the French planter would lose 25 per Cent on his Net Produce, which in his opinion, would ruin the Sugar Islands. That there come to Eustatia every Year 10 or 12 Dutch Ships with Osnabrigs, Cambricks and all sorts of European and East India Manufactures with which all Vessels coming thither are supplied to the great Detriment of the Mother Country. That during the late War the French Trade was entirely stopt except to their sugar Islands, and he thinks it possible to prevent their making any sugar but for themselves. That if the 25 per Cent above mentioned was taken from them a French Sugar Planter must become a Bunkrupt. That he agrees with Admiral Knowles that they can have Lumber no where but from us, for the Reasons set forth by the Admiral; That their Northern Colonies can be of no Use while Nova Scotia is ours, and they have no port to the East. That they cannot be supplied with Lumber from Europe on Account of the Expense. That smuggling is practiced over all the Northern Colonies, particularly at Rhode Island where it is less Expensive, and Goods are landed and exported again as English. That large Quantities of Rum distilled of French Molosses are imported into Ireland and the West of England. That 16 Puncheons from North America are now bonded at the Custom House which is Proof that those Colonies have more than they want. That the Leeward Islands would willingly give their Rum for Lumber.

Mr. Gray said, he had lived Six Years in Jamaica and had supplied 30,000 Soldiers and Sailors (Admiral Vernon's Fleet) with Rum, and that he was of Opinion, that Jamaica would now very nearly furnish a sufficient Quantity for all America; in which Admiral Knowles agreed with him upon computing the number of inhabitants in all the Colonies from the number of those in Newfoundland including the Fishery.

It being late and impossible to close the Evidence to Day their Lordships desired these Gentlemen to attend again to Morrow at 12 o'Clock, and bring with them such Persons of Ability as would ingenuously give the Board their Sentiments as to the proper Remedy to be applied to the Evils complained of.

Friday Dec[r] 7th. The Merchants, Sugar Planters &c. attending as desired, Mr. Sharpe, their Sollicitor, proceeded in producing his Evidence in Support of the Allegation in Support of their Memorial.

Mr. Moore being particularly asked, repeated the assertion he made yesterday that if the Officers of the Customs at New York and Pensilvania did their Duty, the illicit Trade complained of could not be carried on in these Provinces. That four Snows would sufficiently watch the Coast of Rhode Island and New England.

Mr. Gray being asked what quantity of Rum our Islands make, said, that Jamaica makes 12,000 Puncheons per annum of 110 Gallons each; that if the uncultivated parts of that Island were settled it might make 30,000 Puncheons and in three or four Years might make 50,000, if there was demand for it; and that the Encrease of the Demand would be the most effectual Encouragement to settle the uninhabited Parts of the Island.

That if this illicit Trade was effectually prohibited, the Northern Colonies would then be obliged to take from our own Sugar Colonies the Commodities they now take from the French, and that our Islands could very well supply them. That Barbadoes does now make 12,000 Puncheons of 110 Gallons each; to which Mr. Moore added that Island does actually ship off 10,000 Puncheons, but very little or no Molosses.

Mr. Maitland said, that Antigua makes from 10. to 12,000 Puncheons of Rum, St. Christopher's 6,000, Montserrat 1500 Puncheons, That Nevis chiefly sold their Molosses, and made but very little Rum as yet, but are now going into that Trade, and will be capable of making about 1500 Puncheons.

Being asked what is the proportion of Sugar to Molosses, he said that in St. Christophers, the Sugars not being rich yield little Molosses that five Hogsheads of Sugar would yield three of Molosses in that Island; but in Antigua where the Sugars are Richer, three Hogsheads of Sugar yield two of Molosses: That something depends on the Skill of the Distiller but much more on the difference of riches in the Sugar Cane. That 100 Gallons of Jamaica Rum, are equal to 140 Gallons of French Rum, and is Ten per Cent stronger than the Rum of our other Islands.

Mr. Whitaker said that Barbadoes makes about 12000 Puncheons of Rum every Year.

Mr. Sharpe observed to their Lordships, that the above Quantities of Rum, said now to be made in our Sugar Islands, at the most moderate Computation amounted to 41,500 Puncheons or 4,565,000 Gallons to which Quantity the Allowance for Superior Strength in the Jamaica Rum would still make a considerable Addition.

Mr. Gray said that Lumber is of late brought to Jamaica in less Quantities and at a higher Price than in the Year 1733. That between the years 1740 and 1746 he several Times bought Lumber and had offered Rum and Molosses in Exchange for it, but was refused, and obliged to pay Cash. That he also bought flower for the Fleet, for which he was obliged to pay in Specie.

Mr. Maitland said, That he had enquired of many Captains in the North American Trade, who had assured him, That Lumber was double what it was Twenty Years ago; That the Price is now from 20 to 30£ Currency per 1000. Boards or Staves; which could have been bought 10 Years ago for half the first Price, the Reason for which he conceived to be the Quantities of Timber that have been cut down by the new Settlers. That the Want of Lumber would daily encrease by the Continuation of the Northern Colonies to supply the French with it; and that as the Difficulties of getting Necessaries for a Sugar Plantation must greatly discourage a Sugar Planter, the making a less Quantity of Sugar, and of Course the Decline of our Sugar Colonies must be the Consequence of this Illicit Trade, the Prohibition whereof would be the best means of settling the Island of Jamaica of which he said not $\frac{1}{6}$th or $\frac{1}{10}$th Part is at present cultivated. Mr. Maitland confessed that he had never been in North America, but that the Information he had given had been grounded on the frequent Intercourse he had had with the People of that Country. And with regard to what he had asserted on the Article of Lumber, he said he had been concerned in a vessel that went to North America to load Lumber and carry it to St. Christophers, but it was then so scarce that the Captain was obliged to disobey his Orders, and take freight to Cape Breton.

Mr. Penny said, that in the Year 1733, being at Nevis, he had offered Rum or Molosses in Exchange for a Cargo of Lumber, that his Offer being refused, and he still insisting he would only pay in Rum, he was at length obliged to take an Order to pay in Rum to a third Person. That it is the constant Practice of the North American Vessels at Nevis to insist on being paid in Specie.

Mr. Sharpe then desired their Lordships Leave to open a New Head of Evidence, in order to show that this Illicit Trade extends its effects to the Trade of Great Britain, and the great Quantities of Sail Cloth, Linens, Silks, Spices, China &c are illegally carried to the Northern Colonies, whereby this Nation is deprived of the benefit of supplying them: And having called upon Mr. James Warren, a Merchant settled at Marseilles to give an account of the Trade carried on from that Port to North America.

Mr. Warren acquainted their Lordships, That he has had Cargoes of Pitch, Tar, Logwood and Timber from Piscataway, and others of Spars, Staves, and Planks, sometimes from Boston, sometimes from Carolina, That he has also received small Parcels of Rum and Furs from Canceau That he has been settled at Marseilles for 30 Years and these consignments have come four Times in some Years That it is not an accidental but a settled Trade carried on by Commission That the Returns made from Marseilles consisted of Silks, Velvets, Gold and Silver Wire Buttons Silk Stockings Gold and Silver Lace Twist Gloves Olives Capers Wine and Earthen Ware That the Ships concerned in this Trade go from Marseilles to Cadiz where they load Salt and then proceed to Boston.

Captain Cole said That he has known of several of these Cargoes consigned to Mr. Warren to another Merchant of the same Name and other Houses at Marseilles as much or more than to Mr. Warren That he remembers the same Trade carried on at Toulon That Lions Hatts were sent in great Quantities to North America in Exchange for Furrs That the Governors of Mississipi have told him there was very little Timber in that Country That they never sent any Lumber from thence to their own Colonies and very little from St. Lawrence on Account of the difficult Navigation of that River And that he was therefore persuaded the French could not be supplied with Lumber but by our Northern Colonies That it is easy to stop this Trade at New York and Pensilvania there being so few Ports in those Provinces and that four Sloops of War would effectually watch the Coasts of Rhode Island and New England but that Cruizers on the Contraband Vessels would do better

Being asked whether Spars brought in as above to Toulon have not been sold to the French King's Arsenal he answered that Spars Blocks and other Things of the like Nature had been so sold and Mr. Warren said that he sold a Cargoe of Timber and Plank from Piscataway to the Arsenal at Marseilles.

Mr. Bellowes Register to the Court of Admiralty acquainted their Lordships That he had in his Custody by Virtue of his Office several Letters and Papers which were taken by a Man of War in Scotland some of which confirm the Evidence given Yesterday by Admiral Knowles, and others prove a regular Trade between Messieurs Quinsies of Boston and Messieurs

Hops of Amsterdam in which Trade several Persons at Newcastle and Cowes were also concerned Mr. Bellowes then read the following Letters Vizt.

> A Letter from Thomas and Adrian Hop to Edward and Josias Quinsey dated the 23d August 1745 giving advice of a Cargoe of Naval Stores Arms Powder &c shipped for their Account.
>
> A Letter of Ralph Car to the said Josias Quinsey dated at Newcastle Septr 6th 1745.

Their Lordships desired Mr. Bellowes to prepare and lay before them an Abstract of all Letters in his Custody relating to this Illicit Trade that they may the more readily turn to such as are material Evidence in the Matters now under their Consideration.

Mr. Sharpe then read the following Papers.

> The Proceedings in the Court of Admiralty against the Elizabeth of Eustatia Joseph Blake sole Owner James Johnson Master taken by Capt. Philpot of his Majestys Ship the Woolwich She was bound to the Mole a French Island laden with Oznabrigs Glasses &ca
>
> Proceedings in the said Court against the Sloop Endeavour Lawrence Paine Master which Vessel sailed from New York 31st January 1746/7 as a Flagg of Truce with Twelve French Prisoners to Cape Francois She carried with her a Cargoe of Fish Flower Beef Bread and Beer which she sold there received back nine Prisoners and took in a Cargoe of 36 Hogsheads Molosses and 200 Weight of Loaf Sugar which she was to carry back to New York but was taken by Captain Hughes and carried to Boston where both Ship and Cargoe were acquitted
>
> Mr. Clinton's Instructions to said Flagg of Truce
>
> Proceedings in the Court of Admiralty at Boston against the Victory Brigantine which sailed as a Flagg of Truce from Newport 12th January 1746/7 with five French Prisoners to Cape Francois she carried thither a Cargoe of 300 Quintals of Cod Fish some Onions &c with

the Produce whereof she bought 174 Casks of Molosses of different Sizes which were to have been delivered to Jos: Whipple Esqr of Newport her Owner But on the 23^{d} May 1747 she was taken off Lock Island [Block Island ?] by the Hind Sloop Capt. Hughes and carried into Boston where the Ship was acquited but the Cargoe was condemned

The Sentence passed by the Admiralty Judge of Boston on the said Ship and Cargoe.

The Case of the Sloop Mary of Bermuda which sailed in August 1747 as a Flagg of Truce with one French Prisoner to Leogan she carried out 200 and odd Barrels of Flower for which she had a Clearance to Jamaica and took in at Leogan 89 Hogsheads of Sugar 8 Hogsheads and 2 Barrels of Indigo and 48 Loaves of Refined Sugar and 6 English Prisoners Her Owners were Francis Jones and George Forbes Esq^{rs} and Company of Bermuda She was taken the 1st Octr 1747 and had when taken a French Pass.

Mr. Sharpe would have proceeded to offer some Observations on the Evidence that had been laid before their Lordships but it being late the Gentlemen were desired to attend again on Monday next at 12 o'clock.

. . . . .

Monday Decr 10th. The Merchants Sugar Planters &c with their Sollicitor Mr. Sharpe and the Agents of the Northern Colonies with their Sollicitor Mr. Paris attending as desired by the Minutes of Friday last Mr. Sharpe recapitulating what had been given in Evidence at the two Hearings on the Subject observed to their Lordships

That he had in the opening of the Case stated the Act of 1733 and insisted that the Sense of the Legislature had thereby determined the illicit Trade complained of to be in its Consequences destructive and had intended that Act to operate as a Prohibition That he had proved the Northern Colonies and particularly Rhode Island to have been guilty of carrying on this Trade. . . . [He recapitulates. As a remedy he proposes in behalf of the sugar planters a prohibition of all trade between North America and the French Colonies. Then turning to the Northern demand for West Indian products he stated:] he had estimated it (and in the Opinion of Good Judges reasonably at 10,000 Hogsheads of Rum and 16000 Hogsheads Molosses per Annum which several Quantities may be had from our own Islands That the Duties of these several Quantities under the Act of 1733 were

			£
For 10,000	Hheads	Sugar	25,000
12,000		Rum	45,000
16,000		Molosses	40,000
		Per Annum	110,000.

That this Sum for 16 Years to 1749 amounts to £1,760,000 and whatever the Duties that appear by the Custom House Accounts to have been actually paid fall short of that Sum so much are the Duties of which the Crown has been defrauded And if the Northern Colonies should make their annual Demand to exceed the above Estimate the Arrear of Duties due from them would increase in proportion That upon the whole the Consequences of this Illicit Trade were so pernitious in every respect that the putting an effectual Stop to it is of so great importance to the very being of the Trade of the Sugar Islands and to the benefit that ought to accrue to Great Britain from the Northern Colonies that he hoped and entreated their Lordships Assistance in obtaining an Act of Parliament for the total Prohibition of all Trade between the Colonies of North America and all foreign Settlements.

To this Proposal Mr. Paris very strongly objected as prejudicial to the Interest of North America and a means of hindering the settling of it That the Consequences of such a Prohibition if determined on would be so fatal to the Northern Colonies that he entreated that time might be allowed them to answer the several Articles of which they were accused.

Mr. Bollen Agent for the Massachusetts Bay said that the Proposal made by Mr. Sharpe tended to the Destruction of all the North American Trade whether legal or illegal That several of the Persons who had given Evidence were not sufficiently acquainted with matters they had spoken of and that the point in question was of such Consequence that he prayed their Lordships not to determine hastily but to grant time for the Northern Colonies to make answer according to the Request made by their respective Agents on the 13th. of last Month.

Mr. Abercromby Agent for North Carolina and Mr. Charles Agent for the Province of New York urged the same Request . . .

Mr. Tomlinson as a Sugar Colony Merchant (not an agent) objected to the Prohibition proposed and said that he should look on the Contrivers of this application as the greatest enemies to the Sugar Colonies That Mr. Sharpe had sufficiently proved the Fact he had asserted and there were still other instances of illicit Trade of which he himself knew But that the Act of 1732 [*sic*] if effectually executed was sufficient Prohibition and if it had not been found effectual the Collectors and Officers of the Customes were chiefly to blame . . . [He denies the truth of Mr. Maitland's evidence on the price of lumber.] That the Traders of New England have desired to take Molasses but that the Sugar Islands will not part with a Drop. That these are the true Reasons why Specie is insisted on by the North America Traders That he believed the Calculation of Rum made in our Sugar Islands to be right but that the Consumption was so great that they had none to spare . . . if this Affair should come before Parliament he expected the Wisdom of the Legislature would as the best Remedy for the Evils complained of Repeal part of the Act of 1733.

. . . . .

Jan. 10th 1750/1 . . . [Mr. Paris, representing the Northern Colonies, said] his Principal had forbidden him to enter into Arguments upon the Matter without Instructions from his Constituents he therefore renewed his former Motion for a Copy of the Memorial to transmit the same to the Colonies for their Answer thereto in which Motion he was seconded by all the other Agents of the several Northern Colonies respectively.

Mr. Sharpe then acquainted their Lordships that pursuant to

their Directions a General Meeting of the Merchants Sugar Planters &c. had been held and they had agreed to withdraw their former Proposal of a total prohibition and to propose a limited one instead vizt. That the Northern Colonies should be permitted to trade with foreign Settlements provided they did not take their Sugar Rum and Molosses in Return He then delivered to their Lordships a Paper[3] containing the Proposals agreed on at the said Meeting which paper being read and shewn to the Agents of the Northern Colonies they declared [themselves] apprehensive of the same bad Consequences from this as from the former Proposal of a total prohibition but desired copies of the said Paper and further time to consider it.
Ordered.

That Copies of the said Paper be accordingly made and given to the said Agents of the Northern Colonies.

[3] The document is in C.O. 323: 13, O 87.

INDEX

INDEX

For Product Safety Concerns and Information please contact our EU representative GPSR@taylorandfrancis.com
Taylor & Francis Verlag GmbH, Kaufingerstraße 24, 80331 München, Germany

www.ingramcontent.com/pod-product-compliance
Lightning Source LLC
LaVergne TN
LVHW011101300926
849160LV00008B/242
9780367142599